Arid and semi-arid regions are defined as areas where water is at its most scarce. The hydrological regime in these areas is extreme and highly variable, where flash floods from a single large storm can exceed the total runoff from a sequence of years. Globally, these areas face the greatest pressures to deliver and manage freshwater resources. Problems are further exacerbated by population growth, increasing domestic water use, expansion of agriculture, pollution, and the threat of climate change. However, there is little guidance on the hydrology of arid areas, and none on the decision support tools that are needed to underpin flood and water resource management.

As a result, UNESCO initiated the Global Network for Water and Development Information for Arid Lands (G-WADI), and arranged a workshop of the world's leading experts to discuss the hydrological modelling tools required to support water management in these areas. This book presents chapters from contributors to the workshop. It includes case studies from the world's major arid regions, including Africa, the Middle East, the USA, India, and Australia, to demonstrate model applications. It contains web links to tutorials and state-of-the-art modelling software. This volume will be valuable for researchers and engineers working on the water resources of arid and semi-arid regions.

HOWARD WHEATER is Head of Environmental and Water Resource Engineering in the Department of Civil and Environmental Engineering at Imperial College London, and co-chair of G-WADI.

SOROOSH SOROOSHIAN is Distinguished Professor of Civil and Environmental Engineering and Director of the Centre for Hydrometeorology and Remote Sensing at the Henry Samueli School of Engineering, University of California at Irvine.

K. D. SHARMA is Director of the National Institute of Hydrology, India, and a member of the G-WADI Steering Committee. He is also a Visiting Fellow at the Chinese Academy of Sciences and the Winand Staring Centre, the Netherlands.

INTERNATIONAL HYDROLOGY SERIES

The **International Hydrological Programme** (IHP) was established by the United Nations Educational, Scientific and Cultural Organization (UNESCO) in 1975 as the successor to the International Hydrological Decade. The long-term goal of the IHP is to advance our understanding of processes occurring in the water cycle and to integrate this knowledge into water resources management. The IHP is the only UN science and educational programme in the field of water resources, and one of its outputs has been a steady stream of technical and information documents aimed at water specialists and decision-makers.

The **International Hydrology Series** has been developed by the IHP in collaboration with Cambridge University Press as a major collection of research monographs, synthesis volumes and graduate texts on the subject of water. Authoritative and international in scope, the various books within the series all contribute to the aims of the IHP in improving scientific and technical knowledge of fresh-water processes, in providing research know-how and in stimulating the responsible management of water resources.

Hydrological Modelling in Arid and Semi-Arid Areas

Howard Wheater

Imperial College of Science, Technology and Medicine, London

Soroosh Sorooshian

University of California, Irvine

K. D. Sharma

National Institute of Hydrology, Roorkee, India

CAMBRIDGE
UNIVERSITY PRESS

University Printing House, Cambridge CB2 8BS, United Kingdom

One Liberty Plaza, 20th Floor, New York, NY 10006, USA

477 Williamstown Road, Port Melbourne, VIC 3207, Australia

314-321, 3rd Floor, Plot 3, Splendor Forum, Jasola District Centre, New Delhi - 110025, India

79 Anson Road, #06-04/06, Singapore 079906

Cambridge University Press is part of the University of Cambridge.

It furthers the University's mission by disseminating knowledge in the pursuit of education, learning and research at the highest international levels of excellence.

www.cambridge.org
Information on this title: www.cambridge.org/9781108460415

© Cambridge University Press 2008

First published 2008
First paperback edition 2018

A catalogue record for this publication is available from the British Library

ISBN 978-0-521-86918-8 Hardback
ISBN 978-1-108-46041-5 Paperback

Contents

Contributors

S. Ahmed
National Geophysical Research Institute,
Indo-French Centre for Groundwater Research,
Hyderabad, India

B. F. W. Croke
Integrated Catchment Assessment and
Management Centre, Centre for Resource and Environmental
Studies, The Australian National University,
Canberra, Australia

G. de Marsily
Université Pierre et Marie Curie – Paris VI,
UMR CNRS Sisyphe,
Paris, France

D. C. Goodrich
USDA Agricultural Research Service,
Southwest Watershed Research Center,
Tucson, Arizona, USA

L. E. Hay
USGS, WRD,
Denver, Colorado, USA

Y. Hong
Department of Civil and Environmental Engineering,
University of California,
Irvine, California, USA

K.-L. Hsu
Department of Civil and Environmental Engineering
University of California, Irvine, California, USA

D. A., Hughes
Institute for Water Research, Rhodes University,
Grahamstown, South Africa

B. Imam
Department of Civil and Environmental Engineering,
University of California, Irvine, California, USA

A. J. Jakeman
Integrated Catchment Assessment and Management Centre,
Centre for Resource and Environmental Studies,
The Australian National University,
Canberra, Australia

G. Leavesley
USGS, WRD,
Denver, Colorado, USA

E. Ledoux
Ecole Nationale Supérieure des Mines de Paris,
UMR CNRS Sisyphe, Fontainebleau, France

J.-C. Maréchal
Bureau de Recherches Géologiques et Minières,
Montpellier, France

S. L. Markstrom
USGS, WRD,
Denver, Colorado, USA

N. McIntyre
Department of Civil & Environmental Engineering,
Imperial College,
London, UK

S. N. Miller
University of Wyoming,
Department of Natural Resources,
Laramie, Wyoming, USA

D. J. Semmens
USEPA/ORD/NERL,
Landscape Ecology Branch,
Las Vegas, Nevada, USA

K. D. Sharma
National Institute of Hydrology,
Jalvigyan Bhawan,
Roorkee, India

R. D. Singh
National Institute of Hydrology
Jalvigyan Bhawan,
Roorkee, India

R. E. Smith
USDA Agricultural Research Service,
Fort Collins, Colorado, USA

S. Sorooshian
Center for Hydrometeorology and Remote Sensing (CHRS),
Department of Civil and Environmental Engineering,
University of California,
Irvine, California, USA

C. L. Unkrich
USDA Agricultural Research Service,
Southwest Watershed Research Center,
Tucson, Arizona, USA

R. J. Viger
USGS, WRD,
Denver, Colorado, USA

T. Wagener
Civil and Environmental Engineering,
The Pennsylvania State University,
University Park, Pennsylvania, USA

H. S. Wheater
Department of Civil and Environmental Engineering,
Imperial College,
London, UK

D. A. Woolhiser
USDA Agricultural Research Service,
Fort Collins, Colorado, USA

P. Young
Centre for Research on Environmental Systems & Statistics,
CRES/IEBS,
Lancaster University, Lancaster, UK

Preface

This book is the product of an international workshop supported by UNESCO and co-sponsors under the G-WADI initiative. G-WADI is UNESCO's Global Network for Water and Development Information for Arid Lands. It has the strategic objective of strengthening global capacity to manage water resources in arid and semi-arid areas and seeks to provide a global forum for the exchange of experience, information, and tools. Its specific objectives include:

- improved understanding of the special characteristics of hydrological systems and water management needs in arid areas;
- capacity building of individuals and institutions;
- broad dissemination of understanding of water in arid zones to the user community and the public;
- sharing data and exchanging experience to support research and sound water management;
- raising awareness of advanced technologies for data provision, data assimilation, and system analysis;
- promoting integrated basin management and the use of appropriate decision support tools.

Information on G-WADI products and a news-watch service can be found on the G-WADI web-site (www.g-wadi.org).

Hydrological modelling is playing an increasingly important role in the management of catchments with respect to floods, water resources, water quality, and environmental protection. G-WADI identified a particular gap in the information available to support hydrological modelling for arid and semi-arid areas and hence designed an international workshop, bringing together some of the world's leading specialists in data products, modelling and arid-zone hydrology, to provide state-of-the-art material to workshop participants from arid regions world-wide, including South America, the Middle East, North Africa, Southern Africa, Australia and, particularly, Asia, where the workshop was hosted. This book is a product of that workshop, held in Roorkee, India in March 2005, and the material, comprising state-of-the-art reviews

and case studies, is intended to provide insight and tools to help practitioners world-wide. The focus of the workshop was on the modelling of surface water systems, and a specialist workshop on groundwater modelling is planned for 2007. However, in response to workshop requests, a chapter on groundwater modelling is included in this book for completeness.

The structure of the book is as follows:

Chapter 1 provides a review of some of the special hydrological features of arid areas and an introduction to modelling concepts, and Chapter 2 introduces new data products, focusing on satellite-derived estimates of precipitation.

Experience of hydrological modelling in southern Africa is reported in Chapter 3, and in Australia in Chapter 4, together with an introduction to the IHACRES software. In Chapter 5 the USDA KINEROS model – one of the few models specifically designed to represent arid-zone processes – is presented, in its current GIS format, with applications from the arid United States. In Chapter 6 ephemeral flow and sediment modelling is discussed, based on Indian experience. Chapter 7 introduces the USGS Modular Modelling System, which incorporates a varied suite of models and support systems, with applications to North Africa and China, and in Chapter 8 tool-boxes for stochastic analysis are discussed, together with the problem of model regionalization – i.e., the application of models to ungauged catchments.

In Chapters 9 and 10 the focus is on the problem of forecasting floods in real time. The current state-of-the art of time-series models is presented in Chapter 9, with an example from a semi-arid Australian catchment. In Chapter 10 Indian flooding problems are reviewed and the Indian flood forecasting experience is reported, largely based on traditional methods, but rapidly being updated with more modern modelling methods and communications technology.

Issues of groundwater modelling are addressed in Chapter 11, with examples drawn from India, and the book concludes with a summary of web-site access to data products, modelling tools, and tutorials.

Acknowledgements

The editors particularly wish to thank the contributors to the book for their enthusiastic input to both the workshop and the book, the sponsors of the workshop, without whom none of this activity would have been possible, and the workshop attendees, who provided an informed audience and helpful feedback. Financial support for the contributors, regional representatives and international organization was provided by UNESCO's International Hydrology Programme and the UK Government's Department for International Development. We are indebted to the National Institute of Hydrology, Roorkee, who provided local organization and superb hospitality, together with their sister institution IIT Roorkee. Support for participants from the Asian region to attend the workshop was provided by UNESCO's regional offices in Delhi and Tehran.

1 Modelling hydrological processes in arid and semi-arid areas: an introduction to the workshop

H. S. Wheater

1.1 INTRODUCTION

In the arid and semi-arid regions of the world, water resources are limited, and under severe and increasing pressure due to expanding populations, increasing per capita water use and irrigation. Point and diffuse pollution, increasing volumes of industrial and domestic waste, and over-abstraction of groundwater provide a major threat to those scarce resources. Floods are infrequent, but extremely damaging, and the threat from floods to lives and infrastructure is increasing, due to urban development. Ecosystems are fragile, and under threat from groundwater abstractions and the management of surface flows. Added to these pressures is the uncertain threat of climate change. Clearly, effective water management is essential, and this requires appropriate decision support systems, including modelling tools.

Modelling methods have been widely used for over 40 years for a variety of purposes, but almost all modelling tools have been primarily developed for humid area applications. Arid and semi-arid areas have particular challenges that have received little attention. One of the primary aims of this workshop is to bring together world-wide experience and some of the world's leading experts to provide state-of-the-art guidance for modellers of arid and semi-arid systems.

The development of models has gone hand-in-hand with developments in computing power. While event-based models originated in the 1930s and could be used with hand calculation, the first hydrological models for continuous simulation of rainfall-runoff processes emerged in the 1960s, when computing power was sufficient to represent all of the land-phase processes in a simplified, "conceptual" way. Later, in the 1970s and 1980s, increases in power enabled "physically based" hydrological models to be developed, solving a coupled set of partial differential equations to represent overland, in-stream, and subsurface flow and transport processes, together with evaporation from land and water surfaces. And currently, global climate models are able to represent the global hydrological cycle with simplified physics-based models. In parallel, recent developments in computer power provide the ability to use increasingly powerful methods for the analysis of model performance and to specify the uncertainty associated with hydrological simulations. There have, as a result, been important developments in our understanding of modelling strengths and limitations. The workshop will present a range of modelling approaches and introduce methods of uncertainty analysis.

The relationship between models and data is fundamental to the modelling task. Current technology and computing power can provide powerful pre- and post-processors for hydrological models through Geographic Information Systems, linking with digital data sets to provide a user-friendly modelling environment. Some of these methods will be demonstrated here, and an important issue for discussion is the extent to which such methods are applicable to data-sparse environments, and for countries where the underlying digital data may be hard to obtain. Global developments in remote sensing, coupled with modelling and data assimilation, are providing new sources of information. For example, precipitation estimates for mid-latitudes are now available in near real-time; remote sensing of water body elevation is approaching the point where resolution is useful for real-time hydrological modelling. Again, the workshop will illustrate new data products and discuss their applicability (see Chapter 2 by Sorooshian *et al.*).

This introductory chapter aims to set the scene with a perspective on the strengths and weaknesses of alternative modelling approaches, the special features of arid areas, and the consequent modelling challenges.

1.2 RAINFALL-RUNOFF MODELLING

The book presupposes a basic understanding of modelling, and for those requiring more introductory material, the text book by Beven (2000) provides an excellent introduction, and several

Hydrological Modeling in Arid and Semi-Arid Areas, ed. Howard Wheater, Soroosh Sorooshian, and K. D. Sharma. Published by Cambridge University Press.

recent advanced texts are also available (e.g., Wagener *et al.*, 2004; Duan *et al.*, 2003; Singh and Frevert, 2002a,b.). Nevertheless a brief introduction to modelling terminology and issues is included here, to provide a common framework for subsequent discussion.

A model is a *simplified* representation of a real-world system, and consists of a set of simultaneous equations or a logical set of operations contained within a computer program. Models have *parameters*, which are numerical measures of a property or characteristics that are constant under specified conditions. A *lumped* model is one in which the parameters, inputs, and outputs are spatially averaged and take a single value for the entire catchment. A *distributed* model is one in which parameters, inputs, and outputs vary spatially. A semi-distributed model may adopt a lumped representation for individual subcatchments. A model is *deterministic* if a set of input values will always produce exactly the same output values, and *stochastic* if, because of random components, a set of input values need not produce the same output values. An *event-based* model produces output only for specific time periods, whereas a *continuous* model produces continuous output.

The tasks for which rainfall-runoff models are used are diverse, and the scale of applications ranges from small catchments, of the order of a few hectares, to that of global models. Typical tasks for hydrological simulation models include:

- modelling existing catchments for which input–output data exist, e.g., extension of data series for flood design of water resource evaluation, operational flood forecasting, or water resource management;
- runoff estimation on ungauged basins;
- prediction of effects of catchment change, e.g., land use change, climate change;
- coupled hydrology and geochemistry, e.g., nutrients, acid rain
- coupled hydrology and meteorology, e.g., Global Climate Models

Clearly, the modelling approach adopted will, in general, depend on the required scale of the problem (space-scale and time-scale), the type of catchment, and the modelling task. Some of the tasks pose major challenges, and it is helpful to consider a basic classification of model types, after Wheater *et al.* (1993), and their strengths and weaknesses.

1.2.1 Metric models

At the simplest level, all that is required to reproduce the catchment-scale relationship between storm rainfall and stream response to climatic inputs, is a volumetric loss, to account for processes such as evaporation, soil moisture storage, and ground-water recharge, and a time-distribution function, to represent the various dynamic modes of catchment response. This is the basis of the unit hydrograph method, developed in the 1930s, which, in its basic form, represents the stream response to individual storm events by a non-linear loss function and linear transfer function. The simplicity of the method provides a powerful tool for data analysis. Once a set of assumptions has been adopted (separating fast and slow components of the streamflow hydrograph and allocating rainfall losses), rainfall and streamflow data can be readily analyzed, and a unique model determined.

This analytic capability has been widely used in regional analysis. In the UK, for example, the 1975 Flood Studies Report (NERC, 1975) used data from 138 UK catchments to define regression relationships between the model parameters, and storm and catchment characteristics for the rainfall loss and transfer functions. This lumped, event-based model provides the basic tool for current UK flood design, and, through the regional regression relationships, a capability to model flow on ungauged catchments (the regional relationships were updated in the 1999 Flood Estimation Handbook (Institute of Hydrology, 1999) through the replacement of manual by digital map-based characteristics).

The unit hydrograph is also widely adopted internationally in the form of the US Soil Conservation Service model, available within the US Corps of Engineers HEC1 model. For an application to flood protection in Jordan, see Al-Weshah and El-Khoury (1999). Synthetic unit hydrographs can readily be generated based on default model parameters, which is particularly helpful in data-scarce situations. However, relatively little work has been done to evaluate the associated uncertainty with these estimates.

This data-based approach to hydrological modelling has been defined as metric modelling (Wheater *et al.*, 1993). The essential characteristic of metric models is that they are based primarily on observations and seek to characterise system response from those data. In principle, such models are limited to the range of observed data, and effects such as catchment change cannot be directly represented. In practice, the analytical power of the method has enabled some effects of change to be quantified; the UK regional analysis found the degree of urban development to be an important explanatory variable, and this is used in design to mitigate impacts of urbanization.

The unit hydrograph is a simple, event, model with limited performance capability. However methods of time-series analysis can be used to identify more complex model structures for event or continuous simulation. These are typically based on parallel linear stores, and provide a capability to represent both fast- and slow-flow components of a streamflow hydrograph (see for example Chapter 4 by Croke and Jakeman). These provide a powerful set of tools for use, with updating techniques, in real-time flood forecasting (see Chapter 9 by Young).

1.2.2 Conceptual models

The most common class of hydrological model in general application incorporates prior information in the form of a conceptual representation of the processes perceived to be important. The model form originated in the 1960s, when computing power allowed, for the first time, integrated representation of the terrestrial phase of the hydrological cycle, albeit using simplified relationships, to generate continuous flow sequences. These conceptual models are characterized by parameters that usually have no direct, physically measurable identity. The Stanford Watershed Model (Crawford and Linsley, 1966) is one of the earliest examples, and, with some 16–24 parameters, one of the more complex. To apply these models to a particular catchment, the model must be calibrated, i.e., fitted to an observed data set to obtain an appropriate set of parameter values, using either a manual or automatic procedure. Many of the models presented in the workshop (e.g., by Hughes (Chapter 3), Sharma (Chapter 6), Leavesley *et al.* (Chapter 7), and Wheater *et al.* (Chapter 8)) fall into this category.

The problem arises with this type of model that the information content of the available data is limited, particularly if a single performance criterion (objective function) is used (see Kleissen *et al.* 1990) and hence in calibration the problem of non-identifiability arises, defined by Beven (1993) as "equifinality." For a given model, many combinations of parameter values may give similar performance (for a given performance criterion), as indeed may different model structures. This has given rise to two major limitations. If parameters cannot be uniquely identified, then they cannot be linked to catchment characteristics, and there is a major problem in application to ungauged catchments. Similarly, it is difficult to represent catchment change if the physical significance of parameters is ambiguous.

Developments in computing power, linked to an improved understanding of modelling limitations, have led to some important theoretical and practical developments for conceptual modelling. Firstly, recognizing the problem of parameter ambiguity, appropriate methods to analyze and represent this have been developed. The concept of generalized sensitivity analysis was introduced (Spear and Hornberger, 1980), in which the search for a unique best fit parameter set for a given data set is abandoned; parameter sets are classified as either "behavioral" (consistent with the observed data) or "non-behavioral" according to a defined performance criterion. An extension of this is the generalized likelihood uncertainty estimation (GLUE) procedure (Beven and Binley, 1992; Freer *et al.*, 1996). Using Monte Carlo simulation, parameter values are sampled from the feasible parameter space (conditioned on prior information, as available). Based on a performance criterion, a "likelihood" measure can be evaluated for each simulation. Non-behavioral simulations can be rejected (based on

a pre-selected threshold value), and the remainder assigned rescaled likelihood values. The outputs from the runs can then be weighted and ranked to form a cumulative distribution of output time-series, which can be used to represent the modelling uncertainty. This formal representation of uncertainty is an important development in hydrological modelling practice, although it should be noted that the GLUE procedure lumps together various forms of uncertainty, including data error, model structural uncertainty and parameter uncertainty. More generally, Monte Carlo analysis provides a powerful set of methods for evaluating model structure, parameter identifiability, and uncertainty. For example, in a recent refinement (Wagener *et al.*, 2003a,b), parameter identifiability is evaluated using a moving window to step through the output time-series, thus giving insight into the variability of model performance with time.

A second development is a recognition that much more information is available within an observed flow time-series than is indicated by a single performance criterion, and that different segments of the data contain information of particular relevance to different modes of model performance (Wheater *et al.*, 1986). This has long been recognised in manual model calibration, but has only recently been used in automatic methods. A formal methodology for multi-criterion optimization has been developed for rainfall-runoff modelling (e.g., Gupta *et al.*, 1998; Wagener *et al.*, 2000, 2002). Provision of this additional information reduces the problem of equifinality (although the extent to which this can be achieved is an open research issue), and provides new insights into model performance. For example, if one parameter set is appropriate to maximize peak flow performance, and a different set to maximize low flow performance, this may indicate model structural error, or in particular that different models apply in different ranges. Modelling tool-kits for model building and Monte Carlo analysis are currently available, which include GLUE and other associated tools for analysis of model structure, parameter identifiability, and prediction uncertainty (Lees and Wagener, 1999; Wagener *et al.*, 1999).

An important reason for detailed analysis of model structure and parameter identifiability is to explore the trade-off between identifiability and performance to produce an optimum model (or set of models) for a particular application. Thus for regionalization, the focus would be on maximizing identifiability (i.e., minimizing parameter uncertainty), so that parameters can be related to catchment characteristics.

In several senses, therefore, current approaches to parsimonious conceptual modelling represent an extension of the metric concept (and have thus been termed hybrid metric–conceptual models). There has been a progressive recognition that the 1960s first-generation conceptual models, while seeking a comprehensive and integrated representation of the component processes, are non-identifiable. The current generation of stochastic analysis tools allows detailed investigation of model structure and parameter uncertainty, leading to parameter-efficient models that seek to extract the maximum information from the available data. They also allow formal recognition of uncertainty in model parameters, and provide the capability to produce confidence limits on model simulations.

1.2.3 Physics-based modelling

An alternative approach to hydrological modelling is to seek to develop "physics-based models," i.e., models explicitly based on the best available understanding of the physics of hydrological processes. Such models are based on a continuum representation of catchment processes and the equations of motion of the constituent processes are solved numerically using a grid, of course discretized relatively crudely in catchment-scale applications. They first became feasible in the 1970s when computing power became sufficient to solve the relevant coupled partial differential equations (Freeze and Harlan, 1969; Freeze, 1972). The models are thus characterized by parameters that are, in principle, measurable and have a direct physical significance. An important theoretical advantage is that if the physical parameters can be determined a priori, such models can be applied to ungauged catchments, and the effects of catchment change can be explicitly represented. However, whether this theoretical advantage is achievable in practice is an open question at present.

One of the best known models is the *Système Hydrologique Européen* (SHE) model (Abbott *et al.*, 1986a,b), originally developed as a multi-national European research collaboration. In the UK this has been the subject of progressive development by the University of Newcastle-upon-Tyne, and is known as the SHETRAN model (now including TRANsport of solutes and sediments). A recent description is reported by Ewen *et al.* (2000). The catchment is discretized on a grid square basis for the representation of land surface and subsurface processes, creating a column of finite difference cells, which interact with cells from adjacent columns to represent lateral flow and transport. River networks are modelled as networks of stream links, with flow again represented by finite difference solutions of the governing equations. The resulting model is complex, computationally demanding and data intensive. Ewen *et al.* (2000) note that a one-year simulation typically has a two hour run time on an advanced UNIX system.

In practice two fundamental problems arise with such models. The underlying physics has been (necessarily) derived from small-scale, mainly laboratory-based, process observations. Hence, firstly, the processes may not apply under field conditions and at field-scales of interest. There is, for example, numerical evidence that the effects of small-scale heterogeneity may not be captured by effective, spatially aggregated, properties (Binley and Beven,

1989). Secondly, although the parameters may be measurable at small-scale, they may not be measurable at the scales of interest for application. An obvious example of both is the representation of soil water flow at hillslope-scale. Field soils are characterized by great heterogeneity and complexity. Macropore flow is ubiquitous, yet neglected in physics-based models, for lack of relevant theory and supporting data; the Richards' equation commonly used for unsaturated flow depends on strongly non-linear functional relationships to represent physical properties, for which there is no measurement basis at the spatial-scales of practical modelling interest. And field studies such as those of Pilgrim *et al.* (1978) demonstrate that the dominant modes of process response cannot be specified a priori. For more detailed discussion see, for example, Beven (1989).

There is, therefore, a need for fundamental research to address issues such as the appropriate process representation and parameterization at a given scale. For groundwater flow and transport, significant progress has been made; new theoretical approaches to the representation of heterogeneity have been developed (Dagan, 1986), and stochastic numerical methods have been developed to represent explicitly the uncertainty associated with heterogeneous properties (e.g., Wheater *et al.*, 2000) and to incorporate conditioning on field observations. Extension to the more complex problems of field-scale hydrology is urgently needed, but severely constrained by data availability.

Most of the complexity of physically based models, and the associated problems discussed above, arise from the representation of subsurface flows, and the inherent lack of observability of subsurface properties. The situation often met in arid areas is that overland flow is the dominant runoff mechanism, and surface properties are, in principle, much more readily obtained. It was therefore argued by Woolhiser 30 years ago (Woolhiser, 1971), that it is in this environment that physics-based models are most likely to be successful. The well-known KINEROS model is an outstanding example, and is presented in its latest form in Chapter 5 by Semmens *et al.*

1.3 HYDROLOGICAL PROCESSES IN ARID AREAS

Despite the critical importance of water in arid and semi-arid areas, hydrological data have historically been severely limited. It has been widely stated that the major limitation of the development of arid-zone hydrology is the lack of high quality observations (McMahon, 1979; Nemec and Rodier, 1979; Pilgrim *et al.*, 1988). There are many good reasons for this. Populations are usually sparse and economic resources limited; in addition, the climate is harsh and hydrological events infrequent, but damaging. However, in the general absence of reliable long-term data and experimental

Table 1.1 *Summary of Muscat rainfall data (1893–1959)[a]*

Monthly rainfall (mm)	Jan	Feb	Mar	Apr	May	June	July	Aug	Sept	Oct	Nov	Dec
Mean	31.2	19.1	13.1	8.0	0.38	1.31	0.96	0.45	0.0	2.32	7.15	22.0
Standard deviation	38.9	25.1	18.9	20.3	1.42	8.28	4.93	2.09	0.0	7.62	15.1	35.1
Max.	143.0	98.6	70.4	98.3	8.89	64.0	37.1	14.7	0.0	44.5	77.2	171.2
Mean number of raindays	2.03	1.39	1.15	0.73	0.05	0.08	0.10	0.07	0.0	0.13	0.51	1.6
Max. daily fall (mm)	78.7	57.0	57.2	51.3	8.9	61.5	30.0	10.4	0.0	36.8	53.3	57.2
Number of years record	63.0	64	62	63	61	61	60	61	61	60	61	60

[a] After Wheater and Bell, 1983

research, there has been a tendency to rely on humid-zone experience and modelling tools, and data from other regions. At best, such results will be highly inaccurate. At worst, there is a real danger of adopting inappropriate management solutions which ignore the specific features of dryland response.

Despite the general data limitations, there has been some substantial and significant progress in development of national data networks and experimental research. This has given new insights and we can now see with greater clarity the unique features of arid zone hydrological systems and the nature of the dominant hydrological processes. This provides an important opportunity to develop methodologies for flood and water-resource management which are appropriate to the specific hydrological characteristics of arid areas and the associated management needs, and hence to define priorities for research and hydrological data. The aim here is to review this progress and the resulting insights, and to consider some of the implications.

1.3.1 Rainfall

Rainfall is the primary hydrological input, but rainfall in arid and semi-arid areas is commonly characterized by extremely high spatial and temporal variability. The temporal variability of point rainfall is well known. Although most records are of relatively short length, a few are available from the nineteenth century. For example, Table 1.1 presents illustrative data from Muscat (Sultanate of Oman) (Wheater and Bell, 1983), which shows that a wet month is one with one or two raindays. Annual variability is marked and observed daily maxima can exceed annual rainfall totals.

For spatial characteristics, information is much more limited. Until recently, the major source of detailed data has been from the South West USA, most notably the two, relatively small, densely instrumented basins of Walnut Gulch, Arizona ($150 \, km^2$) and Alamogordo Creek, New Mexico ($174 \, km^2$), established in the 1950s (Osborn *et al.*, 1979). The dominant rainfall for these basins is convective; at Walnut Gulch 70% of annual rainfall occurs from purely convective cells, or from convective cells developing along weak, fast-moving cold fronts, and falls in the period

July to September (Osborn and Reynolds, 1963). Raingauge densities were increased at Walnut Gulch to give improved definition of detailed storm structure and are currently better than one per $2 \, km^2$. This has shown highly localized rainfall occurrence, with spatial correlations of storm rainfall of the order of 0.8 at 2 km separation, but close to zero at 15–20 km spacing. Osborn *et al.* (1972) estimated that to observe a correlation of $r^2 = 0.9$, raingauge spacings of 300–500 m would be required.

Recent work has considered some of the implications of the Walnut Gulch data for hydrological modelling. Michaud and Sorooshian (1994) evaluated problems of spatial averaging for rainfall-runoff modelling in the context of flood prediction. Spatial averaging on a 4 km × 4 km pixel basis (consistent with typical weather radar resolution) gave an underestimation of intensity and led to a reduction in simulated runoff of on average 50% of observed peak flows. A sparse network of raingauges (one per $20 \, km^2$), representing a typical density of flash flood warning system, gave errors in simulated peak runoff of 58%. Evidently there are major implications for hydrological practice, and we will return to this issue, below.

The extent to which this extreme spatial variability is characteristic of other arid areas has been uncertain. Anecdotal evidence from the Middle East underlays comments that spatial and temporal variability was extreme (FAO, 1981), but data from south-west Saudi Arabia obtained as part of a five-year intensive study of five basins (Saudi Arabian Dames and Moore, 1988), undertaken on behalf of the Ministry of Agriculture and Water, Riyadh, have provided a quantitative basis for assessment. The five study basins range in area from 456 to $4930 \, km^2$ and are located along the Asir escarpment (Fig. 1.1), three draining to the Red Sea, two to the interior, towards the Rub al Khali. The mountains have elevations of up to 3000 m asl, hence the basins encompass a wide range of altitude, which is matched by a marked gradient in annual rainfall, from 30 to 100 mm on the Red Sea coastal plain to up to 450 mm at elevations in excess of 2000 m asl.

The spatial rainfall distributions are described by Wheater *et al.* (1991a). The extreme spottiness of the rainfall is illustrated for the $2869 \, km^2$ Wadi Yiba by the frequency distributions of the number

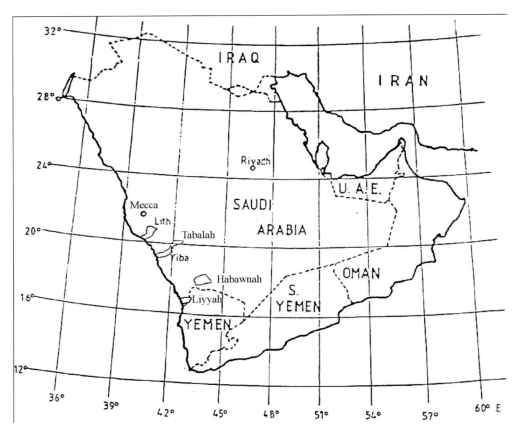

Figure 1.1 Location of Saudi Arabian study basins

of gauges at which rainfall was observed given the occurrence of a catchment rainday (Table 1.2). Typical inter-gauge spacings were 8–10 km, and on 51 % of raindays only one or two raingauges out of 20 experienced rainfall. For the more widespread events, subdaily rainfall showed an even more spotty picture than the daily distribution. An analysis of relative probabilities of rainfall occurrence, defined as the probability of rainfall occurrence for a given hour at Station B given rainfall at Station A, gave a mean value of 0.12 for Wadi Yiba, with only 5 % of values greater that 0.3. The frequency distribution of rainstorm durations shows a typical occurrence of one or two-hour duration point rainfalls, and these tend to occur in mid-late afternoon. Thus rainfall will occur at a few gauges and die out, to be succeeded by rainfall in other locations. This is illustrated for Wadi Lith in Fig. 1.2, which shows the daily rainfall totals for the storm of May 16 1984 (Fig. 1.2a), and the individual hourly depths (Figs. 1.2b–1.2e). In general, the storm patterns appear to be consistent with the results from the south-west USA and area reduction factors were also generally consistent with results from that region (Wheater *et al.*, 1989).

The effects of elevation were investigated, but no clear relationship could be identified for intensity or duration. However, a strong relationship was noted between the frequency of raindays and elevation. It was thus inferred that once rainfall occurred, its point properties were similar over the catchment, but occurrence was more likely at the higher elevations. It is interesting to note that a similar result has emerged from a recent analysis of rainfall in Yemen (UNDP, 1992), in which it was concluded that daily rainfalls observed at any location are effectively samples from a population that is independent of position or altitude.

It is dangerous to generalize from samples of limited record length, but it is clear that most events observed by those networks are characterized by extremely spotty rainfall, so much so that in the Saudi Arabian basins there were examples of wadi flows generated from zero observed rainfall. However, there were also some indications of a small population of more wide-spread rainfalls, which would obviously be of considerable importance in terms of surface flows and recharge. This reinforces the need for long-term monitoring of experimental networks to characterize spatial variability.

For some other arid or semi-arid areas, rainfall patterns may be very different. For example, data from arid New South Wales, Australia have indicated spatially extensive, low intensity rainfalls

Table 1.2 *Wadi Yiba raingauge frequencies and associated conditional probabilities for catchment rainday occurrence*

Number of gauges	Occurrence	Probability
1	88	0.372
2	33	0.141
3	25	0.106
4	18	0.076
5	10	0.042
6	11	0.046
7	13	0.055
8	6	0.026
9	7	0.030
10	5	0.021
11	7	0.030
12	5	0.021
13	3	0.013
14	1	0.004
15	1	0.004
16	1	0.005
17	1	0.004
18	1	0.004
19	0	0.0
20	0	0.0
TOTAL	235	1.000

(Cordery *et al.*, 1983), and recent research in the Sahelian zone of Africa has also indicated a predominance of widespread rainfall. This was motivated by concern to develop improved understanding of land-surface processes for climate studies and modelling, which led to a detailed (but relatively short-term) international experimental programme, the HAPEX-Sahel project based on Niamey, Niger (Goutorbe *et al.*, 1997). Although designed to study land surface/atmosphere interactions, rather than as an integrated hydrological study, it has given important information. For example, Lebel *et al.* (1997) and Lebel and Le Barbe (1997) note that a 100 raingauge network was installed and report information on the classification of storm types, spatial and temporal variability of seasonal and event rainfall, and storm movement. Of total seasonal rainfall, 80 % was found to fall as widespread events which covered at least 70 % of the network. The number of gauges allowed the authors to analyze the uncertainty of estimated areal rainfall as a function of gauge spacing and rainfall depth.

Recent work in southern Africa (Andersen *et al.*, 1998; Mocke, 1998) has been concerned with rainfall inputs to hydrological models to investigate the resource potential of the sand rivers of north-east Botswana. Here, annual rainfall is of the order of

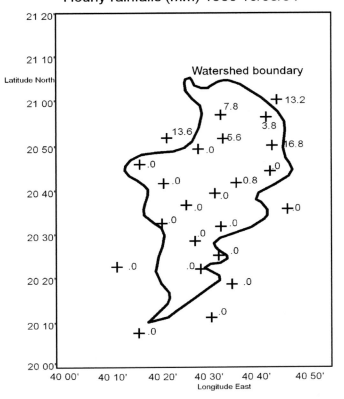

Figure 1.2 (a)–(e) Spatial distribution of daily and hourly rainfall, Wadi Al-Lith

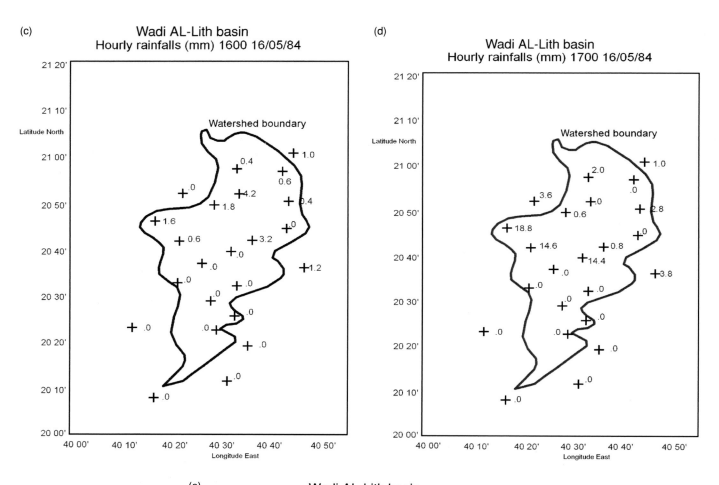

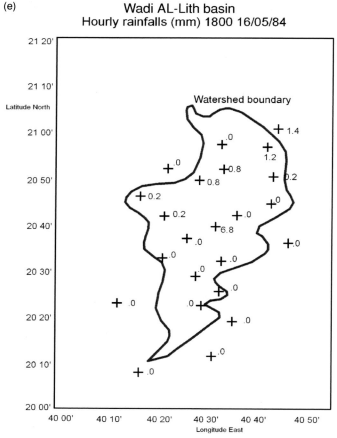

Figure 1.2 (*cont.*)

600 mm, and available rainfall data are spatially sparse, and apparently highly variable, but of poor data quality. Investigation of the representation of spatial rainfall for distributed water resource modelling showed that use of convential methods of spatial weighting of raingauge data, such as Theissen polygons, could give large errors. Large subareas had rainfall defined by a single, possibly inaccurate gauge. A more robust representation resulted from assuming catchment-average rainfall to fall uniformly, but the resulting accuracy of simulation was still poor.

1.3.2 Rainfall-runoff processes

The lack of vegetation cover in arid and semi-arid areas removes protection of the soil from raindrop impact, and soil crusting has been shown to lead to a large reduction in infiltration capacity for bare soil conditions (Morin and Benyamini, 1977). Hence infiltration of catchment soils can be limited. In combination with the high intensity, short duration convective rainfall discussed above, extensive overland flow can be generated. This overland flow, concentrated by the topography, converges on the wadi channel network, with the result that a flood flow is generated. However, the runoff generation process due to convective rainfall is likely to be highly localized in space, reflecting the spottiness of the spatial rainfall fields, and to occur on only part of a catchment, as illustrated above.

Linkage between inter-annual variability of rainfall, vegetation growth, and runoff production may occur. Our modelling in Botswana suggests that runoff production is lower in a year which follows a wet year, due to enhanced vegetation cover, which supports observations reported by Hughes (1995).

Commonly, flood flows move down the channel network as a flood wave, moving over a bed that is either initially dry or has a small initial flow. Hydrographs are typically characterized by extremely rapid rise times, of as little as 15–30 minutes (Fig. 1.3). However, losses from the flood hydrograph through bed infiltration are an important factor in reducing the flood volume as the flood moves downstream. These transmission losses dissipate the flood, and obscure the interpretation of observed hydrographs. It is not uncommon for no flood to be observed at a gauging station, when further upstream a flood has been generated and lost to bed infiltration.

As noted above, the spotty spatial rainfall patterns observed in Arizona and Saudi Arabia are extremely difficult, if not impossible, to quantify using conventional densities of raingauge network. This, taken in conjunction with the flood transmission losses, means that conventional analysis of rainfall–runoff relationships is problematic, to say the least. Wheater and Brown (1989) present an analysis of Wadi Ghat, a 597 km^2 subcatchment of Wadi Yiba, one of the Saudi Arabian basins discussed above. Areal rainfall was estimated from five raingauges and a classical unit hydrograph analysis was undertaken. A striking illustration of the ambig-

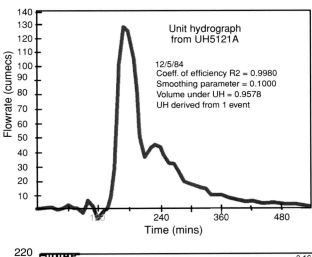

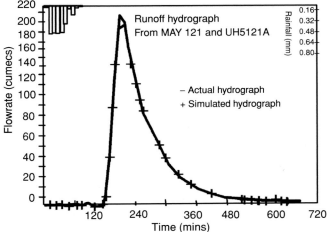

Figure 1.3 Surface water hydrographs, Wadi Ghat May 12, 1984: observed hydrograph and unit hydrograph simulation

uity in observed relationships is the relationship between observed rainfall depth and runoff volume (Fig. 1.4). Runoff coefficients ranged from 5.9 to 79.8 %, and the greatest runoff volume was apparently generated by the smallest observed rainfall! Goodrich *et al.* (1997) show that the combined effects of limited storm areal coverage and transmission loss give important differences from more humid regions. Whereas generally basins in more humid climates show increasing linearity with increasing scale, the response of Walnut Gulch becomes more non-linear with increasing scale. It is argued that this will give significant errors in application of rainfall depth–area-frequency relationships beyond the typical area of storm coverage, and that channel routing and transmission loss must be explicitly represented in watershed modelling.

The transmission losses from the surface water system are a major source of potential groundwater recharge. The characteristics of the resulting groundwater resource will depend on the underlying geology, but bed infiltration may generate shallow water tables, within a few metres of the surface, which can sustain

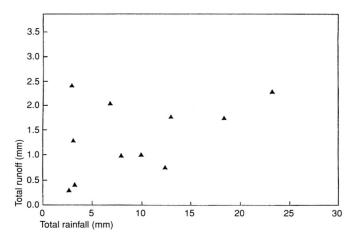

Figure 1.4 Storm runoff as a function of rainfall, Wadi Ghat

supplies to nomadic people for a few months (as in the Hesse of the north of South Yemen), or recharge substantial alluvial aquifers with potential for continuous supply of major towns (as in northern Oman and south-west Saudi Arabia).

The balance between localized recharge from bed infiltration and diffuse recharge from rainfall infiltration of catchment soils will vary greatly depending on local circumstances. However, soil moisture data from Saudi Arabia (Macmillan, 1987) and Arizona (Liu *et al.*, 1995), for example, show that most of the rainfall falling on soils in arid areas is subsequently lost by evaporation. Methods such as the chloride profile method (e.g., Bromley *et al.*, 1997) and isotopic analyses (Allison and Hughes, 1978) have been used to quantify the residual percolation to groundwater in arid and semi-arid areas.

In some circumstances runoff occurs within an internal drainage basin, and fine deposits can support widespread surface ponding. A well-known large-scale example is the Azraq Oasis in north-east Jordan, but small-scale features (Qaas) are widespread in that area. Small-scale examples were found in the HAPEX-Sahel study (Desconnets *et al.*, 1997). Infiltration from these areas is, in general, not well understood, but may be extremely important for aquifer recharge. Desconnets *et al.* report aquifer recharge of between 5 and 20 % of basin precipitation for valley bottom pools, depending on the distribution of annual rainfall.

The characteristics of the channel bed infiltration process are discussed in the following section. However, it is clear that the surface hydrology generating this recharge is complex and extremely difficult to quantify using conventional methods of analysis.

1.3.3 Wadi-bed transmission losses

Wadi bed infiltration has an important effect on flood propagation, but also provides recharge to alluvial aquifers. The balance

between distributed infiltration from rainfall and wadi-bed infiltration is obviously dependant on local conditions, but soil moisture observations from south-west Saudi Arabia imply that, at least for frequent events, distributed infiltration of catchment soils is limited, and that increased near-surface soil moisture levels are subsequently depleted by evaporation. Hence wadi-bed infiltration may be the dominant process of groundwater recharge. As noted above, depending on the local hydrogeology, alluvial groundwater may be a readily accessible water resource. Quantification of transmission loss is thus important, but raises a number of difficulties.

One method of determining the hydraulic properties of the wadi alluvium is to undertake infiltration tests. Infiltrometer experiments give an indication of the saturated hydraulic conductivity of the surface. However, if an infiltration experiment is combined with measurement of the vertical distribution of moisture content, for example using a neutron probe, inverse solution of a numerical model of unsaturated flow can be used to identify the unsaturated hydraulic conductivity relationships and moisture characteristic curves. This is illustrated for the Saudi Arabian Five Basins Study by Parissopoulos and Wheater (1992a).

In practice, spatial heterogeneity will introduce major difficulties to the up-scaling of point profile measurements. The presence of silt lenses within the alluvium was shown to have important effects on surface infiltration as well as subsurface redistribution (Parissopoulos and Wheater, 1990), and subsurface heterogeneity is difficult and expensive to characterize. In a series of two-dimensional numerical experiments it was shown that "infiltration opportunity time," i.e., the duration and spatial extent of surface wetting, was more important than high flow stage in influencing infiltration, that significant reductions in infiltration occured once hydraulic connection was made with a water table, and that hysteresis effects were generally small (Parissopoulos and Wheater, 1992b). Also sands and gravels appeared effective in restricting evaporation losses from groundwater (Parissopoulos and Wheater, 1991).

Additional process complexity arises, however. General experience from the Five Basins Study was that wadi alluvium was highly transmissive, yet observed flood propagation indicated significantly lower losses than could be inferred from *in situ* hydraulic properties, even allowing for subsurface heterogeneity. Possible causes are air entrapment, which could restrict infiltration rates, and the unknown effects of bed mobilization and possible pore blockage by the heavy sediment loads transmitted under flood flow conditions.

A commonly observed effect is that in the recession phase of the flow, deposition of a thin (1–2 mm) skin of fine sediment on the wadi bed occurs, which is sufficient to sustain flow over an unsaturated and transmissive wadi bed. Once the flow has ceased, this skin dries and breaks up so that the underlying alluvium is exposed for subsequent flow events. Crerar *et al.* (1988) observed from laboratory experiments that a thin continous silt layer was formed at low velocities. At higher velocities no such layer occurred, as the bed surface was mobilized, but infiltration to the bed was still apparently inhibited. It was suggested that this could be due to clogging of the top layer of sand due to silt in the infiltrating water, or formation of a silt layer below the mobile upper part of the bed.

Further evidence for the heterogeneity of observed response comes from the observations of Hughes and Sami (1992) from a $39.6\,\mathrm{km^2}$ semi-arid catchment in South Africa. Soil moisture was monitored by neutron probe following two flow events. At some locations immediate response (monitored one day later) occurred throughout the profile, at others, an immediate response near-surface was followed by a delayed response at-depth. Away from the inundated area, delayed response, assumed due to lateral sub-surface transmission, occurred after 21 days.

The overall implication of the above observations is that it is not possible at present to extrapolate from *in situ* point profile hydraulic properties to infer transmission losses from wadi channels. However, analysis of observed flood flows at different locations can allow quantification of losses, and studies by Walters (1990) and Jordan (1977), for example, provide evidence that the rate of loss is linearly related to the volume of surface discharge.

For south-west Saudi Arabia, the following relationships were defined:

$$\mathrm{LOSSL} = 4.56 + 0.02216\,\mathrm{UPSQ} - 2034\,\mathrm{SLOPE} + 7.34$$
$$\mathrm{ANTEC}\ (\mathrm{s.e.}\ 4.15),$$
$$\mathrm{LOSSL} = 3.75 \times 10^{-5}\,\mathrm{UPSQ}^{0.821}\,\mathrm{SLOPE}^{-0.865}\,\mathrm{ACWW}^{0.497} \quad (1.1)$$
$$(\mathrm{s.e.}\ 0.146\ \mathrm{log\ units}\ (+34\,\%)),$$
$$\mathrm{LOSSL} = 5.7 \times 10^{-5}\,\mathrm{UPSQ}^{0.968}\,\mathrm{SLOPE}^{-1.049}$$
$$(\mathrm{s.e.}\ 0.184\ \mathrm{log_e\ units}\ (+44\,\%)),$$

where:

LOSSL	= Transmission loss rate $(1000\,\mathrm{m^3/km})$	(O.R. $1.08-87.9$),
UPSQ	= Upstream hydrograph volume $(1000\,\mathrm{m^3})$	(O.R. $69-3744$),
SLOPE	= Slope of reach (m/m)	(O.R. $0.001-0.011$),
ANTEC	= Antecedent moisture index	(O.R. $0.10-1.00$),
ACWW	= Active channel width (m)	(O.R. $25-231$),
and O.R.	= Observed range.	

$$(1.2)$$

However, generalization from limited experience can be misleading. Wheater *et al.* (1997) analyzed transmission losses between two pairs of flow gauges on the Walnut Gulch catchment for a ten-year sequence and found that the simple linear model of transmission loss as proportional to upstream flow was inadequate. Considering the relationship:

$$V_x = V_0(1-\alpha)^x, \quad (1.3)$$

where V_x is flow volume $(\mathrm{m^3})$ at distance x downstream of flow volume V_0 and α represents the proportion of flow lost per unit distance, then α was found to decrease with discharge volume:

$$\alpha = 118.8(V_0)^{-0.71}. \quad (1.4)$$

The events examined had a maximum value of average transmission loss of $4076\,\mathrm{m^3/km}$ in comparison with the estimate of Lane *et al.* (1971) of $4800-6700\,\mathrm{m^3/km}$ as an upper limit of available alluvium storage.

The role of available storage was also discussed by Telvari *et al.* (1998), with reference to the Fowler's Gap catchment in Australia. Runoff plots were used to estimate runoff production as overland flow for a $4\,\mathrm{km^2}$ basin. It was inferred that $7000\,\mathrm{m^3}$ of overland flow becomes transmission loss and that once this alluvial storage is satisfied, approximately two-thirds of overland flow is transmitted downstream.

A similar concept was developed by Andersen *et al.* (1998) at larger scale for the sand rivers of Botswana, which have alluvial beds of $20-200\,\mathrm{m}$ width and $2-20\,\mathrm{m}$ depth. Detailed observations of water table response showed that a single major event after a seven-week dry period was sufficient to fully satisfy available alluvial storage (the river bed reached full saturation within ten hours). No significant drawdown occurred between subsequent events and significant resource potential remained throughout the dry season. It was suggested that two sources of transmission loss could be occurring, direct losses to the bed, limited by available storage, and losses through the banks during flood events.

It can be concluded that transmission loss is complex, that where deep unsaturated alluvial deposits exist, the simple linear model as developed by Jordan (1977) and implicit in the results of Walters (1990), may be applicable, but that where alluvial storage is limited, this must be taken into account.

1.3.4 Groundwater recharge from ephemeral flows

The relationship between wadi-flow transmission losses and groundwater recharge will depend on the underlying geology. The effect of lenses of reduced permeability on the infiltration process has been discussed and illustrated above, but once infiltration has taken place, the alluvium underlying the wadi bed is effective in minimizing evaporation loss through capillary rise (the coarse structure of alluvial deposits minimizes capillary effects). Thus Hellwig (1973), for example, found that dropping the water table below 60 cm in sand with a mean diameter of 0.53 mm effectively prevented evaporation losses, and Sorey and Matlock (1969) reported that measured evaporation rates from streambed sand were lower than those reported for irrigated soils.

Parrisopoulos and Wheater (1991) combined two-dimensional simulation of unsaturated wadi-bed response with Deardorff's (1977) empirical model of bare soil evaporation to show that evaporation losses were not, in general, significant for the water balance or water table response in short-term simulation (i.e., for periods up to ten days). However, the influence of vapor diffusion was not explicitly represented, and long-term losses are not well understood. Andersen *et al.* (1998) show that losses are high when the alluvial aquifer is fully saturated, but are small once the water table drops below the surface.

Sorman and Abdulrazzak (1993) provide an analysis of groundwater rise due to transmission loss for an experimental reach in Wadi Tabalah, south-west Saudi Arabia and estimate that on average 75 % of bed infiltration reaches the water table. There is, in general, little information available to relate flood transmission loss to groundwater recharge, however. The differences between the two are expected to be small, but will depend on residual moisture stored in the unsaturated zone and its subsequent drying characteristics. But if water tables approach the surface, relatively large evaporation losses may occur.

Again, it is tempting to draw over-general conclusions from limited data. In the study of the sand-rivers of Botswana, referred to above, it was expected that recharge of the alluvial river beds would involve complex unsaturated-zone response. In fact, observations showed that the first flood of the wet season was sufficient to fully recharge the alluvial river-bed aquifer. This storage was topped up in subsequent floods, and depleted by evaporation when the water table was near-surface, but in many sections sufficient water remained throughout the dry season to provide adequate sustainable water supplies for rural villages. And, as noted above, Wheater *et al.* (1997) showed for Walnut Gulch and Telvari *et al.* (1998) for Fowler's Gap, that limited river bed storage affected transmission loss. It is evident that surface water/groundwater interactions depend strongly on the local characteristics of the underlying alluvium and the extent of its connection to, or isolation from, other aquifer systems.

Very recent work at Walnut Gulch (Goodrich *et al.*, 2004) has investigated ephemeral channel recharge using a range of experimental methods, combined with modelling. These included a reach-water balance method, including estimates of near-channel evapotranspiration losses, geochemical methods, analysis of changes in groundwater levels and microgravity measurements, and unsaturated zone flow and temperature analyses. The conclusions were that ephemeral channel losses were significant as an input to the underlying regional aquifer, and that the range of methods for recharge estimation agreed within a factor of three (reach-water balance methods giving the higher estimates).

An important requirement for recharge estimation has arisen in connection with the proposal for a repository for high-level nuclear waste at Yucca Mountain, Nevada. Flint *et al.* (2002) review a wide range of methods, including analysis of physical data from unsaturated zone profiles of moisture and heat, environmental tracers, and watershed modelling. The results indicate extreme variability in space and time, with watershed modelling giving a range from zero to several hundred mm/year, depending on spatial location. The high values arise due to flow focusing in ephemeral channels, and subsequent channel-bed infiltration.

1.4 HYDROLOGICAL MODELLING AND THE REPRESENTATION OF RAINFALL

The preceding discussion illustrates some of the particular characteristics of arid areas which place special requirements on hydrological modelling, for example for flood management or water-resources evaluation. One evident area of difficulty is rainfall, especially where convective storms are an important influence. The work of Michaud and Sorooshian (1994) demonstrated the sensitivity of flood-peak simulation to the spatial resolution of rainfall input. This obviously has disturbing implications for flood modelling, particularly where data availability is limited to conventional raingauge densities. Indeed, it appears highly unlikely that suitable raingauge densities will ever be practicable for routine monitoring. However, the availability of 2 km resolution radar data in the USA can provide adequate information and radar could be installed elsewhere for particular applications. Morin *et al.* (1995) report results from a radar located at Ben-Gurion airport in Israel, for example.

One way forward is to develop an understanding of the properties of spatial rainfall based on high density experimental networks and/or radar data, and represent those properties within a spatial rainfall model for more general application. It is likely that this would have to be done within a stochastic modelling framework in which equally likely realizations of spatial rainfall are produced, possibly conditioned by sparse observations.

Some simple empirical first steps in this direction were taken by Wheater *et al.* (1991a,b) for S.W. Saudi Arabia and Wheater *et al.* (1995) for Oman. In the Saudi Arabian studies, as noted earlier, raingauge data were available at approximately 10 km spacing and spatial correlation was low. Hence a multi-variate model was developed, assuming independence of raingauge rainfall. Based on observed distributions, seasonally dependent catchment rain-day occurrence was simulated, dependent on whether the preceding day was wet or dry. The number of gauges experiencing rainfall was then sampled, and the locations selected based on observed occurrences (this allowed for increased frequency of raindays with increased elevation). Finally, start-times, durations, and hourly intensities were generated. Model performance was compared with observations. Rainfall from random selections of raingauges was well reproduced, but when clusters of adjacent gauges were evaluated, a degree of spatial organisation of occurrence was observed, but not simulated. It was evident that a weak degree of correlation was present, which should not be neglected. Hence in extension of this approach to Oman (Wheater *et al.*, 1995), observed spatial distributions were sampled, with satisfactory results.

However, this multi-variate approach suffers from limitations of raingauge density and, in general, a model in continuous space (and continuous time) is desirable. A family of stochastic rainfall models of point rainfall was proposed by Rodriguez-Iturbe *et al.*

Table 1.3 *Performance of the Bartlett–Lewis Rectangular Pulse Model in representing July rainfall at gauge 44, Walnut Gulch*

	Mean	Var	ACF1	ACF2	ACF3	Pwet	Mint	Mno	Mdur
Model	0.103	1.082	0.193	0.048	0.026	0.032	51.17	14.34	1.68
Data	0.100	0.968	0.174	0.040	0.036	0.042	53.71	13.23	2.38

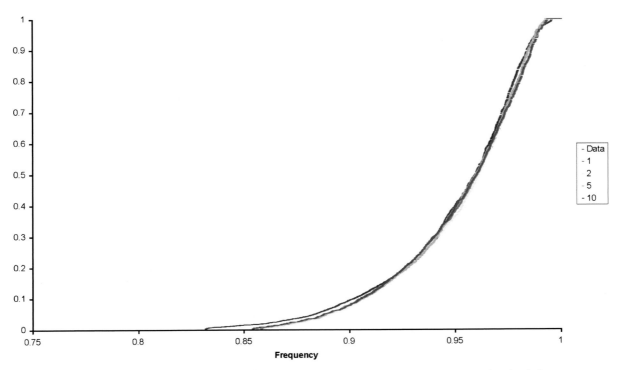

Figure 1.5 (Plate 1) Frequency distribution of spatial coverage of Walnut Gulch rainfall. Observed versus alternative simulations

(1987, 1988) and applied to UK rainfall by Onof and Wheater (1993,1994). The basic concept is that a Poisson process is used to generate the arrival of storms. Associated with a storm is the arrival of raincells, of uniform intensity for a given duration (sampled from specified distributions). The overlapping of these rectangular pulse cells generates the storm intensity profile in time. These models were shown to have generally good performance for the UK in reproducing rainfall properties at different time-scales (from hourly upwards), and extreme values.

Cox and Isham (1988) extended this concept to a model in space and time, whereby the raincells are circular and arrive in space within a storm region. As before, the overlapping of cells produces a complex rainfall intensity profile, now in space as well as time. This model has been developed further by Northrop (1998) to include elliptical cells and storms and is being applied to UK rainfall (Northrop *et al.*, 1999).

Recent work (Samuel, 1999) has been exploring the capability of these models to reproduce the convective rainfall of Walnut

Gulch. In modelling point rainfall, the Bartlett-Lewis Rectangular Pulse Model was generally slightly superior to other model variants tested. Table 1.3 shows representative performance of the model in comparing the hourly statistics from 500 realizations of July rainfall in comparison with 35 years from one of the Walnut Gulch gauges (gauge 44), where Mean is the mean hourly rainfall (mm), Var its variance, ACF1,2,3 the autocorrelations for lags 1,2,3, Pwet the proportion of wet intervals, Mint the mean storm inter-arrival time (h), Mno the mean number of storms per month, and Mdur the mean storm duration (h). This performance is generally encouraging (although the mean storm duration is underestimated), and extreme value performance is excellent.

Work with the spatial–temporal model is still at a preliminary stage, but Fig. 1.5 (Plate l) shows a comparison of observed spatial coverage of rainfall for 25 years of July data from 81 gauges (for different values of the standard deviation of cell radius) and Fig. 1.6 (Plate 2) the corresponding fit for temporal lag-0 spatial correlation. Again, the results are encouraging, and there is promise

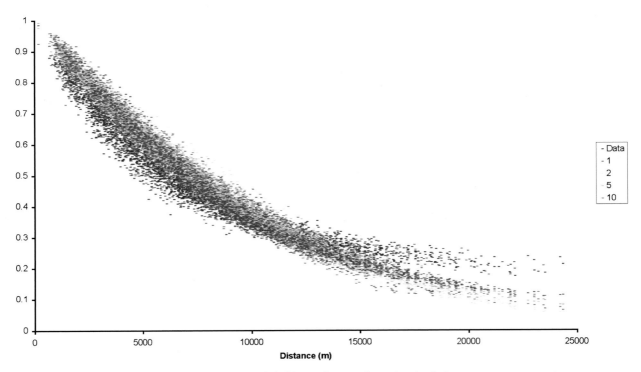

Figure 1.6 (Plate 2) Spatial correlation of Walnut Gulch rainfall. Observed versus alternative simulations

with this approach to address the significant problems of spatial representation for hydrological modelling.

1.5 INTEGRATED MODELLING FOR WATER RESOURCE EVALUATION

Appropriate strategies for water resource development must recognize the essential physical characteristics of the hydrological processes. Surface water storage, although subject to high evaporation losses, is widely used, although temporal variability of flows must be adequately represented to define long term yields. It can be noted that in some regions, for example, the northern areas of southern Yemen, small-scale storage has been developed as an appropriate method to maximize the available resource from spatially localized rainfall. Numbers of small storage units have been developed, some of which fill from localized rainfall. These then provide a short-term resource for a nomadic family and its livestock.

Groundwater is a resource particularly well suited to arid regions. Subsurface storage minimizes evaporation loss and can provide long-term yields from infrequent recharge events. The recharge of alluvial groundwater systems by ephemeral flows can provide an appropriate resource, and this has been widely recognised by traditional development, such as the "*afalaj*" of Oman and elsewhere. There may, however, be opportunities for augmenting

recharge and more effectively managing these groundwater systems. In any case, it is essential to quantify the sustainable yield of such systems, for appropriate resource development.

It has been seen that observations of surface flow do not define the available resource, and similarly observed groundwater response does not necessarily indicate upstream recharge. Figure 1.7 presents a series of groundwater responses from 1985/86 for Wadi Tabalah which shows a downstream sequence of wells 3-B-96, -97, -98, -99 and -100 and associated surface water discharges. It can be seen that there is little evidence of the upstream recharge at the downstream monitoring point.

In addition, records of surface flows and groundwater levels, coupled with ill-defined histories of abstraction, are generally insufficient to define long term variability of the available resource.

To capture the variability of rainfall and the effects of transmission loss on surface flows, a distributed approach is necessary. If groundwater is to be included, integrated modelling of surface water and groundwater is needed. Distributed surface-water models include KINEROS (Wheater and Bell, 1983; Michaud and Sorooshian, 1994) and the model of Sharma (1997, 1998). A distributed approach to the integrated modelling of surface and groundwater response following Wheater *et al.* (1995) is illustrated in Fig. 1.8. This requires the characterization of the spatial and temporal variability of rainfall, distributed infiltration, runoff generation, flow transmission losses, and the ensuing groundwater

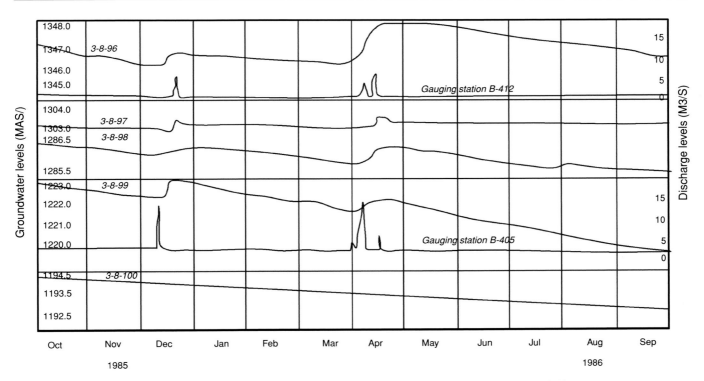

Figure 1.7 Longitudinal sequence of wadi alluvium well hydrographs and associated surface flows, Wadi Tabalah, 1985/6

recharge and groundwater response. This presents some technical difficulties, although the integration of surface and groundwater modelling allows maximum use to be made of available information, so that, for example, groundwater response can feed back information to constrain surface hydrological parameterization. It does, however, provide the only feasible method of exploring the internal response of a catchment to management options.

In a recent application, this integrated modelling approach was developed for Wadi Ghulaji, Sultanate of Oman, to evaluate options for groundwater recharge management (Wheater *et al.*, 1995). The catchment, of area 758 km², drains the southern slopes of Jebal Hajar in the Sharqiyah region of Northern Oman. Proposals to be evaluated included recharge dams to attenuate surface flows and provide managed groundwater recharge in key locations. The modelling framework involved the coupling of a distributed rainfall model, a distributed water-balance model (incorporating rainfall-runoff processes, soil infiltration and wadi flow transmission losses), and a distributed groundwater model (Fig. 1.9).

The representation of rainfall spatial variability presents technical difficulties, since data are limited. Detailed analysis was undertaken of 19 rain gauges in the Sharqiyah region, and of six raingauges in the catchment itself. A stochastic multi-variate temporal–spatial model was devised for daily rainfall, a modified version of a scheme orgininally developed by Wheater *et al.*

(1991a,b). The occurrence of catchment rainfall was determined according to a seasonally variable first-order Markov process, conditioned on rainfall occurrence from the previous day. The number and locations of active raingauges and the gauge depths were derived by random sampling from observed distributions.

The distributed water-balance model represents the catchment as a network of two-dimensional plane and linear channel elements. Runoff and infiltration from the planes was simulated using the SCS approach. Wadi flows incorporate a linear transmission-loss algorithm based on work by Jordan (1977) and Walters (1990). Distributed calibration parameters are shown in Fig. 1.10.

Finally, a groundwater model was developed based on a detailed hydrogeological investigation which led to a multi-layer representation of uncemented gravels, weakly/strongly cemented gravels and strongly cemented/fissured gravel/bedrock, using MOD-FLOW.

The model was calibrated to the limited flow data available (a single event) (Table 1.4), and was able to reproduce the distribution of runoff and groundwater recharge within the catchment through a rational association on loss parameters with topography, geology, and wadi characteristics. Extended synthetic data sequences were then run to investigate catchment water balances under scenarios of different runoff exceedance probabilities (20 %, 50 %, 80 %), as in Table 1.4, and to investigate management options.

Table 1.4 *Annual catchment water balance, simulated scenarios*

Scenario	Rainfall	Evaporation	Groundwater recharge	Runoff	% Runoff
Wet	88	0.372	12.8	4.0	4.6
Average	33	0.141	11.2	3.5	4.0
Dry	25	0.106	5.5	1.7	3.2

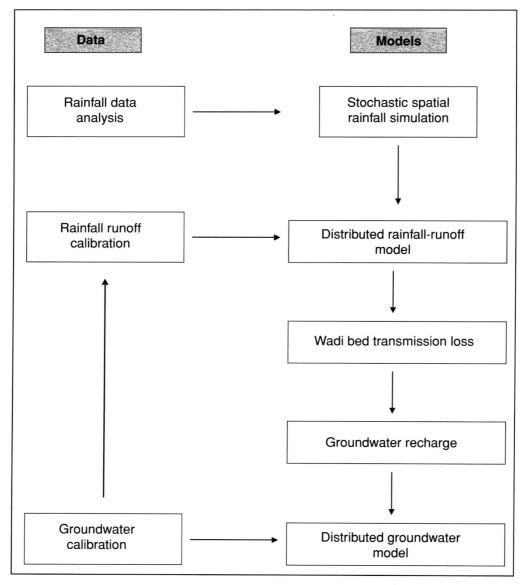

Figure 1.8 Integrated modelling strategy for water resource evaluation

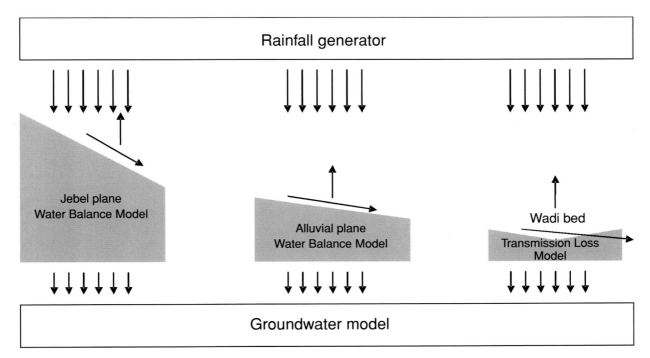

Figure 1.9 Schematic of the distributed water-resource model

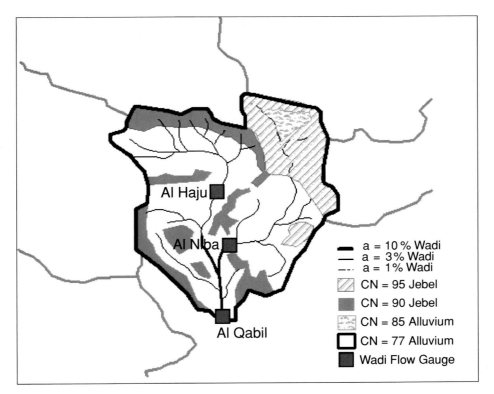

Figure 1.10 Distributed calibration parameters, water balance model, Wadi Ghulaji

1.6 CONCLUSIONS

It has been shown that, for many applications, the hydrological characteristics of arid areas present severe problems for conventional methods of analysis. Recent data are providing new insights. These insights must be used as the basis for development of more appropriate methods for flood design and water-resource evaluation and, in turn, to define data needs and research priorities. Much high-quality research is needed, particularly to investigate processes such as spatial rainfall, and infiltration and groundwater recharge from ephemeral flows.

For developments to maximize the resource potential, define long-term sustainable yields and protect traditional sources, it is argued that distributed modelling is a valuable, if not essential tool. However, this confronts severe problems of characterization of rainfall, rainfall-runoff processes, and groundwater recharge, and of understanding the detailed hydrogeological response of what are often complex groundwater systems. Similarly, new approaches to flood design and management are required which represent the extreme value characteristics of arid areas and recognize the severe problems of conventional rainfall-runoff analysis.

Above all, basic requirements are for high quality data of rainfall, surface-water flows and groundwater response to support regional analyses and the development of appropriate methodologies. Too often, studies focus on either surface or subsurface response without taking an integrated view. Too often, networks are reduced after a few years without recognition that the essential variability of wadi response can only be characterized by relatively long records. Quality control of data is vital, but can easily be lost sight of with ready access to computerized databases.

Superimposed on these basic data needs is the requirement for specific process studies, including sediment transport, surface water/groundwater interactions in the active wadi channel, evaporation processes and consumptive use of wadi vegetation, and the wider issues of groundwater recharge. These are challenging studies, with particularly challenging logistical problems, and require the full range of advanced hydrological experimental methods to be applied, particularly integrating quantity and quality data to deduce system responses, and making full use of remote sensing and geophysical methods to characterize system properties.

It must not be forgotten that, in general, data networks are under threat world-wide, and a major priority for hydrologists must be to promote recognition of the value of data for water management, the importance of long records in a region characterized by high inter-annual variability, and of the particular technical and logistical difficulties in capturing hydrological response in arid areas. The current International Hydrological Programme rightly prioritizes hydrological data as the essential foundation for effective management. The results of both detailed research and regional analyses are required for the essential understanding of wadi hydrology which must underlie effective management.

1.7 REFERENCES

Abbott, M. B., Bathurst, J. C., Cunge, J. A., O'Connell, P. E., and Rasmussen, J. (1986a). An introduction to the European Hydrological System – Système Hydrologique Européen, SHE. 1. History and philosophy of a physically based, distributed modelling system. *J. Hydrol.*, **87**, 45–59.

Abbott, M. B., Bathurst, J. C., Cunge, J. A., O'Connell, P. E., and Rasmussen, J. (1986b). An introduction to the European Hydrological System – Système Hydrologique Européen, SHE. 2. Structure of a physically-based, distributed modelling system. *J. Hydrol.*, **87**, 61–77.

Allison, G. B. and Hughes, M. W. (1978). The use of environmental chloride and tritium to estimate total recharge to an unconfined aquifer. *Aust. J. Soil Res.*, **16**, 181–95.

Al-Weshah, R. A. and El-Khoury, F. (1999). Flood analysis and mitigation for Petra area in Jordan. *ASCE Journal of Water Resources Planning and Management*, **May/June**, 170–77.

Andersen, N. J., Wheater, H. S., Timmis, A. J. H., and Gaongalelwe, D. (1998). Sustainable development of alluvial groundwater in sand rivers of Botswana. In *Hydrology in a Changing Environment*, vol. II, ed. Howard Wheater and Celia Kirby. Chichester: Wiley, 367–76.

Beven, K. J. (1989). Changing ideas in hydrology: the case of physically based models. *J. Hydrol.*, **105**, 157–72.

Beven, K. J. (1993). Prophecy, reality and uncertainty in distributed hydrological modelling. *Adv. Water Resourc.*, **16**, 41–51.

Beven, K. J. (2000). *Rainfall-Runoff Modelling – the Primer*. Chichester: Wiley.

Beven, K. J. and Binley, A. M. (1992). The future of distributed models: model calibration and predictive uncertainty. *Hydrol. Processes*, **6**, 279–98.

Binley, A. M. and Beven, K. J. (1989). A physically based model of heterogeneous hillslopes. 2: Effective hydraulic conductivities. *Water Resour. Res.*, **25(6)** 1227–33.

Bromley, J., Edmunds, W. M., Fellman E. *et al.* (1997). Estimation of rainfall inputs and direct recharge to the deep unsaturated zone of southern Niger using the chloride profile method. *J. Hydrol.*, **188–189**, 139–54.

Cordery, I., Pilgrim, D. H., and Doran, D. G. (1983). Some hydrological characteristics of arid western New South Wales. *The Institution of Engineers, Australia, Hydrology and Water Resources Symp.* Nov, 287–92.

Cox, D. R. and Isham, V. (1988). A simple spatial–temporal model of rainfall, *Proc. Roy. Soc.*, **A415**, 317–28.

Crawford, N. H. and Linsley, R. K. (1966). *Digital simulation in hydrology: Stanford Watershed Model IV*. Tech. Rpt. 39, Stanford University, California.

Crerar, S., Fry, R. G., Slater, P. M., van Langenhove, G. and Wheeler, D. (1988). An unexpected factor affecting recharge from ephemeral river flows in SWA/Namibia. In *Estimation of Natural Groundwater Recharge*, ed. Simmers, I. Dordrecht: D. Reidel Publishing Company, 11–28.

Dagan, G. (1986). Statistical theory of groundwater flow and transport: pore to laboratory; laboratory to formation and formation to regional scale. *Water Resour. Res.*, **22**, 120–35.

Deardorff, J. W. (1977). A parameterization of ground-surface moisture content for use in atmospheric prediction models. *J. Applied. Meteorol.*, **16**, 1182–5.

Desconnets, J. C., Taupin, J. D., Lebel, T., and Leduc, C. (1997). Hydrology of the HAPEX-Sahel Central Super-Site: surface water drainage and aquifer recharge through the pool systems. *J. Hydrol.*, **188–189**, 155–178.

Duan, Q., Gupta, H. V., Sorooshian, S., Rousseau, A. N., and Turcotte, R. (2003). Calibration of watershed models. In *Water Science and Application*, Vol. 6, ed. Qingyun, Duan, Gupta, H. V., Sorooshian, S., Rousseau, A. N., and Turcotte, R., Washington, DC: American Geophysical Union.

Ewen, J., Parkin, G., and O'Connell, P. E. (2000). SHETRAN: distributed river basin flow and transport modeling system. *J. Hydrol. Eng.*, **July**, 250–8.

Flint, A. L., Flint, L. E., Kwicklis, E. M., Fabryka-Martin, J. T., and Bodvarsson, G. S. (2002). Estimating recharge at Yucca Mountain, Nevada, USA: comparison of methods. *Hydro. J.*, **10**, 180–204.

Food and Agriculture Organization (FAO) (1981). *Arid zone hydrology for agricultural development. Irrig. Drain. Pap. 37.* Rome: FAO.

Freer, J., Beven, K., and Abroise, B. (1996). Bayesian uncertainty in runoff prediction and the value of data: an application of the GLUE approach. *Water Resour. Res.*, **32**, 2163–73.

Freeze, R. A. (1972). Role of subsurface flow in generating surface runoff. 2: Upstream source areas. *Water Resour. Res.*, **8(5)** 1272–83.

Freeze, R. A. and Harlan, R. L. (1969). Blueprint for a physically based, digitally simulated hydrologic response model. *J. Hydrol.*, **9**, 237–58.

Goodrich, D. C., Lane, L. J., Shillito, R. M. *et al.* (1997). Linearity of basin response as a function of scale in a semi-arid watershed. *Water Resour. Res.*, **33(12)**, 2951–65.

Goodrich, D. C., Williams, D. G., Unkrich, C. L. *et al.* (2004). Comparison of methods to estimate ephemeral channel recharge, Walnut Gulch, San Pedro River Basin, Arizona. In *Groundwater Recharge in a Desert Environment: The Southwestern United States*, ed. J. F. Hogan, F. M. Phillips, and B. R. Scanlon. Water Science and Application 9. Washington, DC: American Geophysical Union, 77–99.

Goutorbe, J. P., Dolman, A. J., Gash, J. H. C., Kerr, Y. H., Lebel, T., Prince S. D., and Stricker, J. N. M. (eds.) (1997). *HAPEX-Sahel.* The Netherlands: Elsevier (reprinted from *J. Hydrol.* **188–89/1–4**).

Gupta, H. V., Sorooshian, S., and Yapo, P. O. (1998). Towards improved calibration of hydrological models: multiple and non-commensurable measures of information. *Water Resour. Res.* **34(4)**, 751–63.

Hellwig, D. H. R. (1973). Evaporation of water from sand. 3: The loss of water into the atmosphere from a sandy river bed under arid climatic condtions. *J. Hydrol.*, **18**, 305–16.

Hughes, D. A. (1995). Monthly rainfall-runoff models applied to arid and semi-arid catchments for water resource estimation purposes. *Hydr. Sci. J.*, **40(6)**, 751–69.

Hughes, D. A. and Sami, K., (1992). Transmission losses to alluvium and associated moisture dynamics in a semiarid ephemeral channel system in Southern Africa. *Hydrol. Proc.*, **6**, 45–53.

Institute of Hydrology (1999). *Flood Estimation Handbook*, 5 vols. Wallingford UK: Institute of Hydrology.

Jordan, P. R. (1977). Streamflow transmission losses in Western Kansas. *J. Hydr. Div., Am. Soc. Civ. Eng.*, **108**, HY8, 905–19.

Kleissen, F. M., Beck, M. B., and Wheater, H. S. (1990). Identifiability of conceptual hydrochemical models. *Water Resour. Res.*, **26**(2), 2979–92.

Lane, L. J., Diskin, M. H., and Renard, K. G. (1971). Input–output relationships for an ephemeral stream channel system. *J. Hydrol.*, **13**, 22–40.

Lebel, T., Taupin, J. D., and D'Amato, N. (1997). Rainfall monitoring during HAPEX-Sahel. 1: General rainfall conditions and climatology. *J. Hydrol.* **188–9**, 74–96.

Lebel, T. and Le Barbe, L. (1997). Rainfall monitoring during HAPEX-Sahel. 2: Point and areal estimation at the event and seasonal scales. *J. Hydrol.*, **188–189**, 97–122.

Lees, M. J. and Wagener, T. (1999). *A Monte Carlo Analysis Toolbox (MCAT) for Matlab.* User manual. Imperial College, London.

Liu, B., Phillips, F., Hoines, S., Campbell, A. R., and Sharma, P. (1995). Water movement in desert soil traced by hydrogen and oxygen isotopes, chloride, and chlorine-36, southern Arizona. *J. Hydrol.* **168**, 91–110.

Macmillan, L. C. (1987). *Regional evaporation and soil moisture analysis for irrigation application in arid areas.* University of London, MSc thesis, Dept of Civil Engineering, Imperial College.

Mocke, R. (1998). *Modelling the sand-rivers of Botswana: distributed modelling of runoff and groundwater recharge processes to assess the sustainability of rural water supplies.* University of London, MSc thesis, Dept. of Civil Engineering, Imperial College.

Michaud, J. D. and Sorooshian, S. (1994). Effect of rainfall-sampling errors on simulations of desert flash floods. *Water Resour. Res.*, **30(10)**. 2765–75.

Morin, J. and Benyamini, Y. (1977). Rainfall infiltration into bare soils. *Water Resour. Res.*, **13(5)**, 813–17.

Morin, J., Rosenfeld, D., and Amitai, E. (1995). Radar rain field evaluation and possible use of its high temporal and spatial resolution for hydrological purposes. *J. Hydrol.*, **172**, 275–92.

McMahon, T. A. (1979). Hydrological characteristics of arid zones. *Proc. Symposium on the Hydrology of Areas of Low Precipitation,* Canberra, IAHS Publ. No. 128, 105–23.

Natural Environment Research Council (NERC) (1975). *Flood Studies Report,* 5 vols. Wallingford, UK: NERC.

Nemec, J. and Rodier, J. A. (1979). Streamflow characteristics in areas of low precipitation. *Proc. Symposium on the Hydrology of Areas of Low Precipitation.* Canberra: IAHS Publ. No. 128, 125–40.

Northrop, P. J. (1998). A clustered spatial–temporal model of rainfall. *Proc. Roy. Soc.*, **A454**, 1875–88.

Northrop, P. J., Chandler, R. E., Isham, V. S., Onof, C., and Wheater, H. S. (1999). Spatial–temporal stochastic rainfall modelling for hydrological design. In *Hydrological Extremes: Understanding, Predicting, Mitigating,* ed. Gottschalk, L., Olivry, J.-C., Reed, D., and Rosbjerg, D. IAHS Publn. No. 255. Wallingford, UK: IAHS Press, 225–35.

Onof, C. and Wheater, H. S. (1993). Modelling of British rainfall using a random parameter Bartlett–Lewis rectangular pulse model, *J. Hydrol.*, **149**, 67–95.

Onof, C. and Wheater, H. S. (1994). Improvements of the modelling of British rainfall using a modified random parameter Bartlett–Lewis rectangular pulse model, *J. Hydrol.*, **157** 177–95.

Osborn, H. B. and Reynolds, W. N. (1963). Convective storm patterns in the southwestern United States. *Bull. IASH*, **8(3)**, 71–83.

Osborn, H. B., Renard, K. G., and Simanton, J. R., (1979). Dense networks to measure convective rainfalls in the southwestern United States. *Water Resour. Res.*, **15(6)**, 1701–11.

Osborn, H. B., Lane, L. J., and Hondley, J. F. (1972). Optimum gaging of thunderstorm rainfall in southeastern Arizona. *Water Resour. Res.*, **8(1)**, 259–65.

Parissopoulos, G. A. and Wheater, H. S. (1990). Numerical study of the effects of layers on unsaturated-saturated two-dimensional flow. *Water Resour. Mgmt.* **4**, 97–122.

Parissopoulos, G. A. and Wheater, H. S. (1991). Effects of evaporation on groundwater recharge from ephemeral flows. In *Advances in Water Resources Technology*, ed. G. Tsakiris. The Netherlands: A. A. Balkema. 235–45.

Parrisopoulos, G. A. and Wheater, H. S. (1992a). Experimental and numerical infiltration studies in a wadi stream-bed. *J. Hydr. Sci.*, **37**, 27–37.

Parissopoulos, G. A. and Wheater H. S. (1992b). Effects of hysteresis on groundwater recharge from ephemeral flows, *Water Resour. Res.*, **28(11)**, 3055–61.

Pilgrim, D. H., Chapman, T. G., and Doran, D. G. (1988). Problems of rainfall-runoff modelling in arid and semi-arid regions. *Hydrol. Sci. J.*, **33(4)**, 379–400.

Pilgrim, D. H., Huff, D. D., and Steele, T. D. (1978). A field evaluation of subsurface and surface runoff. *J. Hydrol.*, **38**, 319–41.

Rodriguez-Iturbe, I., Cox, D. R., and Isham, V. (1987). Some models for rainfall based on stochastic point processes. *Proc. Roy. Soc.*, **A410**, 269–88.

Rodriguez-Iturbe, I., Cox, D. R., and Isham, V. (1988). A point process model for rainfall: further developments. *Proc. Roy. Soc.*, **A417**, 283–98.

Samuel, C. R. (1999). Stochastic rainfall modelling of convective storms in Walnut Gulch, Arizona. University of London Ph.D. Thesis.

Saudi Arabian Dames and Moore (1988). *Representative Basins Study.* Final Report to Ministry of Agriculture and Water, Riyadh, 84 vols.

Sharma, K. D. (1997). Integrated and sustainable development of water resources of the Luni basin in the Indian arid zone. In *Sustainability of Water Resources under Increasing Uncertainty*, IAHS Publn. No.240. Wallingford, UK: IAHS Press, 385–93.

Sharma, K. D. (1998). Resource assessment and holistic management of the Luni River Basin in the Indian desert. In *Hydrology in a Changing Environment*, Vol II, eds. H. Wheater and C. Kirby. Chichester: Wiley, 387–95.

Singh, V. P. and Frevert, D. K. (eds.) (2002a). *Mathematical Models of Large Watershed Hydrology.* Highlands Ranch, CO: Water Resources Publications, LLC.

Singh, V. P. and Frevert, D. K. (eds.) (2002b). *Mathematical Models of Small Watershed Hydrology and Applications.* Highlands Ranch, CO: Water Resources Publications, LLC.

Sorey, M. L. and Matlock, W. G. (1969). Evaporation from an ephemeral streambed. *J. Hydraul. Div., Am. Soc. Civ. Eng.*, **95**, 423–38.

Sorman, A. U. and Abdulrazzak, M. J. (1993). Infiltration – recharge through wadi beds in arid regions. *Hydr. Sci. J.*, **38(3)**, 173–86.

Spear, R. C. and Hornberger, G. M. (1980). Eutrophication in Peel inlet, II: Identification of critical uncertainties via generalised sensitivity analysis. *Water Resour. Res.*, **14**, 43–9.

Telvari, A., Cordery, I. and Pilgrim, D. H. (1998). Relations between transmission losses and bed alluvium in an Australian arid zone stream. In *Hydrology in a Changing Environment*, ed. H. Wheater and C. Kirby. Chichester: Wiley, Vol II, 361–66.

Travers Morgan (1993). Detailed hydrotechnical recharge studies in Wadi Ghulaji. Final Report to Ministry of Water Resources, Oman.

UNDP (1992). *Surface Water Resources*. Final report to the Government of the Republic of Yemen High Water Council UNDP/DESD PROJECT YEM/88/001 Vol III, June.

Wagener, T., Lees, M. J., and Wheater, H. S. (1999). *A Rainfall-Runoff Modelling Toolbox (RRMT) for Matlab*. User manual. Imperial College, London.

Wagener, T., Boyle, D. P., Lees, M. J., Wheater, H. S., Gupta, H. V., and Sorooshian, S. (2000). A framework for the development and application of hydrological models. *Proc. BHS 7th National Symp., Newcastle-upon-Tyne*, 3.75–3.81.

Wagener, T., Lees, M. J., and Wheater, H. S. (2002). A toolkit for the development and application of parsimonious hydrological models. In *Mathematical Models of Small Watershed Hydrology*, Vol. 2, ed. Singh, Frevert, and Meyer. Highlands Ranch, CO: Water Resources Publications LLC.

Wagener, T., McIntyre, N., Lees, M. J., Wheater, H. S., and Gupta, H. V. (2003a). Towards reduced uncertainty in conceptual rainfall-runoff modelling: dynamic identifiability analysis. *Hydrol. Proc.*, **17**, 455–76.

Wagener, T. Wheater, H. S., and Gupta, H. V. (2003b). Identification and evaluation of conceptual rainfall-runoff models. In *Advances in Calibration of Watershed Models*, ed. Duan, Q., Sorooshian, S., Gupta, H. V., Rousseau, A., and Turcotte, R. AGU Monograph Series USA: American Geophysical Union, 29–47.

Wagener, T., Wheater, H. S., and Gupta, H. V. (2004). *Rainfall-Runoff Modelling in Gauged and Ungauged Catchments*. London: Imperial College: Press.

Walters, M. O. (1990). Transmission losses in arid region. *J. Hydr. Eng.*, **116(1)**, 127–38.

Wheater, H. S. and Bell, N. C. (1983). Northern Oman flood study. *Proc. Instn. Civ. Engrs. Part II*, **75**, 453–73.

Wheater, H. S. and Brown, R. P. C. (1989). Limitations of design hydrographs in arid areas: an illustration from southwest Saudi Arabia. *Proc. 2nd Natl. BHS Symp.* 3.49–3.56.

Wheater, H. S., Bishop, K. H., and Beck, M. B. (1986). The identification of conceptual hydrological models for surface water acidification. *J. Hydrol. Proc.* **1**, 89–109.

Wheater, H. S., Larentis, P., and Hamilton, G. S. (1989). Design rainfall characteristics for southwest Saudi Arabia. *Proc. Inst. Civ. Eng., Part 2*, **87**, 517–38.

Wheater, H. S., Butler, A. P., Stewart, E. J., and Hamilton, G. S. (1991a). A multivariate spatial–temporal model of rainfall in S.W. Saudi Arabia. I: Data characteristics and model formulation. *J. Hydrol.*, **125**, 175–99.

Wheater, H. S., Onof, C., Butler, A. P., and Hamilton, G. S. (1991b). A multivariate spatial–temporal model of rainfall in southwest Saudi Arabia. II: Regional analysis and long-term performance. *J. Hydrol.*, **125**, 201–20.

Wheater, H. S., Jakeman, A. J., and Beven, K. J. (1993). Progress and directions in rainfall-runoff modelling. In *Modelling Change in Environmental Systems*, ed. Jakeman, A. J. Beck, M. B. and McAleer, M. J. Chichester: Wiley, 101–32.

Wheater, H. S., Jolley, T. J., and Peach, D. (1995). A water resources simulation model for groundwater recharge studies: an application to Wadi Ghulaji, Sultanate of Oman. In *Proc. Intnl. Conf. on Water Resources Management in Arid Countries*. Sultanate of Oman, Muscat: Ministry of Agriculture and Water, 502–10.

Wheater, H. S., Woods Ballard, B., and Jolley, T. J. (1997). An integrated model of arid zone water resources: evaluation of rainfall-runoff simulation performance. In *Sustainability of Water Resources Under Increasing Uncertainty*, ed., Rosbjerg, D. Bootayeb, N.–E. Gustard, A. Kundzewicz, Z. W. and Rasmusen, P. F. IAHS Pubn. No. 240, pp. 395–405.

Woolhiser, D. A. (1971). Deterministic approach to watershed modelling. *Nordic Hydrol.* **11**, 146–66.

2 Global precipitation estimation from satellite image using artificial neural networks

S. Sorooshian, K.-L. Hsu, B. Imam, and Y. Hong

2.1 INTRODUCTION

Precipitation is the key hydrologic variable linking the atmosphere with land-surface processes, and playing a dominant role in both weather and climate. The Global Water and Energy Cycle Experiment (GEWEX), recognizing the strategic role of precipitation data in improving climate research, strongly emphasized the need to achieve global measurement of precipitation with sufficient accuracy to enable the investigation of regional to global water and energy distribution. Additionally, many other international research programs have also placed high priority on the development of reliable global precipitation observation.

During the past few decades, satellite-sensor technology has facilitated the development of innovative approaches to global precipitation observations. Clearly, satellite-based technologies have the potential to provide improved precipitation estimates for large portions of the world where gauge observations are limited. Recently many satellite-based precipitation algorithms have been developed (Ba and Gruber, 2001; Huffman *et al.*, 2002; Joyce *et al.*, 2004; Negri *et al.*, 2002; Sorooshian *et al.*, 2000; Tapiador 2002; Turk *et al.*, 2002; Vicente *et al.*, 1998; Weng *et al.*, 2003). These algorithms generate precipitation products consisting of higher spatial and temporal resolution with potential to be used in hydrologic research and water-resources applications. Evaluation of recently developed precipitation products over various regions is ongoing (Ebert, 2004; Kidd, 2004; Janowiak, 2004).

In this chapter, we will introduce one near-global precipitation product generated from the PERSIANN (Precipitation Estimation from Remotely Sensed Information using Artificial Neural Networks) algorithm. PERSIANN is an adaptive, multi-platform precipitation estimation system, which uses artificial neural network (ANN) technology to merge high-quality, sparsely sampled data from NASA, NOAA, and DMSP low altitude polar-orbital satellites (TRMM, DMSP F-13, F-14, and F-15, NOAA-15, -16, -17) with continuously sampled data from geosynchronous satellites (GOES) (Hsu *et al.*, 1997, 1999; Ferraro and Marks, 1995; Janowiak *et al.*, 2000; Sorooshian *et al.*, 2000; Weng *et al.*, 2003). The precipitation product generated from PERSIANN covers 50°S–50°N at 0.25° spatial resolution and hourly temporal resolution.

2.2 NEAR-GLOBAL PERSIANN PRECIPITATION DATA FOR HYDROLOGIC APPLICATIONS

Figure 2.1 (Plate 3) shows the precipitation generation flow from the PERSIANN algorithm. The PERSIANN algorithm provides global precipitation estimation using combined geostationary and low-orbital satellite imagery. Two major stages are involved in processing a satellite image into surface rainfall rates. The algorithm first extracts and classifies local texture features from the long-wave infrared image of geostationary satellites to a number of texture patterns, and then it associates those classified cloud-texture patterns to the surface rainfall rates. PERSIANN generates rainfall rate every 30 minutes. To set up PERSIANN for better capturing the high temporal variation of precipitation, the whole globe is separated into a number of organized subdivisions, while each subdivision consists of an area coverage of 15° × 60°. PERSIANN model parameters in each subdivision are adjusted from passive microwave rainfall estimates processed from low-orbital satellites from NASA, NOAA, and DMSP low altitude polar-orbital satellites (TRMM, DMSP F-13, F-14, and F-15, NOAA-15, -16, -17) (Ferraro and Marks, 1995; Weng *et al.*, 2003). Although other sources of precipitation observation, such as ground-based radar and gauge observations, are potential sources for the adjustment of model parameters, they are not included in the current PERSIANN product generation. Evaluation of the PERSIANN product using gauge and radar measurements is ongoing to ensure

Hydrological Modeling in Arid and Semi-Arid Areas, ed. Howard Wheater, Soroosh Sorooshian, and K. D. Sharma. Published by Cambridge University Press.
© Cambridge University Press 2008.

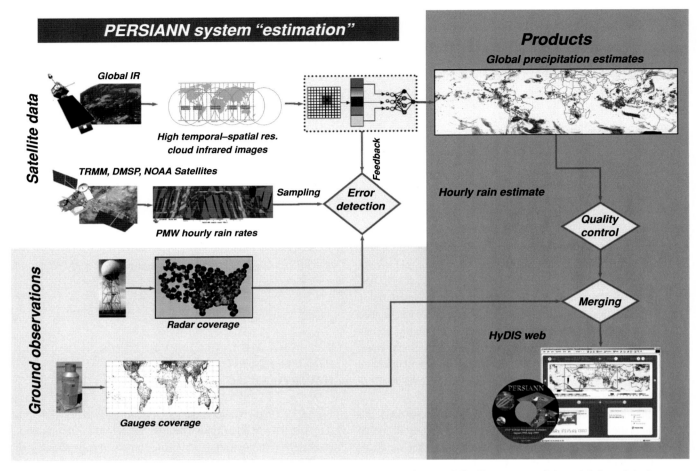

Figure 2.1 (Plate 3) Current operational implementation of the PERSIANN system produces and distributes near-real-time global precipitation products at 0.25° six-hourly resolution.

the quality of generated rainfall data. PERSIANN generates a near-global (50°S–50°N) product at 0.25° spatial resolution and hourly temporal resolution.

In conjunction with the various phases of PERSIANN development, we have also developed the Hydrologic Data and Information System (HyDIS), which has been providing various means to share research quality PERSIANN data with researchers and the general public world-wide. HyDIS is a river basin and country-based web GIS system that includes a global precipitation-mapping server, which provides direct access to near real-time global six-hourly PERSIANN precipitation estimates (http://hydis8.eng.uci.edu/hydis-unesco/). Currently, multiple years of PERSIANN data since year 2000 have been generated. The data are distributed through the HyDIS system to our partners world-wide to conduct model evaluation and data assimilation studies at climate-, meso-, and hydrologic-scales. PERSIANN data visualization and service through the HyDIS demonstration page is listed in Figure 2.2 (Plate 4). The user-friendly

interface of HyDIS enables users to view and collect data in a selected region and at the required accumulated interval period. For those users who would like to receive multiple years of global data, six-hourly global PERSIANN data is added together at the end of month. The data is also available through HyDIS at: hydis8.eng.uci.edu/persiann.

In addition to the interactive map server, which provides the flexibility of regional selection, zooming, and data subsetting, HyDIS tools have been expanded to accommodate UNESCO's Global Water and Development Information (G-WADI) project. The expanded tools include automated generation of country-based aridity mapping and multiple access points to PERSIANN data: (a) original HyDIS Mapserver interface, (b) pull-down menus, which allow the user to select continental region and country, and (c) text listing of countries within each continent. Water resources relevant information was provided through the ability to select from several precipitation accumulation intervals (6 hours, 1, 3, 5, 15, and 30 days). A pull-down menu access-point

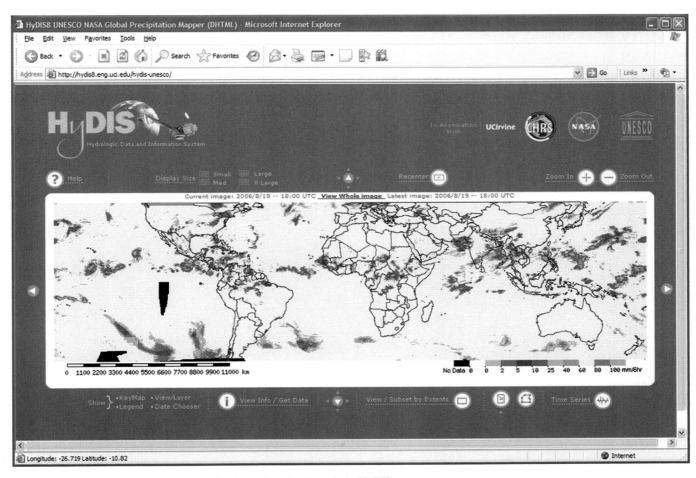

Figure 2.2 (Plate 4) PERSIANN data visualization and service through the HyDIS system

widget has been designed to permit mirror sites to provide HyDIS data access. Access to the text tabulation is provided through a simple image page that allows the user to click on a continental region to access its relevant table. Future updates will include histograms of aridity as well as land-cover classes within political divisions of each country.

2.3 EVALUATION AND APPLICATIONS

Several regional evaluation studies of current satellite high-resolution precipitation products including PERSIANN and several others are ongoing. These regions include: Australia (BMRC precipitation validation page: http://www.bom.gov.au/bmrc/SatRainVal/dailyval.html) and the United States (CPC precipitation page: http://www.cpc.ncep.noaa.gov/products/janowiak/us_web.shtml). Ground gauge and radar data are used in the evaluation to provide an overview of daily as well as seasonal statistics of

satellite and ground-based precipitation observations. Figure 2.3 (Plate 5), which shows a sample BMRC evaluation over Australia, also demonstrates PERSIANN's ability to capture the distribution of daily precipitation in a manner consistent with daily gauge analysis for January 22, 2005. Other validation sites cover part of western Europe (University of Birmingham precipitation validation page: http://kermit.bham.ac.uk/%7Ekidd/ipwg_eu/ipwg_eu.html) and PERSIANN GEWEX Coordinated Enhanced Observation Period (CEOP) sites (http://www.ceop.net/) are under preparation (http://hydis0.eng.uci.edu/CEOP/).

Precipitation is a key forcing variable of the global, regional, and local water and energy cycle. Providing reliable precipitation observation will contribute to improving our understanding of the evolution of convective precipitation during the Monsoon season and the diurnal evolution of the precipitation cycle. Similarly, the product will provide modellers with a unique data set that could be utilized to improve numerical weather predictability as it provides a critical element for data assimilation and ensemble forecasts. For years PERSIANN precipitation products have been used in

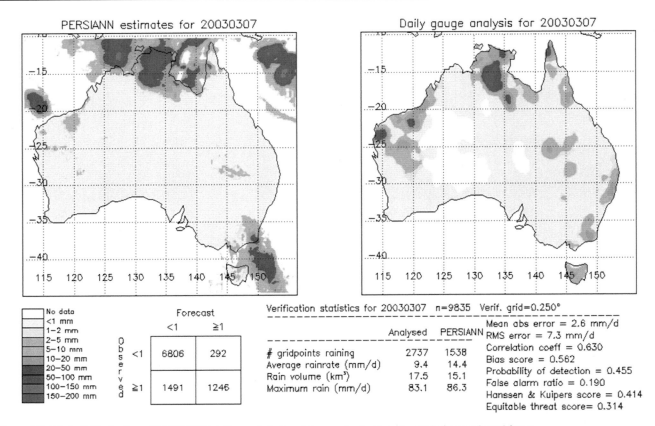

Figure 2.3 (Plate 5) Evaluation of PERSIANN daily precipitation data over Australia region (this figure adapted from: http://www.bom.gov.au/bmrc/SatRainVal/dailyval.html)

a number of hydrologic research and application studies. These studies have included: (1) validation of daily rainfall and diurnal rainfall patterns against observations provided by the TRMM field campaign, (2) evaluation of MM5 numerical weather forecast model estimates over the south-west United States, Mexico, and adjacent oceanic regions, (3) assimilation of PERSIANN data into a Regional Atmospheric Modelling System (RAMS) model to investigate the land-surface hydrologic process, (4) merging of the gauge and PERSIANN systems over the Mexico Region, (5) the use of satellite-based precipitation estimates for runoff prediction in ungauged basins, (6) investigation of the impacts of assimilating satellite rainfall estimates on rainstorm forecast over southwest United States, and (7) analysis of multiple precipitation products and preliminary assessment of their impact on the Global Land Data Assimilation System (GLDAS) land surface states (Gochis *et al.*, 2002; Guevara, 2002; Hong *et al.*, 2005; Li *et al.*, 2003; Xu *et al.*, 2005; Sorooshian *et al.*, 2002; Yi, 2002; Yucel *et al.*, 2002).

Figure 2.4 (Plate 6) shows the diurnal precipitation patterns retrieved from PERSIANN data and NCEP WSR-88 radar data during the summer season (JJA) 2002 around central and northern

America. Note that both sources of data show consistent diurnal precipitation patterns over the land and ocean; the amplitude of the data are similar to one another; in the phase, however, there is approximatly a one-hour lag between PERSIANN and WSR-88 radar estimates. A more detailed discussion of using PERSIANN data in documenting the diurnal precipitation pattern is described in Sorooshian *et al.* (2002).

Figure 2.5 (Plate 7) shows PERSIANN rainfall applied in the streamflow simulation of the Leaf River Basin (1949 km^2) near Collins, Mississippi. In this experiment, more than three years of PERSIANN precipitation, as well as gauge and radar merged rainfall data, were collected and applied to generate daily streamflow using an operational conceptual hydrologic model (the Sacramental Soil Moisture Accounting Model of National Weather Service). Compared to basin daily observation, the daily streamflows generated from PERSIANN rainfall are not significantly different from those generated from gauge and radar merged data. This demonstrates that the satellite-based rainfall measurement is reaching a level potentially suitable as a precipitation data source for basin-scale hydrologic applications, in particular for regions where ground-based observations are lacking.

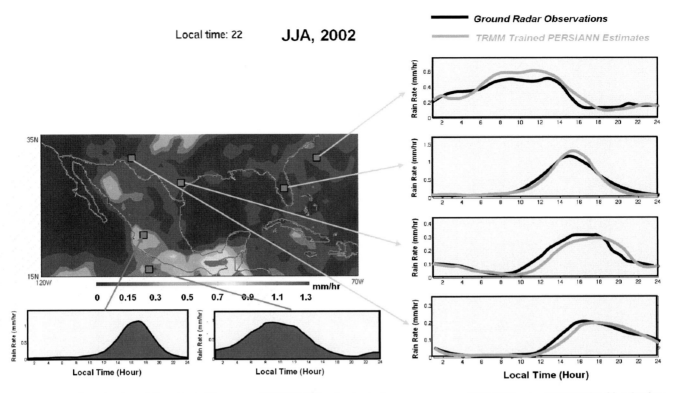

Figure 2.4 (Plate 6) Evaluation of diurnal rainfall pattern of PERSIANN estimates in the summer season (JJA) 2002 using NCEP WSR-88 radar data

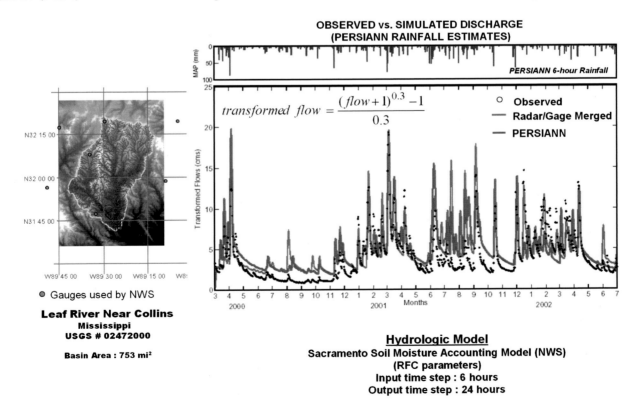

Figure 2.5 (Plate 7) Simulation of daily streamflow using six-hourly accumulated rainfall from (1) satellite-based PERSIANN data and (2) radar and gauge merged data

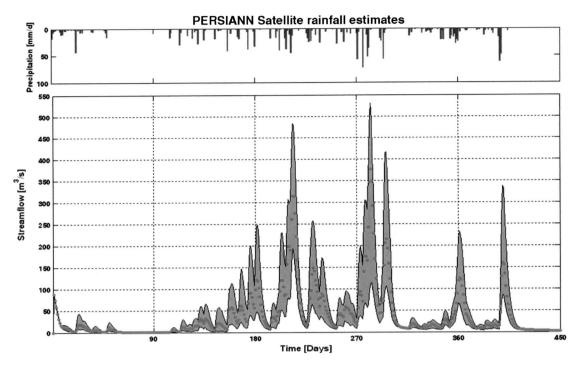

Figure 2.6 (Plate 8) Confidence interval (95 %) of daily streamflow is generated based on the uncertainty of PERSIANN precipitation estimates

The influence of spatial–temporal precipitation errors on hydrologic response is further examined using the stochastic simulation of precipitation error. Figure 2.6 (Plate 8) shows the impact of the PERSIANN estimation error to the generation of streamflow over the Leaf River Basin. In our case, the uncertainty of PERSIANN estimates is evaluated based on radar rainfall measurement. The 95 % confidence interval of simulated streamflow is plotted. It shows that the reliability of estimated streamflow is highly relevant to the quality of precipitation-forcing data. The uncertainty bound is significantly higher during high-flow periods and is lower at low-flow points. The error property of PERSIANN data is under evaluation.

2.4 SUMMARY

In summary, we introduced the operation of the PERSIANN satellite-based algorithm to generate a near global coverage of precipitation data. Currently, multiple years of PERSIANN data are generated and available for public access through the HyDIS data visualization and handling system. With our objective of providing reliable precipitation data for basin-scale hydrologic studies in mind, we have been continuing operation of the PERSIANN algorithm to generate long-term near-global coverage precipitation data, and to improve the quality of data through the dev-

elopment of satellite-based rain retrieval algorithms. In the algorithm development, our recently developed PERSIANN Cloud Classification System (CCS) uses computer image processing and pattern-recognition techniques to process cloud image into rainfall rate (Hong *et al.*, 2004). Preliminary study over northern America shows promise, with the potential improvement of current PERSIANN estimates. Real-time operation of CCS to produce 4 km hourly precipitation over North America is under development (see: http://hydis8.eng.uci.edu/CCS); we will continue to evaluate both PERSIANN and CCS and report our progress.

ACKNOWLEDGEMENTS

Partial support for this research is from NASA-EOS (Grant NA56GPO185), NASA-PMM (grant NNG04GC74G) and NSF STC for "Sustainability of Semi-Arid Hydrology and Riparian Areas" (SAHRA) (Grant EAR-9876800).

REFERENCES

Ba, M. B. and Gruber, A. (2001). GOES multispectral rainfall algorithm (GMSRA), *J. Appl. Meteorol.*, **40**, 1500–14.
Ebert, B. (2004). Monitoring the quality of operational and semi-operational satellite precipitation estimates: the IPWG validation/intercomparison

study. *2nd IPWG Working Group Meeting*, Naval Research Laboratory, Monterey, CA, USA, 25–28, Oct, 2004.

Ferraro, R. R. and Marks, G. F. (1995). The development of SSM/I rainrate retrieval algorithms using ground-based radar measurements. *J. Atmos. Oceanic Technol.*, **12**, 755–770.

Guevara, J. M. (2002). Precipitation estimation over Mexico applying PERSIANN system and gauge data. Masters Thesis, University of Arizona.

Gochis, D. J., Shuttleworth, W. J., and Yang, Z.-L. (2002). Sensitivity of the modeled North American Monsoon regional climate to convective parameterization. *Monthly Weather Rev.*, **130**, 1282–98.

Huffman, G. J., Adler, R. F., Stocker, E. F., Bolvin, D. T., and Nelkin, E. J. (2002). A TRMM-based system for real-time quasi-global merged precipitation estimates. *TRMM International Science Conference*, Honolulu, 22–26 July 2002

Hong, Y., Hsu, K. Sorooshian, S., and Gao, X. (2005). Improved representation of diurnal variability of rainfall retrieval from TRMM-adjusted PERSIANN system, *J. Geophys. Res.*, **110**, D06102.

Hong, Y., Hsu, K., Gao, X., and Sorooshian, S. (2004). Precipitation estimation from remotely sensed imagery using artificial neural network – cloud classification system, *J. Appl. Meteorol.*, **43(12)**, 1834–53

Hsu, K., Gao, X., Sorooshian, S., and Gupta, H. V. (1997). Precipitation estimation from remotely sensed information using artificial neural networks. *J. Appl. Meteorol.*, **36**, 1176–90.

Hsu, K., Gupta, H. V., Gao, X., and Sorooshian, S. (1999). A neural network for estimating physical variables from multi-channel remotely sensed imagery: application to rainfall estimation. *Water Resour. Res.*, **35**, 1605–18.

Janowiak, J. E., Joyce, R. J., and Yarosh, Y. (2000). A real-time global half-hourly pixel resolution infrared dataset and its applications, *Bull. Am. Meteorol. Soc.*, **82**, 205–17.

Janowiak, J. (2004). Validation of satellite-derived rainfall estimates and numerical model forecasts of precipitation over the US. *2nd IPWG Working Group Meeting*, Naval Research Laboratory, Monterey, CA, USA, 25–28, Oct., 2004.

Joyce, R. J., Janowiak, J. E., Arkin, P. A., and Xie, P. (2004). CMORPH: A method that produces global precipitation estimates from passive microwave and infrared data at high spatial and temporal resolution. *J. Hydrometeorol.*, **5**, 487–503.

Kidd, C. (2004) Validation of satellite rainfall estimates over the mid-latitudes. *2nd IPWG Working Group Meeting*, Naval Research Laboratory, Monterey, CA, USA, 25–28, Oct, 2004.

Li, J., Gao, X., Maddox, R. A. Sorooshian, S., and Hsu, K. (2003). Summer weather simulation for the semi-arid lower Colorado River basin: case tests. *Monthly Weather Rev.*, **131(3)**, 521–41.

Negri, A. J., Xu, L., and Adler, R. F. (2002). A TRMM-calibrated infrared rainfall algorithm applied over Brazil. *J. Geophys. Res.*, **107(D20)**, 8048–62.

Sorooshian, S., Hsu, K. Gao, X., Gupta, H. V., Imam, B., and Braithwaite, D. (2000). Evaluation of PERSIANN system satellite-based estimates of tropical rainfall. *Bull. Am. Meteorol. Soc.*, **81**, 2035–46.

Sorooshian, S., X. Gao, K. Hsu *et al.* (2002). Diurnal variability of tropical rainfall retrieved from combined GOES and TRMM satellite information, *J. Clim.*, **15**, 983–1001.

Tapiador, F. J. (2002). A new algorithm to generate global rainfall rates from satellite infrared imagery. *Revista de Teledeteccion*, **18**, 57–61.

Turk, J., Ebert, E., Oh, H.-J., Sohn, B.-J., Levizzani, V., Smith E., and Ferraro, R. (2002). Verification of an operational global precipitation analysis at short time scales. *1st Intl. Precipitation Working Group (IPWG) Workshop*, Madrid, Spain, 23–27 September 2002.

Vicente, G. A., Scofield, R. A., and Menzel, W. P. (1998). The operational GOES infrared rainfall estimation technique. *Bull. Am. Meteorol. Soc.*, **79**, 1883–98.

Weng, F. W., Zhao, L., Ferraro, R., Pre, G., Li, X., and Grody, N. C. (2003). Advanced Microwave Sounding Unit (AMSU) cloud and precipitation algorithms, *Radio Sci.*, **38(4)**, 8068–79.

Xu, J., Gao, X. and Sorooshian, S. (2004). Investigate the impact of assimilating satellite rainfall estimates on rainstrom forecast over Southwest United States, *Geophysical Research Letters*, **31**.

Yi, H. (2002). Assimilation of satellite-derived precipitation into the Regional Atmospheric Modeling System (RAMS) and its impact on the weather and hydrology in the southwest United States. Ph.D. Dissertation, Department of Hydrology and Water Resources, University of Arizona.

Yucel, I., Shuttleworth, W. J., Pinker, R. T., Lu, L., and Sorooshian, S. (2002). Impact of ingesting satellite-derived cloud cover into the regional atmospheric modeling system, *Monthly Weather Rev.*, **130**, 610–28.

3 Modelling semi-arid and arid hydrology and water resources: the southern African experience

D. A. Hughes

3.1 INTRODUCTION

Southern African hydrology is characterized by a high degree of variability with climate zones varying from tropical to extremely arid. Within individual climate zones, and particularly the semi-arid regions, hydrological (rainfall and streamflow) variability is as high as anywhere in the world (McMahon, 1979). While extremes of floods and droughts, and their social or economic consequences, tend to receive a great deal of publicity, it is often the less dramatic components of hydrological variability that present some of the greatest challenges to sustainable water-resource management. There are many basins within the region that have their headwaters in relatively wet and well-watered regions but then pass through much drier regions. The need to understand streamflow loss, as well as streamflow-generation processes increases the complexity of any modelling study. The fact that many of these rivers also cross national boundaries (Orange, Limpopo, Okavango, etc.) adds to the complexity of managing water resources at the regional scale.

The political and socio-economic history of the region has not been conducive to the collection and maintenance of hydrological records. Even today, despite the recognition of the importance of well-managed water resources to the health and socio-economic well-being of the region, the acquisition of the necessary information does not appear high on the agendas of many of the region's government institutions. Similar constraints have limited the development of local capacity within both the hydrological sciences and water-resource management fields. While development aid funding for often quite grandiose schemes has been made available, they have been largely based on expertise hired from outside the region, with little of that experience and expertise remaining within the region at the end of the project. At the same time, financial rewards and incentives for local staff are frequently inadequate and do not encourage participation and further training in the field.

During the colonial era, several countries developed quite detailed water-resource monitoring networks. However, in many cases, war, inadequate economic resources, and shifting social and political priorities have meant that these networks have not been maintained and in some cases the historical data are not readily accessible. Water-resource utilization and land-use changes likely to impact on natural hydrological processes are also less than well documented in many areas. Even where licensing systems are in operation, the amounts of water abstracted or returned to the river can be highly temporally variable. This makes it difficult to calibrate and validate hydrological model results against historical data, even where they exist.

The development and successful application of hydrological models has therefore been seriously hampered by:

- a high degree of spatial and temporal variation in hydrometeorological variables and resulting streamflow;
- a lack of adequately long or continuous records of rainfall (and other hydrometeorological variables) and streamflow;
- a lack of information on land-use changes and both spatial and temporal variations in water utilization;
- a lack of quantitative understanding of the mechanisms of some critical hydrological process (notably channel transmission losses and surface–ground water interactions);
- a lack of capacity in some parts of the region, frequently associated with a lack of political commitment to addressing that lack of capacity.

Despite these limitations, there are a number of examples of the successful development and application of hydrological models within the region. The methods and models that have been applied need to be seen in the context of limited data availability and the water-resource information requirements of the region. They have therefore generally tended to be pragmatic, rather than scientifically ideal, solutions. This paper will review some of the problems

Hydrological Modeling in Arid and Semi-Arid Areas, ed. Howard Wheater, Soroosh Sorooshian, and K. D. Sharma. Published by Cambridge University Press.
© Cambridge University Press 2008.

with the application of hydrological models, some of the developments that have been achieved as well as some of the opportunities that exist, through the integration of existing methods with new technologies.

3.2 CHARACTERISTICS OF SOUTHERN AFRICAN SEMI-ARID DRAINAGE BASINS

As with arid and semi-arid basins worldwide, one of the most important characteristics is the high degree of spatial variability of rainfall inputs during individual storm events. Coupled with relatively complex associations between soil characteristics (depths and hydraulic properties) and topography, this variability suggests that developing generalizations about patterns of runoff generation can be extremely difficult, even at the scale of relatively small catchments (up to $10\,km^2$). At larger scales, additional processes associated with the spatial discontinuity of channel flow, permeable channel beds, high rates of evaporation, and a lack of antecedent baseflow contribute to complex spatial variability in streamflow.

Mostert *et al.* (1993) identified a potentially important process that contributes to the understanding of the inter-annual water balance of some Namibian basins. Wet seasons contribute to the development of improved vegetation cover in these otherwise poorly vegetated regions. In the following season, the improved vegetation cover can lead to improved infiltration, as well as more effective evapotranspiration losses and a reduction in the relative amount of runoff that occurs (compared to similar rainfall after a dry year). This effect appears to last for over three years following a wet season and was incorporated into the NAMRON model (Mostert *et al.*, 1993). While this process has not been studied in detail elsewhere in the semi-arid parts of the region, it is likely that similar non-seasonal vegetation-cover dynamics could play a major role in explaining the high degree of inconsistency in relationships between rainfall and streamflow.

There have been very few direct studies of channel transmission losses in the region (Crerar *et al.*, 1988; Hughes and Sami, 1992; Görgens and Boroto, 2003), despite the fact that this process has been recognized as one of the most important components of the water balance of many of the regions semi-arid basins. At the small scale, runoff generated during relatively small storm events has to satisfy in-channel pool storage before progressing downstream and contributing to more widespread streamflow. Both pool storage and channel flow are subject to seepage into the bed and banks, the amounts highly dependent upon the local nature of the soil or rock material. The process of recharge into alluvial aquifers is well documented at various scales (Crerar *et al.*, 1988; Görgens and Boroto, 2003), but it remains difficult to develop generalized quantitative approaches to the estimation of such losses.

The highly fractured nature of the material underlying some rock-bed channels (which tend to be found along fracture lineations and zones of geological weakness), suggests that losses from non-alluvial rivers can also be substantial. Unfortunately, there is only anecdotal evidence available for this process and there have been no attempts at quantification in the region. The presence of small farm dams further contributes to the spatial discontinuity found in channel flow in medium sized semi-arid basins. Identifying their presence is a relatively simple matter using remote-sensing approaches. However, quantifying their storage capacities is a different matter, as only the larger developments are usually documented.

At the larger scale, transmission losses and the associated groundwater resources of alluvial aquifers play a major role in some of the region's major basins. Many of the tributaries of the Limpopo River are permanently flowing in their headwaters and then pass through much drier regions and become seasonal rivers due to natural losses, as well as abstractions. The situation can be exacerbated by the utilization of the alluvial groundwater resources in the main Limpopo valley. This increases the storage capacity of the alluvial material at the start of the dry season and can delay the onset of channel flow downstream (Görgens and Boroto, 2003). The Namibian experience (Wheeler *et al.*, 1987) suggests that managing transmission losses to alluvial material can be used as an effective alternative water-resource development strategy to conventional surface water storage, which is subject to large evaporative losses.

A thorough understanding of the total water-resource availability of semi-arid basins should include both surface- and groundwater and implies that they should be modelled together. Understanding surface runoff processes on hillslopes, as well as mechanisms of recharge (Sami and Hughes, 1996) to subsurface storages, should be the key to the joint modelling of surface- and groundwater in semi-arid basins. However, there have been very few studies where these have been considered together.

3.3 DATA AVAILABILITY

Ideally, any development of hydrological models should be based on a sound conceptual understanding of the processes being modelled and backed up with quantitative information that can be used to parameterize a model for a specific application. While there exists a relatively sound conceptual understanding of the processes involved, providing sufficient information to quantify the processes is a different matter.

3.3.1 Rainfall data

Rainfall is one of the key driving variables of any hydrological model, regardless of the climate region. In the semi-arid basins of the region, the spatial variability of the occurrence and depth of rainfall (over almost any time-scale, but particularly short periods), coupled with the relatively sparse observation networks, makes it extremely difficult to satisfactorily quantify the main water inputs. Most of the available rainfall records are based on daily observations, precluding the possibility of defining the real intensity characteristics that are of great importance in semi-arid runoff-generation processes.

The poor spatial distribution of rainfall-measuring stations is partly due to the difficulties of access in the sparsely populated semi-arid regions of southern Africa. The data availability situation is made worse by frequent missing data, closure of stations or the complete collapse of hydrological monitoring during periods of social and political upheaval. Satellite-derived rainfall estimates have the potential to provide much more spatial detail in basin rainfall inputs. However, if this relatively new technology is to be used in conjunction with existing historical rainfall records, it is important that the two data sources are checked for consistency. This is not always as simple as it may appear at first sight due to a lack of overlapping data (see Wilk *et al.*, 2006, for an example from the Okavango basin).

Many of the water-resource modelling approaches used in the region are based on monthly time intervals. While aggregating rainfall data into monthly totals reduces the degree of spatial variability, a great deal of intensity information that can be critical to runoff-generation processes in semi-arid areas is lost.

3.3.2   Evaporation data

In general terms, the availability of evaporation data is far worse than for rainfall data. Given that evapotranspiration is the second largest component of the water balance of semi-arid basins this would seem to be a critical issue with respect to hydrological modelling. However, for many rainfall events that occur in semi-arid basins, the generation of runoff is less dependent upon the antecedent moisture storage characteristics (which are dependent on evaporative losses) than on the rainfall intensity characteristics and soil surface conditions.

One area where evaporation data could play a significant role is in the quantification of channel losses through direct evaporation or transpiration from the riparian vegetation (see the study of Orange River losses by McKenzie *et al.*, 1993). However, whether the accurate quantification of potential evaporation demand, or a thorough understanding of the seepage characteristics of the channel banks is of greater importance remains to be seen.

3.3.3   Streamflow data

Many modellers would argue that the future for simulating the hydrology of ungauged catchments lies in the so-called "physically-based" approach and the use of information on spatially distributed catchment properties (Schulze, 2000, for example). However, there seems to be little doubt that testing the validity of any model formulation, as well as the adequacy of the available input data, still relies upon the availability of observed streamflow data.

Earlier comments about the problems of maintaining raingauge networks apply to an equal, if not greater, extent with respect to streamflow data. Both Namibia and South Africa make use of weir or flume structures as part of their national streamflow-monitoring network. These two countries have relatively good networks within arid and semi-arid areas, despite the large costs of construction and maintenance involved (especially on large rivers with substantial sediment movement). However, there are nevertheless a relatively small number of gauges to cover very diverse hydrological conditions. The situation in most of the other countries is not as good and they rely upon rated sections in rivers with quite dynamic bed conditions, suggesting relatively low confidence in the accuracy of some of the historical data. In some countries the resources (financial and human) available to check ratings, service the gauges in the field and process the raw data are inadequate to maintain continuous records.

In terms of extreme flows, which frequently dominate the long-term mean volumes of streamflow from semi-arid catchments, it has to be recognized that few of the gauging approaches are able to quantify these accurately.

3.3.4   Water abstraction and land-use information

Reference has already been made to the influence of small farm dams on the streamflow dynamics of southern African semi-arid basins. While their impacts are generally straightforward to understand and the spatial extent of their occurrence available from analyses of satellite imagery or aerial photography, there is little quantitative information generally available about their storage capacities. Hughes and Sami (1993) illustrated the importance of small farm dams through a study of a $670\,km^2$ basin in the Eastern Cape province of South Africa. Over 50 % of the dams (a total of 364) have a full supply volume of less than $2000\,m^3$ and catchment areas of less than $2\,km^2$. The runoff storage capacity (dam volume divided by catchment area) of 60 % of the dams is less than 2 mm, while 20 % have storage capacities of between 5 and 20 mm. While these capacities may appear quite low, they are more than sufficient to absorb runoff generated in a substantial proportion of storm events.

The situation with respect to information on major dams is far better, although an improved understanding of rates of sedimentation, and hence their medium- to long-term storage dynamics would be an advantage.

The other major source of water in semi-arid basins is groundwater. The impacts on the surface water-resources of semi-arid basins are mainly where abstraction takes place from alluvial aquifers, thereby affecting the dynamics of channel transmission losses. Information is generally available for large alluvial aquifer abstraction schemes (Görgens and Boroto, 2003), but not for smaller, more distributed abstractions.

Land-use changes in the semi-arid basins of the subcontinent are less of an issue than in the wetter parts of the region. However, as many of the larger semi-arid rivers have their headwaters in wetter areas, their channel flow dynamics are still affected. The Sabie–Sand system rises in the Eastern Escarpment of the Mpumalanga Province of South Africa, which has experienced substantial commercial afforestation over the past 60 years or more. It then passes through the relatively arid low veld region, where tributary flow is seasonal. During the dry season, baseflows have been considerably reduced and the relative impacts of transmission losses (largely as a result of transpiration from riparian vegetation) are now greater than under the natural flow regime.

3.4 HYDROLOGICAL MODELLING APPROACHES

A large number of models have been applied within the region, but this discussion will focus on models that have been developed specifically for the region. There is not room here for detailed descriptions of the models and reference should be made to the original material. The focus will therefore be on any model components that have been specifically designed to cater for semi-arid hydrological processes, the perceived advantages and disadvantages of the models and on the successes and failures of application that have been reported.

Most of the models that have been developed within the region have been moderately detailed "conceptual" type models with a relatively large number of parameters. The traditional approach to application has been the manual calibration of the models against observed data and the use of regionalized parameter sets for use in ungauged basins (Midgley *et al.*, 1994, for example). There has therefore been a focus on identifying the associations between model parameter values and measurable basin properties, rather than on the mathematics of parameter interaction and automatic optimization procedures. Most of the models used have been continuous time series (rather than single event) type models designed for water-resource estimation and design purposes. Developments have therefore been driven by the pragmatic requirements of water-resource engineers rather than research orientated scientific understanding.

3.4.1 Pitman monthly time-step model

This model has been more widely applied within the southern African region than any other hydrological model and has also been applied outside the region (Wilk and Hughes, 2002). It was developed in the 1970s (Pitman, 1973) and has undergone a number of revisions since then. It now exits in several forms, each with different additional components added by a range of different developers. However, the core concepts of the original model have been preserved in all the revised versions. The model is an explicit soil-moisture accounting model representing interception, soil moisture and groundwater storages, with model functions to represent the inflows and outflows from these. The Institute for Water Research (IWR) has developed one of these versions and added a number of "refinements" based on assessments throughout the subcontinent as part of the southern Africa FRIEND program (Hughes, 1995, 1997). Subsequently, the IWR has added more explicit groundwater recharge and discharge functions (Hughes, 2004a). Figure 3.1 illustrates the structure of the model.

In semi-arid basin applications the dominant runoff-generating component is a triangular "catchment absorption" function controlled by two variables (ZMIN and ZMAX). The rainfall rate during one time interval of the model is used to determine what proportion of the catchment (the relative area under the triangle between ZMIN and the rainfall rate) will contribute to surface runoff. Hughes (1997) added a third parameter (ZAVE) to allow the triangle to assume an asymmetric shape. The model operates over several iterations (typically four, but user-determined in some versions) and therefore allows the monthly rainfall depth to be subdivided. The original model used a fixed rainfall distribution function, while Hughes (1997) modified this to allow for regional differences in the distribution of rainfall within a month (see Hughes *et al.*, 2003, for a detailed analysis of Zambian rainfall data in relation to establishing a suitable parameter value). These two model components (the rainfall distribution and the absorption function) strongly interact with each other in determining the monthly response of runoff to rainfall in semi-arid applications.

Most of the current versions of the model operate using a semi-distributed, subcatchment scheme, whereby each subarea has its own hydrometeorological inputs and parameter set. The model includes components to allow for abstractions from distributed farm dams, direct from the river, as well as major storage dams at the outlet of each subarea. While the model has no explicit function to estimate channel transmission losses, these have been frequently included in a modelling scheme through the use of "dummy" dams representing the loss storage and evaporating area (Görgens and Boroto, 2003; Hughes *et al.*, 2003). The problem with this approach for perennial rivers flowing through arid areas (e.g., the Lower Okavango) is that the "dummy" reservoir is always full and the losses dependent only on the evaporation rate and the surface area. A new approach, linked to the revised groundwater routines, is being tested on the Okavango. The algorithm is based on two factors: one related to the near channel

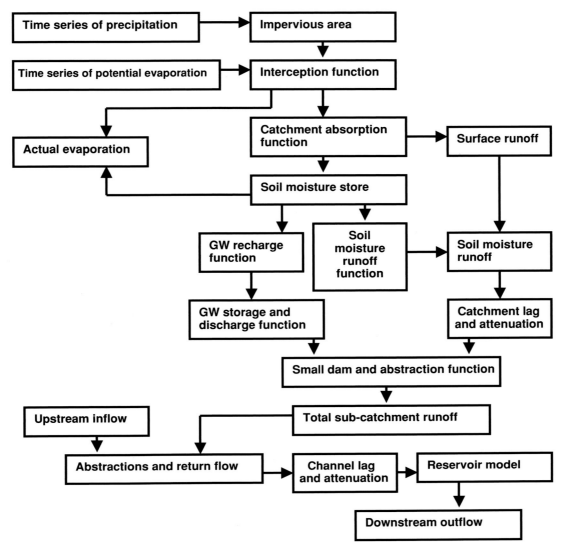

Figure 3.1 Structure of the Pitman model with revised groundwater recharge and discharge routines

groundwater storage level and one to the relative flow rate in the channel. Thus losses increase with increasing channel flow and with lower groundwater storage.

There are 24 model parameters in the modified groundwater version of the model and typically 14 of these are established a priori or through some initial calibration test runs. This leaves ten parameters, which are normally the focus of the calibration effort. The catchment absorption function (three parameters), the maximum size of the soil-moisture storage and the channel-loss parameter frequently dominate the calibration process in semi-arid areas.

One version of the model included a scheme to allow for the "growth and decay" of surface-cover conditions based on antecedent moisture storage conditions (Hughes and Metzler,

1998). The result is that the catchment-absorption and evaporative-loss parameters vary with time to simulate dynamic vegetation cover. This was added to compare the model with the Namrom model (de Bruine *et al.*, 1993) for application in Namibian basins, where this phenomenon has been identified as being very important in the understanding of inter-annual runoff responses (Mostert *et al.*, 1993).

One of the advantages of the Pitman model is the availability of guidelines for parameter estimation provided by the WR90 study (Midgley *et al.*, 1994) that reports on the application of the model to 1946 so-called quaternary catchments covering South Africa, Lesotho, and Swaziland. These guidelines can be used to establish initial parameters for almost any climate region of the subcontinent, which can then be refined through local calibration (where

Table 3.1 *Summary of SA FRIEND project results for the Pitman model (based on monthly flows)*

Country	No. of basins	Range of mean monthly % error	Range of R^2	Range of CE
Namibia	9	−2.1 to 1.5	0.06 to 0.72	−0.84 to 0.68
Botswana	10	−8.7 to 11.8	0.39 to 0.77	0.27 to 0.70
Zimbabwe	14	−8.7 to 2.6	0.45 to 0.90	0.33 to 0.88
Tanzania	5	−7.6 to 3.3	0.33 to 0.81	0.27 to 0.78
Zambia	13	−14.0 to 6.5	0.41 to 0.79	0.29 to 0.78
Malawi	4	−12.4 to 0.9	0.35 to 0.84	0.24 to 0.82
Swaziland	7	−5.2 to 10.0	0.45 to 0.81	0.39 to 0.80
Mozambique	5	−9.0 to 6.1	0.37 to 0.84	0.21 to 0.84

Note: R^2 is the coefficient of determination, while CE is the coefficient of efficiency (Nash and Sutcliffe, 1970)

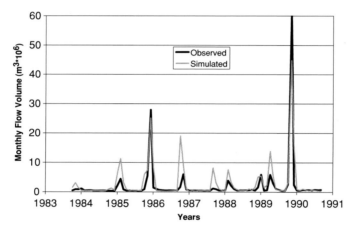

Figure 3.2 Typical Pitman model result for a semi-arid quaternary catchment simulation based on default regional parameters (Little Fish River, Eastern Cape Province)

available data allow). Figure 3.2 illustrates a typical example of a simulation result using the regional parameters (compared with observed data). This study is in the process of being updated and one component of the revision (referred to as WR2005) will be the integration of all recent improvements to the Pitman model and the development of a new "official" version.

There is little doubt that the model is easier to apply and is generally more successful in the humid and temperate parts of the region, rather than the more arid parts (Hughes, 1997 – see Table 3.1 for some summary results). This could be a consequence of the relatively poor definition of the real spatial variations in rainfall input that is typical for many basins, the limitations of the model in terms of the temporal distribution of rainfall within a single month, or the relatively simplistic approach to simulating runoff generation. To isolate which of these influences dominate would require rather more data than are available in most arid

parts of the region. Given that it is not always possible to achieve a very good one-to-one fit with observed streamflow, it is important to satisfactorily reproduce those statistical properties of the streamflow time series that are critical from a water resource planning and management perspective (Hughes and Metzler, 1998).

3.4.2 Namrom model

This model was designed specifically for use in Namibian basins (de Bruine *et al.*, 1993; Mostert *et al.*, 1993) and therefore includes components to simulate processes that were identified as important in this arid region. Specifically, it has been designed to address the issues of dynamic, non-seasonal, surface-cover conditions, as well as transmission losses to alluvial aquifers. While it has been applied with a reasonable degree of success to a number of basins in Namibia, it has not been applied elsewhere and therefore its

Table 3.2 *Summary of the results of applying the NAMROM, Pitman and modified Pitman (NAMPit) to 5 Namibian basins*

Model	Range of mean daily % error	Range of R^2	Range of CE
NAMROM	−4.1 to 2.3	0.39 to 0.90	0.30 to 0.90
Pitman	−2.1 to 0.6	0.23 to 0.72	0.15 to 0.68
NAMPit	−2.5 to 3.5	0.22 to 0.82	0.10 to 0.82

general applicability is largely untested. The basic concepts are sound and other models could possibly benefit from being adapted to include similar approaches (Hughes and Metzler, 1998).

The model is based on a single equation for total effective precipitation, using four parameters: antecedent weighting factor (seasonally varying), initial loss, subcatchment loss factor, and loss exponent. A regression equation is then developed for observed runoff and total effective precipitation. The model is therefore more of a statistical regression-type model with weighting parameters having some perceived physical meaning.

Hughes and Metzler (1998) compared the original Pitman model and a modified version (including a dynamic vegetation-cover process) with the Namron model. In general terms the 20 years of data available for calibration indicated that the Namrom model performed the best with the modified Pitman model a close second (Table 3.2). The simulated means and standard deviations of monthly flow were all very similar. However, when the calibrated models were applied to all the available rainfall data (68 years) and a reservoir yield assessment undertaken, the results for the three models were quite different. For five existing reservoirs and a demand of 40 % MAR, the simulated shortfalls varied substantially for the three models (between 1 and 17 % of the design demand across the five reservoirs). This serves to further illustrate the point made at the end of the previous section and empasizes the issue discussed by Görgens (1983) about the minimum length of calibration data required for semi-arid modelling purposes.

3.4.3 ACRU model

The ACRU model has been developed by the Bioresources and Environmental Engineering Hydrology School of the University of KwaZulu-Natal (Schulze, 1994). It is a daily time-step model designed around a multi-layer soil-moisture accounting scheme and has a large number of parameters that require quantification. It is designed to be used in ungauged basins on the basis of parameters evaluated through default relationships with measurable catchment properties (soils, vegetation, management practices, etc. – see Schulze, 2000). It has been mostly applied in the temperate and humid parts of South Africa and has been frequently used for assessing the impacts of various land-use modifications, specifically commercial afforestation. It has been less widely applied in the drier parts of the region and there appears to be very little documentation of the success of its application in semi-arid to arid basins. It is still very unclear how the model performs with default parameters under different situations of catchment data availability (in terms of spatial resolution, accuracy, etc.).

3.4.4 VTI model

The Variable Time Interval model was developed at the IWR, Rhodes University as part of a detailed study of the catchment response characteristics of a medium sized semi-arid basin (670 km^2) in the Eastern Cape Province of South Africa (Hughes and Sami, 1994). It has subsequently been applied to a wide range of basins (Table 3.3) within the region under the Southern African FRIEND programme (Hughes, 1997). It is essentially a daily model that can use shorter modelling time-intervals during periods of assumed high process activity (based on thresholds of rainfall intensity), given that shorter time interval rainfall data are available. Figures 3.3 and 3.4 illustrate the structure of the model and indicate that the main moisture-accounting routines are quite complex with a number of feedback mechanisms. It has explicit functions for most of the processes recognized as being important in semi-arid catchments and is, as a consequence, relatively

Table 3.3 *Summary of SA FRIEND project results for the VTI model (based on daily flows)*

Country	No. of basins	Range of mean daily % error	Range of R^2	Range of CE
Botswana	3	−3.3 to 2.8	0.40 to 0.49	0.23 to 0.43
Zimbabwe	6	−3.5 to 20.0	0.45 to 0.77	0.32 to 0.70
Tanzania	5	−15.2 to −0.2	0.14 to 0.67	−0.13 to 0.62
Swaziland	7	−6.0 to 8.9	0.14 to 0.63	−0.51 to 0.59
Mozambique	5	−9.2 to 4.5	0.44 to 0.75	0.33 to 0.71

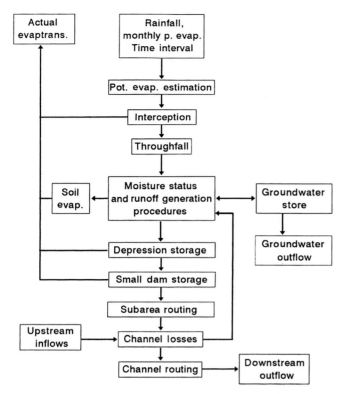

Figure 3.3 Main structure of the variable time interval (VTI) model (from Hughes and Sami, 1994)

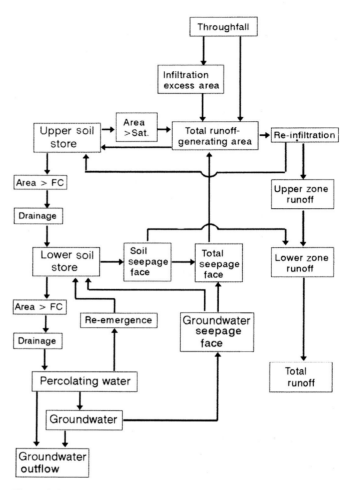

Figure 3.4 Structure of the moisture accounting and main runoff generation components of the VTI model (from Hughes and Sami, 1994)

complex with a large number of parameters. The difficulties of applying the model are associated with the large number of parameter interactions and the fact that many of the parameters are not easy to estimate from known catchment properties. Further details of the parameters and suggestions for the derivation of default values from typically available physical catchment information are provided in Hughes and Sami (1994). In excess of 35 parameters are required for each element (subcatchment) of the spatial distribution system. However, there are between five and eight parameters that are typically modified during calibration, depending on the perceived runoff generation characteristics of a specific catchment. The successful use of the model requires a relatively detailed understanding of its structure and a sound conceptual understanding of the dominant runoff-generation mechanisms of the catchment, as well as good quality input climate data.

The results for several Botswana basins (Hughes, 1997) suggested that, while the model is capable of reproducing daily flow duration curves (Fig. 3.5), the correspondence statistics based on daily flows are quite poor. This may be due to the model structure, but is almost certainly also related to the lack of spatial detail in the available rainfall data. The results for seasonal rivers in Zimbabwe were somewhat better (Figure 3.6), which may be a reflection of the smaller size of the Zimbabwean basins. Results for many of the wetter parts of southern Africa, and where rea-

sonable confidence can be expressed in the available rainfall data, are very encouraging.

3.4.5 Monash model

SMEC (1991) applied the monthly Pitman model as well as the daily Monash model to basins in Botswana. However, the results are presented as monthly summaries and it is therefore difficult to evaluate the model in terms of its ability to simulate daily flows. While the monthly simulations using the Monash model appear to be slight improvements over the monthly Pitman model, no attempts were made to use the "dummy dam" approach to simulating transmission losses for the latter.

3.5 MODELLING ENVIRONMENTS

One of the limitations identified by a SADC (Southern Africa Development Community) report on the implementation of

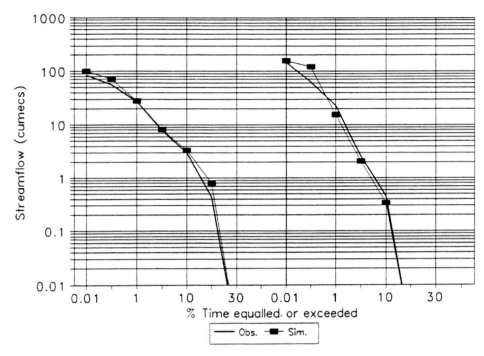

Figure 3.5 Daily flow duration curves for the Ntse River, Botswana (extracted from Hughes, 1997). The left-hand side is the calibration period (1973 to 1979); the right-hand side is the validation period (1980 to 1990)

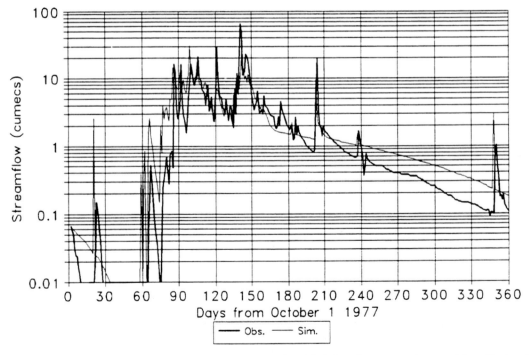

Figure 3.6 Observed and simulated (VTI model) daily flows for the Lumani River, Zimbabwe for one year (extracted from Hughes, 1997)

regional water resource assessments (SADC, 2001) was the lack of access to appropriate models and modelling tools. The need for a common approach and shared methods was also identified as something that could address some of the issues of lack of capacity in hydrological assessment within the southern African region. The SA FRIEND project made use of an integrated modelling environment package (HYMAS – see Hughes *et al.*, 1994 and Hughes, 2004b), but this was developed in a DOS environment and had become outdated and was not very "user friendly." Subsequently, the Institute for Water Research has developed the SPATSIM (Spatial and Time Series Information Modelling) system for Windows (Hughes, 2002). This package makes use of an ESRI Map Objects spatial front end, linked to a database table structure, for data storage and access and includes relatively seamless links to a range of hydrological and water-resource estimation models. It also includes a wide range of data preparation and analysis facilities typically required in hydrological modelling studies. It was originally developed to bring together many of the environmental flow assessment procedures used in South Africa (Hughes, 2004b), but has much wider applicability. The SPATSIM approach has been adopted as the core modelling environment to be used for the update of the South African water-resource information system (WR90 – see Midgley *et al.*, 1994). Further details about the SPATSIM package can be accessed through the IWR web site at http://www.ru.ac.za/institutes/iwr and looking for the "Hydrological Models and Software" link.

The SPATSIM version of the Pitman monthly model has been applied extensively within South Africa, as well as the Okavango basin, Tanzania, Swaziland, Lesotho, and Mozambique. One of the perceived advantages of SPATSIM is that it can be extended to include new modelling approaches quite easily and that these can be developed and added by organizations, other than the IWR, after some initial training.

There are similar packages available internationally, such as Mike Basins, ArcHydro, etc., which are now being introduced into the southern African region. There have been no comparative studies of their relative effectiveness in terms of transferring modelling technology and building capacity in regional water-resource assessment. This issue is important in a region that has limited existing capacity.

3.6 DISCUSSION AND CONCLUSIONS

The previous sections have briefly identified some of the issues that are relevant to the application of hydrological models in the semi-arid and arid regions of southern Africa. This section summarizes the limitations that are related to process understanding, available data, and modelling methods, as well as the opportunities that exist to address these limitations. Some of these opportunities

were highlighted during the G-WADI Workshop held in Roorkee, India during 2005.

3.6.1 Process understanding

There is little doubt that the main limitation to the further development of existing models is the general lack of quantitative understanding of the process of channel transmission loss at various scales. The current lack of research resources, in terms of finances and personnel, suggest that there will only be very limited progress in this regard in the foreseeable future. This is, of course, regrettable in that any future development of model algorithms to account for this important process in the semi-arid catchments of the region will be based on only a very restricted conceptual appreciation of real life.

Establishing networks of groundwater observation boreholes in the large-scale alluvial systems of the region could prove to be a relatively inexpensive and useful approach to quantifying the product of transmission losses (aquifer recharge). However, these observations would need to be coupled with a better understanding of evaporative losses from the alluvial aquifers to be of real benefit.

3.6.2 Available data

Perhaps the main limitation from a data perspective is the lack of adequately representative (certainly in space and frequently in time) rainfall data. There are clear indications that the uncertainty associated with the results of many modelling studies could be reduced if additional information was available about the space–time distribution of rainfall. The prospects for improving the network of ground-based measurements would seem to be poor and therefore the future probably lies in the use of satellite and radar measurements. However, the importance of establishing relationships between new remotely sensed data and historical ground-based measurements has already been noted (Wilk *et al.*, 2006; Hughes *et al.*, 2006). The contribution of Sorooshian *et al.* (2005) in enhancing the awareness of global satellite rainfall-data products is invaluable in this regard. A research priority in the southern African region should be to begin assessing the new data products and establishing their potential for future use.

Obviously, additional streamflow-monitoring stations would be of great value in providing additional calibration data for modelling studies. However, apart from the normal data collection issues associated with a lack of resources in the region, many years are required to accumulate sufficiently representative data in semi-arid regions.

In terms of physical catchment data (such as land use, vegetation, soils, etc.), remote-sensing methods appear to have been applied successfully elsewhere in the world. The priority therefore seems to be to adopt similar approaches in southern Africa and adapt the existing models to be able to make better use of such data.

3.6.3 Model structures and model developments

In a recent review of South African modelling approaches, Hughes (2004c) concluded that the available models have proved to be invaluable for managing water resources in the region, but that the future challenges lay in integrating international developments in model application with local models that have been demonstrated to "work." However, for the semi-arid parts of the region there are still water-resource management issues that the available models are less capable of addressing with confidence (Hughes and Metzler, 1998; Hughes, 2005).

There have been very few studies of the efficiency of the structures of the models used in southern Africa from the point of view of parameter sensitivity and identifiability (Wheater *et al.*, 2005). Recent developments in other parts of the world have indicated that simple models can perform as well as more complex models with more parameters and that the latter are frequently over-parameterized. Several methods were presented by Wheater *et al.* (2005), as well as by Young (2005) that could be used to investigate the structure of models. Many of these can be applied quite easily and might suggest simplifications to the structures of existing models. The success with which the simple daily time-step IHACRES model (Croke and Jakeman, 2005) has been applied in semi-arid basins of Australia and elsewhere, suggests that further trials with simple models of this type could be of value to the southern Africa region.

The data exploration approaches discussed by Young (2005), although not applicable to ungauged basins where much of the simulated data are required, could be used to investigate several modelling issues, such as:

- the extent to which the signals in the available rainfall data can be used to explain the signals in the available streamflow data;
- the "order" of the best model that can be used to explain the streamflow data from the rainfall (and possibly climate data) and the identification of the number of possible flow paths existing in the basin;
- whether the use of time-varying parameters can generate better results and therefore whether there are some non-linear effects that need to be accounted for in the best model structure.

While the models that are developed through Young's (2005) approach may not be applicable to ungauged basins (as they largely rely upon the availability of data), the methods may be very relevant for assessing the current conceptual understanding of basin-scale hydrological processes and therefore the approaches to the application of existing models.

3.6.4 Conclusions

Given the difficulties, referred to above, of addressing the lack of understanding of some processes and of resolving data deficiency problems in the short term, it has not been a straightforward task to identify how existing models should be improved. However, the requirement for more detailed information and more reliable estimates of development, or climate-change impacts suggest that the models currently in use do need to be modified. For example, within South Africa the priority for practical model applications used to be the design of water supply storage reservoirs, where the accurate simulation of dry-season low flows was not all that important. The recent emphasis on environmental flows and ecologically sustainable water-resource development has changed these priorities and the reliable estimation of low flows under natural and modified streamflow regimes is of far greater importance. This does not necessarily mean that the models that have been used in the past need to be replaced. In many cases it is only the approaches to using a model (changing the emphasis of the calibration, for example) that need to be modified. However, there may be real benefits of including a relatively simple daily time-step model to the range of available tools.

Ultimately, the success of any modelling study depends upon the quality and appropriateness of the model, the quality of the data inputs to the model and the experience of the user applying the model and assessing the results. Adopting a pragmatic approach to improving model estimates and focusing on what might be achievable in the future, it is suggested that there is a need for:

- only limited changes to existing models;
- further investigation of the structures of existing models with the intention of developing an improved understanding of the most effective methods of their application in different situations;
- better use of available data (hydrometeorological and spatial basin-property data) and especially satellite rainfall data;
- improved guidelines for model use and training of model users.

REFERENCES

Boroto, R. A. J. and Gorgens, A. H. M. (2003). Estimating transmission losses along the Limpopo River: an overview of alternative methods. Hydrology of the Mediterranean and semi-arid regions. *Proc. Montpellier Conference.* IAHS Publication No 278. Wallingford, UK: IAHS Press, 138–43.

Crerar, S., Fry, R. G., Slater, P. M., van Langenhove, G., and Wheeler, D. (1988). An unexpected factor affecting recharge from ephemeral river flows in SW Africa/Namibia. In *Estimation of Natural Groundwater Recharge*, ed. I. Simmers, Dordrecht: Reidel D. Pub. Co., 11–28.

Croke, B. F. W. and Jakeman, A. J. (2005). *Use of the IHACRES rainfall-runoff model in arid and semi-arid regions.* Paper presented at the GWADI workshop held at the National Institute of Hydrology, Roorkee, India, March 2005.

de Bruine, B. A., Grobler, A. J., and Crerar, S. E. (1993). Evaluation of a specialised rainfall-runoff model for catchments in Nambia. *Proc. 6th South African National Hydrology Symposium*, Pietermaritzburg, Sept. 1993. Pietermaritzburg, South Africa: University of Natal, 225–33.

Görgens, A. H. M. (1983). Reliability of calibration of a monthly rainfall-runoff model: the semiarid case. *Hydrol. Sci. J.*, **28(4)**, 485–98.

Hughes, D. A. (1995). Monthly rainfall-runoff models applied to arid and semi-arid catchments for water resource estimation purposes. *Hydrol. Sci. J.*, **40(6)**, 751–69.

Hughes, D. A. (1997). Southern African FRIEND – the application of rainfall-runoff models in the SADC region. Water Research Commission Report No. 235/1/97. Pretoria, South Africa: Water Research Commission.

Hughes, D. A. (2002). The development of an information modeling system for regional water resource assessments. *Proc. of the 4th International Conference on FRIEND*. IAHS Publ. No. 274, 43–49.

Hughes, D. A. (2004a). Incorporating ground water recharge and discharge functions into an existing monthly rainfall-runoff model. *Hydrol. Sci. J.*, **49(2)**, 297–311.

Hughes, D. A. (ed.) (2004b) SPATSIM, an integrating framework for ecological Reserve determination and implementation. Water Research Commission Report No 1160/1/04. Pretoria, South Africa: Water Research Commission.

Hughes, D. A. (2004c). Three decades of hydrological modelling research in South Africa. *South African J. Sci.*, **100**, 638–42.

Hughes, D. A. (2005). Hydrological issues associated with the determination of environmental water requirements of ephemeral rivers. *River Res. Appl.*, **21(8)**, 899–908.

Hughes, D. A., Andersson, L., Wilk, J., and Savenije, H. H. G. (2006). Regional calibration of the Pitman model for the Okavango River. *J. Hydrology*, **331**, 30–42.

Hughes, D. A. and Metzler, W. (1998). Assessment of three monthly rainfall-runoff models for estimating the water resource yield of semi-arid catchments in Namibia. *Hydrol. Sci. J.*, **43(2)**, 283–97.

Hughes, D. A., Murdoch, K. A., and Sami K. (1994). A hydrological model application system – a tool for integrated river basin management. In *Integrated River Basin Development*, ed. C. Kirby and W. R. White. Chichester, UK: Wiley, 397–406.

Hughes, D. A., Mwelwa, E., Andersson, L., and Wilk, J. (2003). Regional water resources and river flow modelling. In *Southern African FRIEND Phase II 2000–2003*, ed. J. Meigh and M. Fry. Wallingford, UK: Centre for Ecology and Hydrology, 5–30.

Hughes, D. A. and Sami, K. (1992). Transmission losses to alluvium and associated moisture dynamics in a semi-arid ephemeral channel system in southern Africa. *Hydrolog. Proc.*, **6**, 45–53.

Hughes, D. A. and Sami, K. (1993). The Bedford catchments: an introduction to their physical and hydrological characteristics. Unpubl. Report to the Water Research Commission by the Institute for Water Research, Rhodes University, Grahamstown, South Africa.

Hughes, D. A. and Sami, K. (1994). A semi-distributed, variable time interval model of catchment hydrology – structure and parameter estimation procedures. *J. Hydrol.*, **155**, 265–91.

McKenzie, R. S., Roth, C., and Stoffberg, F. (1993). Orange River losses. *Proc. 6th South African National Hydrology Symposium*, Pietermaritzburg, Sept. 1993. Pietermaritzburg, South Africa: University of Natal, 351–8.

McMahon, T. A. (1979). Hydrological characteristics of arid zones. *Symposium on the Hydrology of Areas of Low Precipitation*. IAHS Publ. No 128. Proc. Canberra Symposium, Dec. 1979: 105–23.

Midgley, D. C., Pitman W. V., and Middleton, B. J. (1994). *Surface Water Resources of South Africa 1990*. Vols I–VI. Water Research Commission Reports N0. 298/1.1/94 to 298/6.1/94. Pretoria, South Africa: Water Research Commission.

Mostert, A. C., McKenzie, R. S., and Crerar, S. E. (1993). A rainfall/runoff model for ephemeral rivers in an arid or semi-arid environment. *Proc. 6th South African National Hydrology Symposium*, Pietermaritzburg, Sept 1993. Pietermaritzburg, South Africa: University of Natal, 219–24.

Nash, J. E. and Sutcliffe, J. V. (1970). River flow forecasting through conceptual models. Part I – a discussion of principles. *J. Hydrol.*, **10**, 282–90.

Pitman, W. V. (1973). *A mathematical model for generating monthly flows from meteorological data in South Africa*. Report No. 2/73, Hydrological Research Unit, University of the Witwatersrand, South Africa.

SADC (2001) *Assessment of surface water resources*. Regional Strategic Action Plan for Integrated Water Resource Development and Management in SADC. Project document 14/03. SADC Water Sector Coordinating Unit.

Sami, K. and Hughes, D. A. (1996). A comparison of recharge estimates to a fractured sedimentary aquifer in South Africa from a chloride mass balance and an integrated surface-subsurface model. *J. Hydrol.*, **179**, 111–36.

Schulze, R. E. (1994). *Hydrology and Agrohydrology: A Text to Accompany the ACRU 3.00 Agrohydrological Modelling System*. Water Research Commission Report TT69/95. Pretoria, South Africa: Water Research Commission.

Schulze, R. E. (2000). Modelling hydrological responses to land use and climate change: a southern African perspective. *Ambio*, **29**, 12–22.

SMEC (1991) *Botswana National Water Master Plan Study, Volume 6 – Hydrology*. Prepared by the Snowy Mountains Engineering Corporation. WLPU Consultants and Swedish Geological International AB for the Department of Water Affairs, Botswana.

Sorooshian, S., Hsu, K.-L., Imam, B., and Hong Y. (2005). Global precipitation estimations from satellite image using artificial neural networks. Paper presented at the GWADI workshop held at the National Institute of Hydrology, Roorkee, India, March 2005.

Wheater, H. S., McIntyre, N., and Wagener, T. (2005). Calibration, uncertainty and regional analysis of conceptual rainfall-runoff models. Paper presented at the GWADI workshop held at the National Institute of Hydrology, Roorkee, India, March 2005.

Wilk, J. and Hughes, D. A. (2002). Calibrating a rainfall-runoff model for a catchment with limited data. *Hydrol. Sci. J.*, **47(1)**, 3–17.

Wilk, J., Kniveton, D., Andersson, L., *et al.* (2006). Estimating rainfall and water balance over the Okavango River basin for hydrological applications. *J. Hydrol.*, **331**, 18–29.

Wheeler, D., Crerar, S., Slater, P., and van Langenhove, G. (1987). Investigation of the recharge processes to alluvium in ephemeral rivers. *Proc. 3rd South African National Hydrology Symposium*, Grahamstown, Sept. 1987. Grahamstown, South Africa: Rhodes University, 395–417.

Young, P. C. (2005). Real-time flow forecasting. Paper presented at the GWADI workshop held at the National Institute of Hydrology, Roorkee, India, March 2005.

4 Use of the IHACRES rainfall-runoff model in arid and semi-arid regions

B. F. W. Croke and A. J. Jakeman

Streamflow in arid and semi-arid regions tends to be dominated by rapid responses to intense rainfall events. Such events frequently have a high degree of spatial variability, coupled with poorly gauged rainfall data. This sets a fundamental limit on the capacity of any rainfall-runoff model to reproduce the observed flow. The IHACRES model is a parameterically efficient rainfall-runoff model that has been applied to a large number of catchments covering a diverse range of climatologies. While originally designed for more temperate climates, the model has been successfully applied to a number of ephemeral streams in Australia. The recently released Java-based version of IHACRES (Croke et al., 2005, 2006) includes a modified loss module, as well as a cross-correlation analysis tool, new fit indicators and visualization tools. Additional advances that will be included in future releases of the IHACRES model include: an alternative non-linear loss module which has a stronger physical basis, at the cost of a slightly more complicated calibration procedure; a baseflow filtering approach that uses the SRIV algorithm to estimate the parameter values; a modified routing module that includes the influence of groundwater losses (subsurface outflow and extraction of groundwater); and a technique to estimate the response curve directly from observed streamflow data.

4.1 INTRODUCTION

Successful management of water resources requires qualitative analysis of the effects of changes in climate and land-use practices on streamflow and water quality. While expert knowledge can provide indications of such impacts, detailed analysis requires the use of mathematical models to separate the water balance dynamically (at the temporal scale at which the important processes are operating). This includes separation of incident precipitation into losses to evapotranspiration, runoff to streams, recharge to groundwater systems and changes in short-term catchment storages. Some

of the processes which need to be considered are: evapotranspiration and feedback to the atmosphere; vegetation dynamics; groundwater levels and the resulting effect on soil waterlogging and salinization; reservoir storage capacity reliability; wetland dynamics; urban runoff; flooding; erosion in crop and pasture lands, as well as channel erosion and sedimentation; and aquatic ecosystem functions.

Arid and semi-arid areas tend to be dominated by intense rainfall events with a high degree of spatial variability. This typically leads to a rapid response profile, and in areas without weather radar coverage, poor rain-gauge density prevents accurate estimation of the rainfall depth and spatial distribution for a particular event. Further, if only daily rainfall data are available, then calibrating a rainfall-runoff model at a daily time-step means that most of the information contained in the hydrograph is not used (note that runoff here means total streamflow, not just surface runoff).

Another important consideration for calibration of models for catchments in arid and semi-arid areas is the frequency of events. Such catchments tend to have fewer streamflow events than catchments in wetter climates. This means that longer calibration periods are needed in order to reduce the uncertainty in model parameters. Otherwise, parameter values will tend to relate more to the errors in the data, with a significant decrease in performance in simulation compared with calibration.

4.2 IHACRES RAINFALL-RUNOFF MODEL

4.2.1 Data availability

Typically the available data for catchments (other than heavily instrumented research catchments) is limited to daily rainfall and temperature and, in some cases, stream discharge. Thus the mathematical representation most often used is a rainfall-runoff model.

Hydrological Modeling in Arid and Semi-Arid Areas, ed. Howard Wheater, Soroosh Sorooshian, and K. D. Sharma. Published by Cambridge University Press.
© Cambridge University Press 2008.

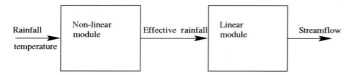

Figure 4.1 Generic structure of the IHACRES model, showing the conversion of climate time series data to effective rainfall using the non-linear module, and the linear module converting effective rainfall to streamflow time series

Rainfall-runoff models fall into several categories: metric, conceptual, and physics-based models (Wheater *et al.*, 1993). Metric models are typically the most simple, using observed data (rainfall and streamflow) to characterize the response of a catchment. Conceptual models impose a more complex representation of the internal processes involved in determining catchment response, and can have a range of complexity depending on the structure of the model. Physics-based models involve numerical solution of the relevant equations of motion.

4.2.2   Model structure

The selection of which model to use should be based on the issue(s) being investigated and the data available. As more complex questions are asked, more complex models may be needed to provide the answers. However, with increasing model complexity comes the cost of increasing uncertainty in the model predictions. The IHACRES model is a hybrid conceptual-metric model, using the simplicity of the metric model to reduce the parameter uncertainty inherent in hydrological models while at the same time attempting to represent more detail of the internal processes than is typical for a metric model. Figure 4.1 shows the generic structure of the IHACRES model. It contains a non-linear loss module which converts rainfall into effective rainfall (that portion which eventually reaches the stream prediction point) and a linear module which transfers effective rainfall to stream discharge. Further modules can be added including one that allows recharge to be an output. The inclusion of a range of non-linear loss modules within IHACRES increases its flexibility in being used to assess the effects of climate and land-use change. The linear module routes effective rainfall to stream through any configuration of stores in parallel and/or in series. The configuration of stores is identified from the time series of rainfall and discharge but is typically either one store only, representing ephemeral streams, or two in parallel, allowing baseflow or slowflow to be represented as well as quickflow. Only rarely does a more complex configuration than this improve the fit to discharge measurements (Jakeman and Hornberger, 1993).

The original structure of the IHACRES model used an exponentially decaying soil-moisture index to convert rainfall into effective

rainfall. Ye *et al.* (1997) adapted this model to improve the performance of the model in ephemeral catchments. This involved introducing a threshold parameter (*l*) and a non-linear relationship (power law with exponent parameter *p*) between the soil-moisture index and the fraction of rainfall that becomes effective rainfall.

The Ye *et al.* (1997) version has been coded within IHACRES_v2.0, reformulated to enable the mass balance parameter c (see below) to be estimated from the gain of the transfer function, and to reduce the interaction between the c and p parameters. The effective rainfall u_k in the revised model is given by:

$$u_k = [c(\phi_k - l)]^p r_k, \tag{4.1}$$

where r_k is the observed rainfall, c, l, and p are parameters (mass balance, soil-moisture index threshold and non-linear response terms, respectively), and ϕ_k is a soil-moisture index given by:

$$\phi_k = r_k + (1 - 1/\tau_k)\phi_{k-1}. \tag{4.2}$$

The drying rate τ_k is given by:

$$\tau_k = \tau_w \exp(0.062 f(T_r - T_k)), \tag{4.3}$$

where τ_w, f, and T_r are parameters (reference drying rate, temperature modulation and reference temperature, respectively). This formulation enables the gain of the transfer function to be directly related to the value of the parameter c, thus simplifying model calibration. This version of the model is more general than the version used within the IHACRES_PC model (Littlewood *et al.*, 1997), which can be recovered by setting parameters l to zero and p to one (with the soil-moisture index in the original model given by $s_k = c\phi_k$). This version of the non-linear module is described in detail in Jakeman *et al.* (1990) and Jakeman and Hornberger (1993). Examples of studies that have used this version of IHACRES (with minor modifications to Equations 4.1 to 4.3) can be found in Hansen *et al.* (1996), Post and Jakeman (1999), Schreider *et al.* (1996), and Ye *et al.* (1997).

The linear module uses exponentially decaying stores to convert rainfall (*U*) into streamflow (*Q*). For a single exponential store:

$$Q_k = -\alpha Q_{k-1} + \beta U_{k-\delta}, \tag{4.4}$$

where δ is the delay between rainfall and streamflow response, α is the storage coefficient (determines decay rate of the store), and β is the fraction of effective rainfall that appears as streamflow in the current time-step. IHACRES_v2.0 uses a second-order transfer function to represent the unit hydrograph response curve, which can be written as:

$$Q_k = \frac{b_0 + b_1 z^{-1} + b_2 z^{-2}}{1 + a_1 z^{-1} + a_2 z^{-2}} U_{k-\delta}, \tag{4.5}$$

where z is the timestep operator (z^{-1} produces a one step shift backwards in time) and a_i and b_i are fitted parameters and are related to the coefficients in Equation (4.4):

$$a_1 = \alpha_q + \alpha_s,$$
$$a_2 = \alpha_q \alpha_s,$$
$$b_1 = \beta_i + \beta_q + \beta_s, \qquad (4.6)$$
$$b_1 = \beta_i(\alpha_q + \alpha_s) + \beta_q \alpha_s + \beta_s \alpha_q,$$
$$b_1 = \beta_i \alpha_q \alpha_s,$$

where the subscripts refer to instantaneous (i), quick (q) and slow (s) components. The volume (V) of each component is given by:

$$V_q = \frac{\beta_q}{1 + \alpha_q}, \qquad V_s = \frac{\beta_s}{1 + \alpha_s}, \qquad V_i = \beta_i, \qquad (4.7)$$

with the sum of the volumes of all components equalling one by definition.

4.2.3 Regionalization

Various versions of the IHACRES model have also been used to address regionalization issues (Post and Jakeman, 1996; Sefton and Howarth, 1998; Kokkonen et al., 2003). These issues require methods for estimating the parameters of models from independent means such as landscape attributes rather than directly from rainfall-discharge time series. The parametric efficiency of IHACRES (often about six parameters) lends itself to regionalization problems, making it easier than complex models to relate its parameters to landscape attributes. The IHACRES model is one of the models to be used by the Top-Down Modelling Working Group (Littlewood et al., 2003) operating as part of the Prediction in Ungauged Basins initiative of the International Association of Hydrological Sciences (http://cee.uiuc.edu/research/pub/).

4.3 NEW VERSION OF IHACRES

There are a number of reasons for the development of a Java-based version of the rainfall-runoff model IHACRES. This includes several recent developments in the model, particularly with respect to the non-linear loss module. One such modification is the development of a catchment-moisture deficit (CMD) accounting system that enables a more process-based determination of the partitioning of rainfall to discharge and evapotranspiration (Croke and Jakeman, 2004). Other enhancements include simulating the effects of retention storages such as farm dams on stream discharge (Schreider et al., 1999) as well as the interaction between groundwater recharge and streamflow by linking a physics-based groundwater discharge model (Sloan, 2000) with the IHACRES model (Croke et al., 2002). The groundwater version of the model was used to assess groundwater recharge in the Jerrabomberra Creek catchment, ACT (Croke et al., 2001). These developments have improved the potential of IHACRES to model the effects

of land use-change on catchment response (e.g., Dye and Croke, 2003), as well as inferring the hydrological response of ungauged catchments.

Further advances in the IHACRES model have been made in the method of calibration. In addition to the current simple refined instrumental variable (SRIV) method of parameter estimation (e.g., Jakeman et al., 1990), a method based on estimating hydrographs directly from streamflow data without the need for rainfall data has been developed (Croke, 2004). This enables higher-resolution streamflow data to be used, reducing the loss of information that occurs when data are binned to a daily time-step. In addition, this calibration method reduces the number of parameters that need to be estimated within the model, thus reducing the parameter uncertainty, while at the same time reducing the time required to calibrate the model.

In order to visualize data such as inputs and model outputs, the new Java-based version of IHACRES makes use of the VisAD library. VisAD (Hibbard, 1998) is a Java component library for interactive visualization and analysis of numerical data. The library is available under the Lesser General Public License (LGPL), which allows the library to be used in commercial applications so long as certain conditions are satisfied. Using VisAD it has been possible to create very sophisticated interactive visualizations of data within IHACRES_v2.0 with a minimum amount of effort. The visualization of data is very important for the calibration and interpretation of models like IHACRES_v2.0, where it is necessary for users to be able to view effective representations of data in order to make appropriate decisions.

A number of changes have been made to the objective functions used in IHACRES_v2.0. The lag 1 correlation coefficients U1 (correlation coefficient of the lagged effective rainfall and model error) and X1 (correlation coefficient of the lagged modelled streamflow and the model error) have been normalized correctly (e.g., a value of +1 corresponds to perfect correlation) to aid in interpretation of these values. Also, a number of objective functions have been added. These are based on the Nash–Sutcliffe model efficiency indicator:

$$R^2 = 1 - \frac{\sum_i (Q_{o,i} - Q_{m,i})^2}{\sum_i (Q_{o,i} - \overline{Q_o})^2}, \qquad (4.8)$$

with the observed flow Q_o, and modelled flow Q_m replaced with the square root (R2_sqrt), logarithm (R2_log) and inverse (R2_inv) of the flow. These objective functions are progressively less biased to peak flows, and more to low flows. In arid and semi-arid catchments, the best objective function is likely to be R2_sqrt as there is rarely a baseflow component in these catchments.

To avoid numerical errors with the logarithmic and inverse versions, the 90 % flow exceedence value (ignoring time-steps with no flow) was added to Q_o. These shift the weighting of the objective

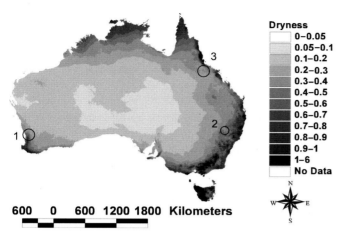

Figure 4.2 Ratio of rainfall to potential evaporation across Australia (rainfall and potential evaporation data from the National Land and Water Resources Audit database (http://adl.brs.gov.au/ADLsearch/). Circles indicate approximate locations of the catchments used in this report: Canning River (1), Mooki River and Coxs Creek (2), and Broughton River (3)

function progressively from the flow peaks to low flows. The logarithmic form, for example, gives a fairly uniform weighting to high and low flows, while the inverse version is heavily biased towards low flows. Subsequently, for studies focused on flood peaks, the traditional Nash–Sutcliffe efficiency should be preferred. However, where simulating low flows are important (e.g., ecological impacts, estimating availability of water for irrigation), the logarithmic or inverse forms of the efficiency indicator should be preferred.

4.4 APPLICATION OF THE MODEL TO AUSTRALIAN CATCHMENTS

A significant fraction of Australia is classed as arid or semi-arid (see Fig. 4.2). Catchments from three different climatic regions are presented here: The Canning River in Western Australia, the Liverpool Plains region of the Namoi River catchment in northern NSW, and the Burdekin River catchment in northern Queensland.

4.4.1 Canning River

The Canning River is a tributary of the Swan River, and is located southeast of Perth in Western Australia (see Fig. 4.2). Data for the gauge at Scenic Drive (616024) have been used to test the IHACRES model. This gauge is upstream of Canning Dam, and has a contributing area of 517 km^2. The data consist of daily rainfall (P), streamflow (Q) and potential evaporation (PE) covering the period from January 1, 1977 to December 31, 1987,

with all data expressed in mm. The catchment has a Mediterranean climate, with about 80 % of the rainfall occurring between May and October. The river is ephemeral, with no flow approximately 54 % of the time. For the 11 years from 1978 to 1987, the mean annual rainfall was 890 mm, the runoff coefficient was 1.8 % and ratio of rainfall to potential evaporation was 0.64 (using the National Land and Water Resources Audit database http://adl.brs.gov.au/ADLsearch/, the ratio of rainfall to potential evaporation was 0.57). Thus, while the catchment is ephemeral, it is just above the threshold for being classed as semi-arid (semi-arid catchments are defined as having P/PE values between 0.2 and 0.5).

The catchment was calibrated on a five-year period from January 1, 1978 to December 31, 1982 (see Fig. 4.3) using the modified Ye et al. (1997) non-linear loss module (Equations 4.1 to 4.3). Since a time series of potential evaporation was used for the variation in the drying rate (rather than the more usual temperature), the reference temperature was set to zero. Calibration runs showed that the model was unable to reproduce the observed flows if the threshold parameter (l) was fixed at zero, and the non-linearity parameter (p) fixed at one. Introducing the threshold parameter resulted in a considerable improvement in the model fit, with R^2 values reaching 0.93 for the calibration period. The calibrated parameter values selected based on the R2_log objective function are shown in Table 4.1.

The calibrated model gave objective function values: $R^2 = 0.90$, R2_sqrt = 0.93, R2_log = 0.95 and R2_inv = 0.93 for the calibration period. For the remainder of the data, the simulated values for the objective functions were $R^2 = 0.87$, R2_sqrt = 0.95, R2_log = 0.97 and R2_inv = 0.96. This shows that, except for the R^2 objective function, the model performed slightly better in the simulation period than in the calibration period. Analysis of the statistics for individual years shows that the model performed very poorly in 1980 ($R^2 = 0.22$) and poorly in 1987 ($R^2 = 0.55$). All other years gave R^2 values of greater than or equal to 0.84 with the exception of 1977 ($R^2 = 0.79$) and 1979 ($R^2 = 0.72$).

4.4.2 Namoi River catchments

The Liverpool Plains region of the Namoi River catchment is located in northern NSW, Australia (see Fig. 4.2). This area is semi-arid, with a ratio of P/PE of between 0.4 and 0.5 for most of the area (except for the Liverpool Range to the south). The main rivers in the region are the Mooki River and Coxs Creek, which drain north from the Liverpool Ranges to the Namoi River.

GAUGE 419034

This gauge is located on the Mooki River at Caroona, and corresponds to a catchment area of 2540 km^2. The catchment was

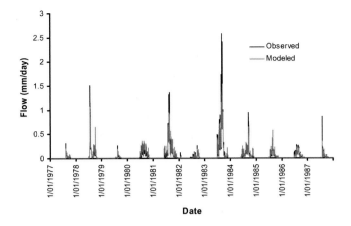

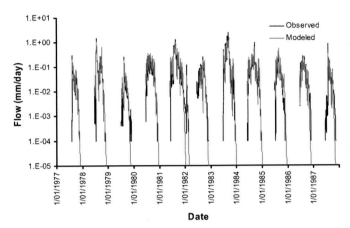

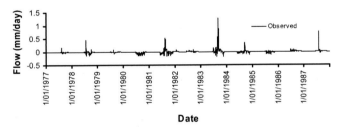

Figure 4.3 Observed and modelled flows for the Canning River at Scenic Drive (gauge 616024)

Table 4.1 *Calibrated parameter values for all catchments (note only* t_w*, f, l and α were free parameters)*

Parameter	616024	419034	419032	120014
c	0.000284	0.002896	0.002465	0.001114
t_w	162	42	42	22
f	2	1	1.5	0.5
T_{ref}	0	20	20	20
l	300	100	100	50
p	1	1	1	1
α	−0.793	−0.323	−0.389	−0.125
V	1	1.744	1	1
Calibration period				
R^2	0.90	0.88	0.86	0.76
R2_sqrt	0.93	0.70	0.86	0.82
R2_log	0.95	0.58	0.77	0.90
R2_inv	0.93	−0.44	0.39	0.91
Simulation period				
R^2	0.87	0.71	0.59	0.72
R2_sqrt	0.95	0.77	0.65	0.78
R2_log	0.97	0.54	0.82	0.87
R2_inv	0.96	−2.99	0.71	0.89

R2_log = 0.54 and R2_inv = −2.99. Thus the model performed slightly better in the simulation period in terms of the R2_sqrt and R2_log objective functions. Analysis of the statistics for individual years shows that the model performed extremely poorly in 8 out of 20 years, with negative R^2 values being recorded. This was balanced with 8 out of the remaining 14 years yielding R^2 values greater than or equal to 0.69.

GAUGE 419052

This gauge is located on the Coxs Creek at Mullaley, and has a catchment area of 2370 km². The calibration period selected was January 1, 1974 to December 31, 1978 (5 years), with the data record extending from December 3, 1972 to January 31, 1989 (a little over 16 years, see Fig. 4.5). The calibrated parameter values were selected based on the R2_sqrt objective function, and are shown in Table 4.1.

For the calibration period, the model yielded objective function values of: $R^2 = 0.86$, R2_sqrt = 0.86, R2_log = 0.77, and R2_inv = 0.39, showing better performance in simulating the low flows in this catchment compared with gauge 419034. For the remainder of the data, the values for the objective functions were $R^2 = 0.59$, R2_sqrt = 0.65, R2_log = 0.82 and R2_inv = 0.71. Thus the model performed slightly better in simulating the low

calibrated on a five-year period from January 1, 1981 to December 31, 1985 (see Fig. 4.4) again using the modified Ye *et al.* (1997) non-linear loss module. The calibrated parameter values selected based on the R2_sqrt objective function are shown in Table 4.1.

For the calibration period, the model yielded objective function values of: $R^2 = 0.88$, R2_sqrt = 0.70, R2_log = 0.58 and R2_inv = −0.44, showing the poor performance in simulating the low flows in this catchment. For the remainder of the data, the values for the objective functions were $R^2 = 0.71$, R2_sqrt = 0.77,

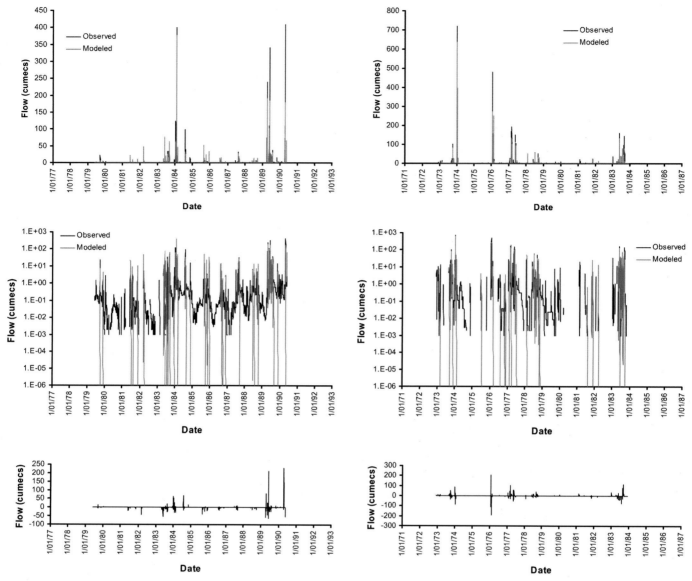

Figure 4.4 Observed and modelled flow for Mooki River (gauge 419034)

Figure 4.5 Observed and modelled flow for Coxs Creek (gauge 419052)

flows for the simulation period (R2_log and R2_inv objective functions). Analysis of the statistics for individual years shows that the model performed extremely poorly in 7 out of 16 years (some of the driest years in the data), with negative R^2 values being recorded. There were four years with R^2 values greater than 0.66, three of these in the calibration period.

The poor R2_inv value for both catchments is a result of not including a slow-flow component in the model. This is evident in the log plots in Figs. 4.4 and 4.5. While these catchments have a slow-flow component, it is not easily identifiable with the parameter estimation method used here. For gauge 419052, the intermit-

tent nature of the slow-flow component is another problem that needs to be addressed.

4.4.3 Burdekin River

The Burdekin River is a large (approximately 130 000 km^2) coastal catchment in northern Queenland (see Fig. 4.2). The catchment has a dry tropical climate, with rainfall dominated by high intensity events. The data used here is for gauge 120014 (Broughton River at Oak Meadows, area 181 km^2). The data

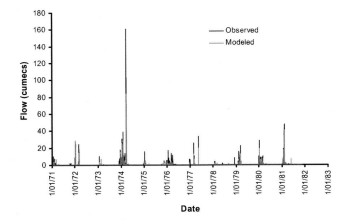

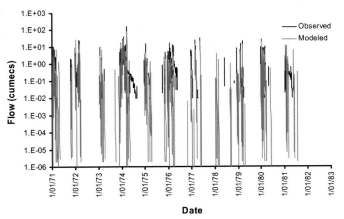

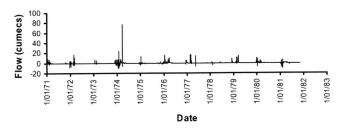

Figure 4.6 Observed and modelled flow for Broughton River (gauge 120014).

extend from November 8, 1970 to December 31, 1987. They consist of rainfall (mm), streamflow (cumecs), and temperature (°C). The mean annual rainfall from 1/1/1980 to 31/12/1999 was 590 mm, and ratio of rainfall to potential-evaporation was 0.32 (derived from rainfall and potential evaporation surfaces obtained from the National Land and Water Resources Audit database http://adl.brs.gov.au/ADLsearch/).

For the calibration period, the model yielded objective function values of: $R^2 = 0.76$, R2_sqrt $= 0.82$, R2_log $= 0.90$ and R2_inv $= 0.91$, showing better performance in simulating the

low flows in this catchment compared with the gauges in the Namoi River catchment (see Fig. 4.6). For the remainder of the data, the values for the objective functions were $R^2 = 0.72$, R2_sqrt $= 0.78$, R2_log $= 0.87$, and R2_inv $= 0.89$. This is the result of the lack of any baseflow component in this catchment. Analysis of the statistics for individual years shows that the model performed extremely poorly in 2 out of 17 years (some of the driest years in the data), with negative R^2 values being recorded. There were seven years with R^2 values greater than 0.6, three of these in the calibration period.

4.5 CONCLUSION

The IHACRES_v2.0 software is a considerable enhancement of the IHACRES_PC software. The software can be used on any platform that has the appropriate Java runtime environment. In addition, the functionality of the software has been increased through the inclusion of additional non-linear modules and alternative calibration techniques, as well as improved visualization of data and modelled results. The model can be applied to arid and semi-arid catchments, though the length of the calibration period should be increased to accommodate the lower frequency of streamflow events.

4.6 ACKNOWLEDGEMENTS

The authors wish to thank Felix Andrews for coding the model, Ian Littlewood for discussions on the development of this version of IHACRES, and the beta testers for help in testing the software.

4.7 REFERENCES

Croke, B. F. W. (2004). A technique for deriving the average event unit hydrograph from streamflow-only data for quick-flow-dominant catchments. *Adv. Water Resour.*, **29**, 493–502 (doi:10.1016/j.advwatres.2005.06.005).

Croke, B. F. W., Andrews, F., Jakeman, A. J., Cuddy, S. M., and Luddy, A. (2005). Redesign of the IHACRES rainfall-runoff model. *29th Hydrology and Water Resources Symposium, Water Capital*. Canberra: Engineers Australia, 21–23 February, 2005 (ISBN 085 825 8439).

Croke, B. F. W., Andrews, F., Jakeman, A. J., Cuddy, S. M., and Luddy, A. (2006). IHACRES Classic Plus: a redesign of the IHACRES rainfall-runoff model. *Environ. Mod. Software*, **21**, 426–7.

Croke, B. F. W. and Jakeman, A. J. (2004). A catchment moisture deficit module for the IHACRES rainfall-runoff model. *Environ. Mod. Software*, **19**, 1–5.

Croke, B. F. W., Smith, A. B., and Jakeman, A. J. (2002). A one-parameter groundwater discharge model linked to the IHACRES rainfall-runoff model. In *Proceedings of the 1st Biennial Meeting of the International Environmental Modelling and Software Society*, A. Rizzoli and A. Jakeman (eds.), Vol. I. University of Lugano, Switzerland, 428–33.

Croke B. F. W., Evans, W. R., Schreider, S. Y., and Buller, C. (2001). Recharge estimation for Jerrabomberra Creek catchment, the Australian Capital Territory. In *MODSIM2001, International Congress on Modelling and Simulation*, Canberra, 10–13 December 2001, F. Ghassemi, P. Whetton, R. Little and M. Littleboy (eds.), Vol. 2., ISBN 0 86740 525 2, 555–60.

Dye P. J. and Croke, B. F. W. (2003). Evaluation of streamflow predictions by the IHACRES rainfall-runoff model in two South African catchments. *Environ. Mod. Software*, **18**, 705–12.

Hansen, D. P., Ye, W., Jakeman, A. J., Cooke, R., and Sharma, P. (1996). Analysis of the effect of rainfall and streamflow data quality and catchment dynamics on streamflow prediction using the rainfall-runoff model IHACRES. *Environ. Software*, **11**, 193–202.

Hibbard, W. (1998). VisAD: connecting people to computations and people to people. *Comp. Graphics*, **32**, 10–12.

Jakeman, A. J. and Hornberger, G. M. (1993). How much complexity is warranted in a rainfall-runoff model? *Water Resour. Res.*, **29**, 2637–49.

Jakeman, A. J., Littlewood, I. G., and Whitehead, P. G. (1990). Computation of the instantaneous unit hydrograph and identifiable component flows with application to two small upland catchments. *J. Hydrol.*, **117**, 275–300.

Kokkonen, T., Jakeman, A. J., Young, P. C., and Koivusalo, H. J. (2003). Predicting daily flows in ungauged catchments: model regionalization from catchment descriptors at the Coweeta Hydrologic Laboratory, North Carolina. *Hydrolog. Proc.*, **17**, 2219–38.

Littlewood, I. G., Croke, B. F. W., Jakeman, A. J., and Sivapalan, M. (2003). The role of "top-down" modelling for Prediction in Ungauged Basins (PUB). *Hydrolog. Proc.*, **17**, 1673–9.

Littlewood, I. G., Down, K., Parker, J. R., and Post, D. A. (1997). *IHACRES – Catchment-scale rainfall-streamflow modelling (PC version) Version 1.0 – April 1997.* Canberra: The Australian National University, Institute of Hydrology and Centre for Ecology and Hydrology.

Post, D. A. and Jakeman, A. J. (1996). Relationships between catchment attributes and hydrological response characteristics in small Australian mountain ash catchments. *Hydrolog. Proc.*, **10**, 877–92.

Post, D. A. and Jakeman, A. J. (1999). Predicting the daily streamflow of ungauged catchments in S.E. Australia by regionalising the parameters of a lumped conceptual rainfall-runoff model. *Ecolog. Modelling*, **123**, 91–104.

Schreider S. Y., Jakeman, A. J., Letcher, R. A., *et al.* (1999). *Impacts and Implications of Farm Dams on Catchment Yield*, iCAM working paper 1999/3. ISSN 1442–3707.

Schreider, S. Yu., Jakeman, A. J., and Pittock, A. B. (1996). Modelling rainfall-runoff from large catchment to basin scale: The Goulburn Valley, Victoria. *Hydrolog. Proc.*, **10**, 863–76.

Sefton, C. E. M. and Howarth, S. M. (1998). Relationships between dynamic response characteristics and physical descriptors of catchments in England and Wales. *J. Hydrol.*, **211**, 1–16.

Sloan, W. T. (2000). A physics-based function for modelling transient groundwater discharge at the watershed scale. *Water Resour. Res.*, **36**, 225–41.

Wheater, H. S., Jakeman, A. J., and Beven, K. J. (1993). Progress and directions in rainfall-runoff modelling. In *Modelling Change in Environmental Systems*, ed. A. J. Jakeman, M. B. Beck and M. J. McAleer Chichester: Wiley, 101–32.

Ye, W., Bates, B. C., Viney, N. R., Sivapalan, M. and Jakeman, A. J. (1997). Performance of conceptual rainfall-runoff models in low-yielding ephemeral catchments. *Water Resour. Res.*, **33**, 153–66.

5   KINEROS2 and the AGWA modelling framework

D. J. Semmens, D. C. Goodrich, C. L. Unkrich, R. E. Smith, D. A. Woolhiser, and S. N. Miller

5.1   INTRODUCTION

This chapter describes the conceptual model, mathematical model, and numerical methods underpinning the Kinematic Runoff and Erosion Model, KINEROS2. The performance of KINEROS2 and its numerous components has been evaluated in numerous studies, which were described in detail by Smith *et al.* (1995a). Here we provide an overview of the geospatial interface for KINEROS2, including data requirements and the major steps and methods used to derive model inputs. An example is provided illustrating how KINEROS2 can be used via AGWA for multi-scale watershed assessment. We conclude with a description of current and planned research and development that is designed to improve both KINEROS2 and AGWA and their usability for environmental management and planning.

5.1.1   KINEROS2

KINEROS2 is a distributed, physically based, event model describing the processes of interception, dynamic infiltration, surface runoff, and erosion from watersheds characterized by predominantly overland flow. The watershed is conceptualized as a cascade of planes and channels, over which flow is routed in a top-down approach using a finite difference solution of the one-dimensional kinematic wave equations. KINEROS2 may be used to evaluate the effects of various artificial features such as urban developments, detention reservoirs, circular conduits, or lined channels on flood hydrographs and sediment yield.

KINEROS2 originated at the US Department of Agriculture (USDA) Agricultural Research Service's (ARS) Southwest Watershed Research Center (SWRC) in the late 1960s as a model that routed runoff from hillslopes represented by a cascade of one-dimensional overland-flow planes contributing laterally to channels (Woolhiser, *et al.*, 1970). Rovey (1974) coupled interactive infiltration to this model and released it as KINGEN (Rovey

et al., 1977). After significant validation using experimental data, KINGEN was modified to include erosion and sediment transport as well as a number of additional enhancements, resulting in KINEROS (KINematic runoff and EROSion), which was released in 1990 (Woolhiser *et al.*, 1990) and described in some detail by Smith *et al.* (1995a). Subsequent research with, and application of, KINEROS has led to additional model enhancements and a more robust model structure, which have been incorporated into the latest version of the model: KINEROS2 (hereafter referred to as K2). K2 is open-source software that is distributed freely via the Internet, along with associated model documentation (www.tucson.ars.ag.gov/kineros).

5.1.2   AGWA

A geographic information system (GIS) user interface for K2, the Automated Geospatial Watershed Assessment (AGWA) tool, facilitates parameterization and calibration of the model. AGWA uses internationally available spatial datasets to delineate the watershed, subdivide it into model elements, and derive all necessary parameter inputs for each model element. AGWA also enables the spatial visualization and comparison of model results, and thus permits the assessment of hydrologic impacts associated with landscape change. The utilization of a GIS further provides a means of relating model results to other spatial information.

Spatially distributed data are required to develop inputs for K2, and the subdivision of watersheds into model elements and the assignation of appropriate parameters are both time-consuming and computationally complex. To apply K2 on an operational basis, there was thus a critical need for automated procedures that could take advantage of widely available spatial datasets and the computational power of geographic information systems (GIS). The AGWA GIS interface for K2 was developed in 2002 (Miller *et al.*, 2002) by the USDA-ARS, US Environmental Protection

Hydrological Modeling in Arid and Semi-Arid Areas, ed. Howard Wheater, Soroosh Sorooshian, and K. D. Sharma. Published by Cambridge University Press.
© Cambridge University Press 2008.

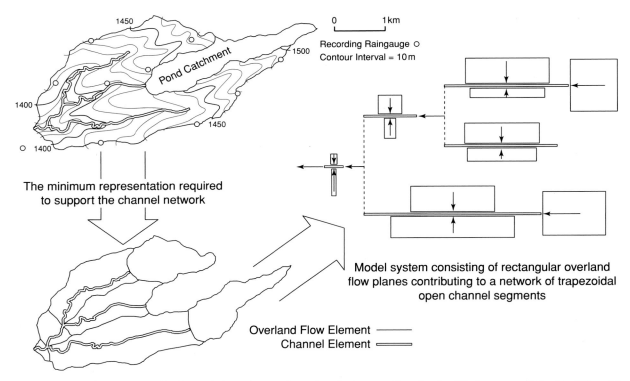

Walnut Gulch Subwatershed No. 11 showing the watershed boundary and primary channel network (the pond catchment is a noncontributing area).

Figure 5.1 Illustration of how topographic data and channel network topology is abstracted into the simplified geometry defined by K2 model elements. Note that overland-flow planes are dimensioned to preserve average flow length, and therefore planes contributing laterally to channels generally do not have widths that match the channel length.

Agency (EPA) Office of Research and Development (ORD), and the University of Arizona (UA) to address this need.

AGWA is an extension for the Environmental Systems Research Institute's ArcView versions 3.X (ESRI, 2001), a widely used and relatively inexpensive PC-based GIS software package (trade names are mentioned solely for the purpose of providing specific information and do not imply recommendation or endorsement by the US EPA or USDA). The GIS framework of AGWA is ideally suited for watershed-based analysis in which landscape information is used for both deriving model input, and for visualization of the environment and modelling results. AGWA is distributed freely via the Internet as a modular, open-source suite of programs (www.tucson.ars.ag.gov/agwa or www.epa.gov/nerlesd1/land-sci/agwa).

5.2   KINEROS2 MODEL DESCRIPTION

5.2.1   Conceptual model elements

In K2, the watershed being modelled is conceptualized as a collection of spatially distributed model elements, of which there can be several types. The model elements effectively abstract the watershed into a series of shapes, which can be oriented so that one-dimensional flow can be assumed. A typical subdivision, from topography to model elements, of a small watershed in the USDA-ARS Walnut Gulch Experimental is illustrated in Fig. 5.1. Further subdivision can be made to isolate hydrologically distinct portions of the watershed if desired (e.g., large impervious areas, abrupt changes in slope, soil type, or hydraulic roughness, etc.). As currently implemented, the computational order of the K2 model simulation must proceed from upslope/upstream elements to downstream elements. This is required to ensure that upper boundary conditions for the element being processed are always defined. Attributes for each of the model-element types are summarized in Table 5.1, and followed by more detailed descriptions in the text.

5.2.1.1   OVERLAND-FLOW ELEMENTS

Overland-flow elements are abstracted as regular, planar, rectangular surfaces with uniform parameter inputs. Non-uniform surfaces, such as converging or diverging contributing areas, or major breaks in slope, may be represented using a cascade of overland-flow elements, each with different parameter inputs.

Table 5.1 *KINEROS2 model-element types and attributes*

Model element type	Attributes
Overland flow	Planes; cascade allowed with varied lengths, widths, and slopes; microtopography
Urban overland	Mixed infiltrating/impervious with runoff-runon
Channels	Simple and compound trapezoidal
Detention Structures	Arbitrary shape, controlled outlet – discharge f (stage)
Culverts	Circular with free surface flow
Injection	Hydrographs and sedigraphs injected from outside the modelled system, or from a point discharge (e.g., pipe, drain)

Microtopographic relief on upland surfaces can be represented by specifying the microtopographic relief height and mean spacing.

5.2.1.2 URBAN ELEMENTS

The urban element represents a composite of up to six overland-flow areas (Fig. 5.2), including various combinations of pervious and impervious surfaces contributing laterally to a paved, crowned street. This model element was originally conceived as a single residential or commercial lot; however, a contiguous series of similar lots along the same street can be combined into a single urban element. The aggregate model representation is offered instead of attempting to describe each roof, driveway, lawn, sidewalk, etc., as individual model elements. The urban element can receive upstream inflow (into the street), but not lateral inflow from adjacent urban or overland-flow elements. The relative proportions of the six overland-flow areas are specified as fractions of the total element area. It is not required to have all six types, but intervening connecting areas must be present if the corresponding indirectly connected area is specified. The element is modelled as rectangular.

5.2.1.3 CHANNEL ELEMENTS

Channels are defined by two trapezoidal cross-sections at the upstream and downstream ends of each reach. Geometric and hydrologic parameters can be uniform, or vary linearly along a reach. If present, base flow can be represented with a constant inflow rate. Compound trapezoidal channels (Fig. 5.3) can be represented as a parallel pair of channels, each with its own hydraulic and infiltrative characteristics. For each channel, the geometric relations for cross-sectional area of flow A and wetted perimeter P are expressed in terms of the same depth, h, whose zero value corresponds to the level of the lowermost channel segment (Fig. 5.3). Note that the wetted perimeters do not include the interface where the two sections join, i.e., this constitutes a frictionless boundary (dotted vertical line). There is no need to explicitly account for mass transfer between the two channels, as it is implicit in the common depth (level water surface) requirement. However, for exchange of suspended sediment, a net transfer rate q_t is recovered via a mass balance after computation of h at the advanced time-step.

5.2.1.4 POND ELEMENTS

In addition to surface and channel elements, a watershed may contain detention storage elements, which receive inflow from up to ten upstream elements and two lateral elements and produce outflow from an uncontrolled outlet structure. This type of element can be used to represent a pond, or a flume or other flow-measuring structure with backwater storage. User-defined rating information is required to parameterize these elements. Infiltration, or seepage, is computed using a constant, user-defined saturated hydraulic conductivity.

5.2.1.5 CULVERT ELEMENTS

In an urban environment, circular conduits must be used to represent storm sewers. To apply the kinematic model, there must be no backwater, and the conduit is assumed to maintain free surface-flow conditions at all times – there can be no pressurization. A schematic drawing of a partially full circular section is shown in Fig. 5.4. There is assumed to be no lateral inflow. The upper boundary condition is a specified discharge as a function of time. No infiltration is computed for culvert elements.

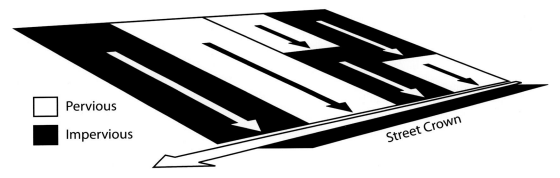

Figure 5.2 Diagram illustrating the layout of an urban element and all six possible contributing areas.

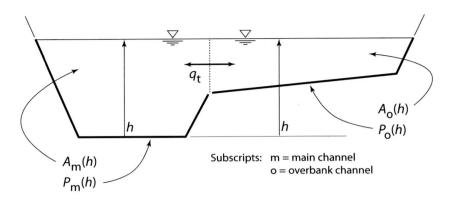

Figure 5.3 Diagram illustrating basic compound channel cross-section geometry

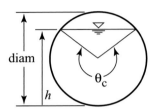

Figure 5.4 Diagram illustrating basic culvert geometry

5.2.1.6 INJECTION ELEMENTS

Injection elements provide a convenient means of introducing water and sediment from sources other than rainfall-derived runoff or base flows. Examples would include effluent from water treatment or industrial sources, or agricultural return flows. Data are provided as a text file listing time (min) and discharge (m^3/s) pairs plus up to five columns of corresponding sediment concentrations by particle class.

5.2.2 Mathematical and numerical model

5.2.2.1 RAINFALL

Rainfall data are entered as time-accumulated depth or time-intensity breakpoint pairs. A time–depth pair simply defines the total rainfall accumulated up to that time. A time–intensity pair defines the rainfall rate until the next data pair. If data are available as time–depth breakpoints, there is no advantage in converting them to intensity as the program must convert intensity to accumulated depth. Rainfall is modelled as spatially uniform over each element, but varies between elements if there is more than one rain gauge.

The spatial and temporal variability of rainfall is expressed by interpolation from rain gauge locations to each plane, pond or urban element (and optionally channels). An element's location is represented by a single pair of x, y coordinates, such as its aerial centroid. The interpolator attempts to find the three closest rain gauges that enclose the element's coordinates. If such a configuration does not exist, it looks for the two closest gauges for which the element's coordinates lie within a strip bounded by two (parallel) lines that pass through the gauge locations and are perpendicular to the line connecting the two points. Finally, if two such points do not exist, the closest gauge alone is used.

If three points are used for the interpolation, the depth at any breakpoint time is represented by a plane passing through the depths above the three points for a given time-step, and the interpolated depth for the element is the depth above its coordinates (Fig. 5.5). For two points, a plane is defined by the two parallel lines, which are considered to be lines of constant depth.

Once the configuration is determined and the spatial interpolating coefficients are computed, an extended set of breakpoint times is constructed as the union of all breakpoint times from the two or three gauges. Final breakpoint depths are computed using the extended set of breakpoint times, interpolating depths within each set of gauge data when necessary. If initial soil saturation is specified in the rainfall file, it will be interpolated using the same spatial interpolation coefficients.

5.2.2.2 INTERCEPTION

As implemented in K2, interception is the portion of rainfall that initially collects and is retained on vegetative surfaces. The effect of interception is controlled by two parameters: the interception depth and the fraction of the surface covered by intercepting vegetation. The interception-depth parameter reflects the average depth of rainfall retained by the particular vegetation type or mixture of vegetation types present on the surface. Rainfall rate is reduced by the cover fraction (i.e., a cover fraction equal to 0.50 gives a 50 % reduction) until the amount retained reaches the interception depth.

5.2.2.3 INFILTRATION

The conceptual model of soil hydrology in K2 represents a soil of either one or two layers, with the upper layer of arbitrary depth,

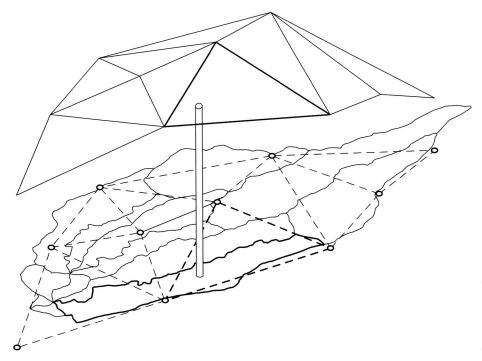

Figure 5.5 Diagrammatic representation of the K2 rainfall interpolation procedure

exhibiting log-normally distributed values of saturated hydraulic conductivity, K_S. The surface of the soil exhibits microtopographic variations that are characterized by a mean micro-rill spacing and height. This latter feature is significant in the model, since one of the important aspects of the K2 hydrology is an explicit interaction of surface flow and infiltration. Infiltration may occur from either rainfall directly on the soil or from ponded surface water created from previous rainfall excess. Also involved in this interaction, as discussed below, is the small-scale random variation of K_S. All of the facets of K2 infiltration theory are presented in much greater detail in Smith *et al.* (2002).

Basic infiltrability

Infiltrability, f_c, is the rate at which soil will absorb water (vertically) when there is an unlimited supply at the surface. Infiltration rate, f, is equal to rainfall, $r(t)$, until this limit is reached. K2 uses the Parlange 3-Parameter model for this process (Parlange *et al.*, 1982), in which the models of Green and Ampt (1911) and Smith and Parlange (1978) are included as the two limiting cases. A scaling parameter, γ, is the third parameter in addition to the two basic parameters K_S and capillary length scale, G. Most soils exhibit infiltrability behavior intermediate to these two models, and K2 uses a weighting "γ" value of 0.85. The state variable for infiltrability is the initial water content, in the form of the soil saturation deficit, $G\Delta\theta_i$, defined as the saturated water content minus the initial water content. In terms of these variables, the basic model is:

$$f_c = K_S \left[1 + \frac{\gamma}{\exp\left(\dfrac{\gamma I}{G\Delta\theta_i}\right) - 1} \right]. \qquad (5.1)$$

The K2 infiltration model employs the *infiltrability depth approximation* (IDA) from Smith *et al.* (2002) in which f_c is described as a function of infiltrated depth I. This approach derives from the "time compression" approximation earlier suggested by Reeves and Miller (1975): time is not compressed, but I is a surrogate for time as an independent variable. This form of infiltrability model eliminates the separate description of ponding time and the decay of f after ponding.

Small-scale spatial variability

The infiltrability model of K2 incorporates the coefficient of variation of K_S, CV_K, as described by Smith and Goodrich (2000). Assuming that K_S is distributed log-normally, there will for all normal values of rain intensity r be some portion of the surface for which $r < K_S$. Thus for that area there will be no potential runoff. Smith and Goodrich (2000) simulated ensembles of distributed point infiltration and arrived at a function for infiltrability

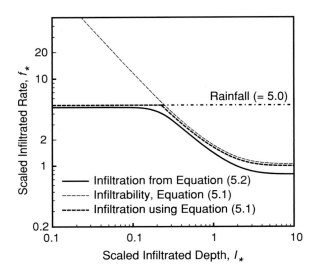

Figure 5.6 Graph showing a comparison of the infiltrability function with and without consideration of randomly varying K_S

which closely describes this ensemble infiltration behavior:

$$f_{e^*} = 1 + (r_{e^*} - 1) \left\{ 1 + \left[\frac{(r_{e^*} - 1)}{\gamma} (e^{\gamma I_{e^*}} - 1) \right]^c \right\}^{-1/c} ,_{r_{e^*} > 1,} \quad (5.2)$$

in which f_{e^*} and r_{e^*} are infiltrability and rain rate scaled on the ensemble effective asymptotic K_S value. This effective ensemble K_e is the appropriate K_S parameter to use in the infiltrability function for an ensemble, and is a function of CV_K and r_{e^*}; the ratio of r to ensemble mean of K_S defined as $\xi(K)$. Smith and Goodrich (2000) describe how effective K_e drops significantly below $\xi(K)$ for low relative rain rates and high relative values of CV_K.

Equation (5.2) also scales I by the parameter pair $G\Delta\theta_i$. The additional parameter c is a function only of CV_K and the value of r. There is evidence in watershed runoff measurements (Smith and Goodrich, 2000) that this function is more appropriate for watershed areas than the basic (uniform K_S) relation of Equation (5.1). Figure 5.6 compares Equation (5.2) for $CV_K = 0.8$ to Equation (5.1), in which CV_K is implicitly zero. Note that Equation (5.2) does not have a ponding point, but rather exhibits a gradual evolution of runoff, and thus Equation (5.2) describes infiltration rate rather than infiltrability.

Infiltration with two-layer soil profiles

For a soil with two layers, either layer can be flow limiting and thus can be the infiltration control layer, depending on the soil properties, thickness of the surface layer, and the rainfall rate. There are several possibilities, most of which have been discussed by Corradini *et al.* (2000) and Smith *et al.* (1993). K2 attempts to model all cases in a realistic manner, including the redistribution

of soil water during periods when r is less than K_S and thus runoff is not generated from rainfall.

Upper soil control: For surface soil layers that are sufficiently deep, this case [$r > K_{S1}$] resembles a single soil profile. However, when the wetting front reaches the layer interface, the capillary drive parameter and the effective value of K_S for Equation (5.1) must be modified. The effective parameters for this case were discussed by Smith (1990). The effective K_S parameter, K_4, is found by solving the steady unsaturated flow equation with matching values of soil capillary potential at the interface.

Lower soil control: When the condition $K_{S1} > r > K_{S2}$ occurs, the common runoff mechanism called saturation runoff may occur. K2 treats the limitation of flow through the lower soil by application of Equation (5.1) or (5.2) to flow through the layer interface, and when that water which cannot enter the lower layer has filled the available pore space in the upper soil, runoff is considered to begin. The available pore space in the upper soil is the initial deficit $\Delta\theta_{1i}$ less rainwater in transit through the upper soil layer. For reasonably deep surface-soil layers, it is possible for control to shift from the lower to the upper if the rainfall rate increases to sufficiently exceed K_{S1} before the surface layer is filled from flow limitations into the lower layer.

An example of runoff generation from single and two-layer soil profiles is illustrated in Fig. 5.7. Note that in both profiles the top soils have identical porosity and saturated hydraulic conductivity. The shallow top layer in the two-layer case has significantly less available pore space to store and transmit infiltrated water to the lower, less permeable, soil layer. The burst of rainfall occurring at roughly 850 minutes into the event produces identical Hortonian runoff from both profiles for approximately 40 minutes. The upper soil layer is controlling in both profiles and runoff is produced by infiltration excess. The long, low-intensity period of rainfall between 950 and 1850 minutes is fully absorbed by both soil profiles, but is effectively filling the available pore space in the shallow upper layer of the two-layer profile. When the rainfall intensity increases at approximately 1850 minutes to around 5 mm/hr ($r < K_S$ of the upper soil layer), runoff is generated from the shallow profile as the lower soil layer in the two-layer systems is now controlling and runoff generation occurs via saturation excess. The single layer profile again generates runoff via infiltration excess when the rainfall intensity increases (at ~2010 minutes) above the infiltrability of the soil.

Redistribution and initial wetting

Rainfall patterns of all types and rainfall rates of any value should be accommodated realistically in a robust infiltration model. This includes the effect on runoff potential of an initial storm period of very low rainfall rates, and the reaction of the soil infiltrability to periods within the storm of low or zero rainfall rates. K2 simulates the wetting-zone changes due to these conditions with

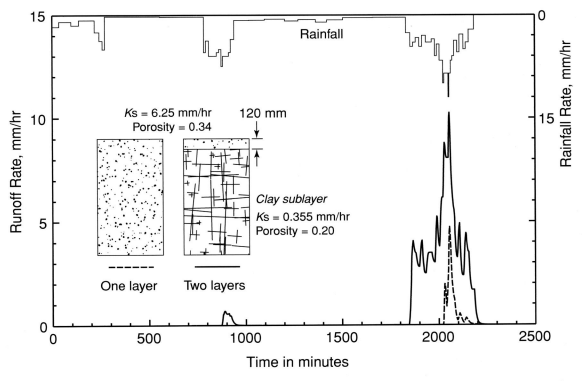

Figure 5.7 Graph of computed runoff rate for simulations based on one- and two-layer soil profiles illustrating infiltration- and saturation-excess runoff generation

an approximation described by Smith *et al.* (1993) and Corradini *et al.* (2000). Briefly, the wetting profile of the soil is described by a water-balance equation in which the additions from rainfall are balanced by the increase in the wetted zone value of two and the extension of the wetted zone depth due to the capillary drive of the wetting front. The soil wetted shape is treated as a similar shape of depth Z with volume $\exists Z(\Delta\theta_0 - \Delta\theta_I)$, where $\exists$ is a constant scale factor defined in Smith *et al.* (1993). Space does not permit detailed description here, but the method is applicable to prewetting of the soil as well as the decrease in $\Delta\theta_0$ during a storm hiatus. It is also applicable, with modification, to soils with two layers.

5.2.2.4 OVERLAND FLOW
The appearance of free water on the soil surface, called ponding, gives rise to runoff in the direction of the local slope (Fig. 5.8). Rainfall can produce ponding by two mechanisms, as outlined in the infiltration section. The first mechanism involves a rate of rainfall, which exceeds the infiltrability of the soil at the surface. The second mechanism is soil filling, when a soil layer deeper in the soil restricts downward flow and the surface layer fills its available porosity. In the first mechanism, the surface-soil water-pressure head is not more than the depth of water, and decreases

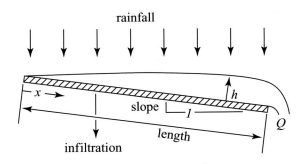

Figure 5.8 Definition sketch for overland flow

with depth, while in the second mechanism, soil water-pressure head increases with depth until the restrictive layer is reached.

Viewed at a very small scale, overland flow is an extremely complex three-dimensional process. At a larger scale, however, it can be viewed as a one-dimensional flow process in which flux is related to the unit area storage by a simple power relation:

$$Q = \alpha h^m, \qquad (5.3)$$

where Q is discharge per unit width and h is the storage of water per unit area. Parameters α and m are related to slope, surface roughness, and flow regime. Equation (5.3) is used in conjunction with the equation of continuity:

$$\frac{\partial h}{\partial t} + \frac{\partial Q}{\partial x} = q(x, t), \qquad (5.4)$$

where t is time, x is the distance along the slope direction, and q is the lateral inflow rate. For overland flow, Equation (5.3) may be substituted into Equation (5.4) to obtain:

$$\frac{\partial h}{\partial t} + \alpha m h^{m-1} \frac{\partial h}{\partial x} = q(x, t). \qquad (5.5)$$

By taking a larger-scale, one-dimensional approach it is assumed that Equation (5.5) describes normal flow processes; it is not assumed that overland-flow elements are flat planes characterized by uniform depth sheet flow.

The kinematic-wave equations are simplifications of the de Saint Venant equations, and do not preserve all of the properties of the more complex equations, such as backwater and diffusive-wave attenuation. Attenuation does occur in kinematic routing from shocks or from spatially variable infiltration. The kinematic routing method, however, is an excellent approximation for most overland-flow conditions (Woolhiser and Liggett, 1967; Morris and Woolhiser, 1980).

Boundary conditions
The depth or unit storage at the upstream boundary must be specified to solve Equation (5.5). If the upstream boundary is a flow divide, the boundary condition is

$$h(0, t) = 0. \qquad (5.6)$$

If another surface is contributing flow at the upper boundary, the boundary condition is

$$h(0, t) = \left[\frac{\alpha_u h_u(L, t)^{m_u} W_u}{\alpha W} \right]^{\frac{1}{m}}, \qquad (5.7)$$

where subscript u refers to the upstream surface, W is width and L is the length of the upstream element. This merely states an equivalence of discharge between the upstream and downstream elements.

Recession and microtopography
Microtopographic relief can play an important role in determining hydrograph shape (Woolhiser *et al.*, 1997). The effect is most pronounced during recession, when the extent of soil covered by the flowing water determines the opportunity for water loss by infiltration. K2 provides for treatment of this relief by assuming the relief geometry has a maximum elevation, and that the area covered by surface water varies linearly with elevation up to this maximum (Fig. 5.9). The geometry of microtopography is completed by specifying a relief scale, which geometrically represents the mean spacing between relief elements.

Given the conceptual relation presented in Figure 5.9, the effective mean hydraulic depth, h_m, is computed as the cross-sectional

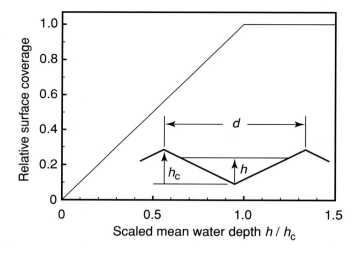

Figure 5.9 Graph showing the assumed relation of covered surface area to scaled mean water depth. Parameter h_c is the microtopographic relief height and d is the mean microtopographic spacing.

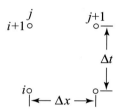

Figure 5.10 Notation for space and time dimensions of the finite difference grid

area of flow divided by the width of the element. The relation here is scaled, and the maximum topographic relief, h_c, is a parameter that can be user-defined. Infiltration from the portion of the surface covered by water proceeds at the infiltrability rate, and the remaining area will have a value of f determined by the rainfall rate. Thus infiltration proceeds during receding flows depending on the microtopography.

Numerical solution
KINEROS2 solves the kinematic-wave equations using a four-point implicit finite difference method. The finite difference form for Equation (5.5) is

$$h_{j+1}^{i+1} - h_{j+1}^i + h_j^{i+1} - h_j^i + \frac{2\Delta t}{\Delta x} \left\{ \theta_w \left[\alpha_{j+1}^{i+1} \left(h_{j+1}^{i+1} \right)^m \right. \right.$$
$$\left. - \alpha_j^{i+1} \left(h_j^{i+1} \right)^m \right] + (1 - \theta_w) \left[\alpha_{j+1}^i \left(h_{j+1}^i \right)^m - \alpha_j^i \left(h_j^i \right)^m \right] \right\}$$
$$- \Delta t (\bar{q}_{j+1} + \bar{q}_j) = 0, \qquad (5.8)$$

where θ_w is a weighting parameter (usually 0.6 to 0.8) for the x derivatives at the advanced time-step. The notation for this method is shown in Fig. 5.10.

A solution is obtained by Newton's method (sometimes referred to as the Newton–Raphson technique). While the solution is unconditionally stable in a linear sense, the accuracy is highly dependent on the size of Δx and Δt values used. The difference scheme is nominally of first-order accuracy.

Roughness relationships

Two options for α and m in Equation (5.5) are provided in K2:

(1) The Manning hydraulic resistance law may be used. In this option

$$\alpha = 1.49 \frac{S^{\frac{1}{2}}}{n} \quad \text{and} \quad m = \frac{5}{3}, \qquad (5.9)$$

where S is the slope, n is a Manning's roughness coefficient for overland flow, and English units are used.

(2) The Chezy law may be used. In this option,

$$\alpha = C S^{\frac{1}{2}} \quad \text{and} \quad m = \frac{3}{2}, \qquad (5.10)$$

where C is the Chezy friction coefficient.

5.2.2.5 CHANNEL FLOW

Unsteady, free-surface flow in channels is also represented by the kinematic approximation to the unsteady, gradually varied flow equations. Channel segments may receive uniformly distributed but time-varying lateral inflow from overland-flow elements on either or both sides of the channel, from one or two channels at the upstream boundary, and/or from an upland area at the upstream boundary. The dimensions of overland-flow elements are chosen to completely cover the watershed, so rainfall on the channel is not considered directly.

The continuity equation for a channel with lateral inflow is:

$$\frac{\partial A}{\partial t} + \frac{\partial Q}{\partial x} = q_\mathrm{c}(x, t), \qquad (5.11)$$

where A is the cross-sectional area, Q is the channel discharge, and $q_\mathrm{c}(x, t)$ is the net lateral inflow per unit length of channel. Under the kinematic assumption, Q can be expressed as a unique function of A, and Equation (5.11) can be rewritten as:

$$\frac{\partial A}{\partial t} + \frac{\partial Q}{\partial A} \frac{\partial A}{\partial x} = q_\mathrm{c}(x, t). \qquad (5.12)$$

The kinematic assumption is embodied in the relationship between channel discharge and cross-sectional area such that:

$$Q = \alpha R^{m-1} A, \qquad (5.13)$$

where R is the hydraulic radius. If the Chezy relationship is used, $\alpha = C S^{1/2}$ and $m = 3/2$. If the Manning equation is used, $\alpha = 1.49 S^{1/2}/n$ and $m = 5/3$. Channel cross-sections may be approximated as trapezoidal or circular, as shown in Figs. 5.3 and 5.4.

Compound channels

K2 contains the ability to route flow through channels with a significant overbank region. The channel may in this case be composed of a smaller channel incised within a larger flood plane or swale. The compound channel algorithm is based on two independent kinematic equations (one for the main channel and one for the overbank section), which are written in terms of the same datum for flow depth. In writing the separate equations, it is explicitly assumed that no energy transfer occurs between the two sections, and upon adding the two equations the common datum implicitly requires the water-surface elevation to be equal in both sections (Fig. 5.3). However, flow may move from one part of the compound section to another. Such transfer will take with it whatever the sediment concentration may be in that flow when sediment routing is simulated. Each section has its own set of parameters describing the hydraulic roughness, bed slope, and infiltration characteristics. A compound-channel element can be linked with other compound channels or with simple trapezoidal channel elements. At such transitions, as at other element boundaries, discharge is conserved and new heads are computed downstream of the transition.

Base flow

K2 allows the user to specify a constant base flow in a channel, which is added at a fractional rate at each computational node along the channel to produce the designated flow at the downstream end of the reach. This feature allows simulation of floods that occur in excess of an existing base discharge, but requires foreknowledge of where those flows originate and at what rate.

Channel infiltration

In arid and semi-arid regions, infiltration into channel alluvium may significantly affect runoff volume and peak discharge. If the channel infiltration option is selected, Equation (5.2) is used to calculate accumulated infiltration at each computational node, beginning either when lateral inflow begins or when an advancing front has reached that computational node. Because the trapezoidal channel simplification introduces significant error in the area of channel covered by water at low flow rates (Unkrich and Osborn, 1987), an empirical expression is used to estimate an "effective wetted perimeter." The equation used in K2 is:

$$p_\mathrm{e} = \min \left[\frac{h}{0.15\sqrt{\mathrm{BW}}}, 1 \right] p, \qquad (5.14)$$

where p_e is the effective wetted perimeter for infiltration, h is the depth, BW is the bottom width, and p is the channel wetted perimeter at depth h. This equation states that p_e is smaller than p until a threshold depth is reached, and at depths greater than the threshold depth, p_e and p are identical. The channel loss rate is obtained by multiplying the infiltration rate by the effective wetted perimeter. A two-layer soil representation is also allowed in channels.

Culverts

The most general discharge relationship and the one often used for flow in pipes is the Darcy–Weisbach formula:

$$S_f = \frac{f_D}{4R} \frac{u^2}{2g}, \qquad (5.15)$$

where S_f is the friction slope, f_D is the Darcy–Weisbach friction factor, and u is the velocity (Q/A). Under the kinematic assumption, the conduit slope S may be substituted for S_f in Equation (5.15), so that

$$u = 2\sqrt{\frac{2g}{f_D} RS}. \qquad (5.16)$$

Discharge is computed by using Equation (5.16), and a specialized relationship between channel discharge and cross-sectional area:

$$Q = \frac{\alpha A^m}{p^{m-1}} \qquad (5.17)$$

where p is the wetted perimeter, α is $[8gS / f_D]^{1/2}$, and $m = 3/2$. Geometric relationships for partially full conduits are discussed further in the original KINEROS documentation (Woolhiser *et al.*, 1990).

Numerical method for channels

The kinematic equations for channels are solved by a four-point implicit technique similar to that for overland-flow surfaces, except that A is used instead of h, and the geometric changes with depth are considered.

5.2.2.6   POND FLOW

In-channel detention structures are modelled as simple reservoirs. Outflow is assumed to be solely a function of water depth, so the dynamics of the storage are described by the mass-balance and outflow equations:

$$\frac{dV}{dt} = q_I - q_O - A_p f_c, \qquad (5.18)$$

in which

V = storage volume [L^3],
q_I = inflow rate [L^3/T],
q_O = outflow rate [L^3/T],
A_p = pond surface area [L^2]
f_c = pond infiltration loss rate [L/T].

Reservoir geometry is described by a user-defined rating table for V, A_p, and q_O. Equation (5.18) is written in finite difference form over a time interval t and the stage at time $t + \Delta t$ is determined using a hybrid Newton–Raphson/bisection method. For a given V, A_p and q_O are estimated using log–log interpolation.

5.2.2.7   EROSION AND SEDIMENTATION

As an optional feature, K2 can simulate the movement of eroded soil in addition to the movement of surface water. K2 accounts separately for erosion caused by raindrop energy (splash erosion), and erosion caused by flowing water (hydraulic erosion). Erosion is computed for upland, channel, and pond elements.

The general equation used to describe the sediment dynamics at any point along a surface flow path is a mass-balance equation similar to that for kinematic water flow (Bennett, 1974):

$$\frac{\partial(AC_s)}{\partial t} + \frac{\partial(QC_s)}{\partial x} - e(x,t) = q_s(x,t), \qquad (5.19)$$

in which

C_s = sediment concentration [L^3/L^3],
Q = water discharge rate [L^3/T],
A = cross-sectional area of flow [L^2],
e = rate of erosion of the soil bed [L^2/T],
q_s = rate of lateral sediment inflow for channels [L^3/T/L].

For upland surfaces, it is assumed that e is composed of two major components – production of eroded soil by splash of rainfall on bare soil, and hydraulic erosion (or deposition) due to the interplay between the shearing force of water on the loose soil bed and the tendency of soil particles to settle under the force of gravity. Thus e may be positive (increasing concentration in the water) or negative (deposition). Net erosion is a sum of splash erosion rate as e_s and hydraulic erosion rate as e_h:

$$e = e_s + e_h. \qquad (5.20)$$

Splash erosion

Based on limited experimental evidence, the splash erosion rate can be approximated as a function of the square of the rainfall rate (Meyer and Wischmeier, 1969). This relationship is used in K2 to estimate the splash erosion rate as follows:

$$e_s = c_f k(h) r^2, \quad q > 0$$
$$e_s = 0, \quad q < 0, \qquad (5.21)$$

in which c_f is a constant related to soil and surface properties, and $k(h)$ is a reduction factor representing the reduction in splash erosion caused by increasing depth of water. The function $k(h)$ is 1.0 prior to runoff and its minimum is zero for very deep flow; it is given by the empirical expression:

$$k(h) = \exp(-c_h h) \qquad (5.22)$$

The parameter c_h represents the damping effectiveness of surface water, and does not vary widely. Both c_f and $k(h)$ are always positive, so e_s is always positive when there is rainfall and a positive rainfall excess (q).

Hydraulic erosion

The hydraulic erosion rate e_h represents the rate of exchange of sediment between the flowing water and the soil over which it flows, and may be either positive or negative. K2 assumes that for any given surface-water flow condition (velocity, depth, slope, etc.), there is an equilibrium concentration of sediment that can be carried if that flow continues steadily. The hydraulic erosion rate (e_h) is estimated as being linearly dependent on the difference between the equilibrium concentration and the current sediment concentration. In other words, hydraulic erosion/deposition is modelled as a kinetic transfer process:

$$e_h = c_g(C_m - C_s)A, \qquad (5.23)$$

in which C_m is the concentration at equilibrium transport capacity, $C_s = C_s(x, t)$ is the current local sediment concentration, and c_g is a transfer-rate coefficient $[T^{-1}]$. Clearly, the transport capacity is important in determining hydraulic erosion, as is the selection of transfer-rate coefficient. Conceptually, when deposition is occurring, c_g is theoretically equal to the particle-settling velocity divided by the hydraulic depth, h. For erosion conditions on cohesive soils, the value of c_g must be reduced, and v_s/h is used as an upper limit for c_g.

Microtopography

The treatment of infiltration and microtopography also interacts with erosion as the effective mean hydraulic depth and related velocity drive the hydraulic portion of erosion. In addition it affects the percentage of the surface that is subject to splash erosion from raindrop impact. This conceptualization of microtopography does not directly define a rill–inter-rill region in terms of erosion treatment. Both "rill" and "interrill" processes can occur simultaneously in shallow flow as splash erosion can occur on the sides of rills and in rills when flow is sufficiently shallow so that raindrop momentum is transmitted to the soil surface (Smith *et al.*, 1995b).

Transport capacity

Transport capacity is determined using the relation of Engelund and Hansen (1967):

$$C_m = \frac{0.05uu_*^3}{g^2dh(\gamma_s - 1)^2}, \qquad (5.24)$$

in which

u is velocity [L/T],
u_* is shear velocity, defined as $\sqrt{ghS}$,
d is particle diameter [L],
γ_s is suspended specific gravity of the particles, $\gamma_s - 1$,
h is water depth [L].

Particle settling velocity

Settling velocity is calculated from particle size and density, assuming the particles have drag characteristics and terminal-fall velocities similar to those of spheres (Fair and Geyer, 1954). This relation is:

$$v_s^2 = \frac{4}{3}\frac{g(\rho_s - 1)}{C_D} \qquad (5.25)$$

in which C_D is the particle drag coefficient. The drag coefficient is a function of particle Reynolds number:

$$C_D = \frac{24}{R_n} + \frac{3}{\sqrt{R_n}} + 0.34, \qquad (5.26)$$

in which R_n is the particle Reynolds number, defined as:

$$R_n = \frac{v_sd}{v}, \qquad (5.27)$$

where v is the kinematic viscosity of water $[L^2/T]$. Settling velocity of a particle is found by solving Equations (5.25), (5.26), and (5.27) for v_s.

Treating a range of particle sizes

Erosion relations are applied to each of up to five particle-size classes, which are used to describe the range of particle sizes found in typical soils. Our experimental and theoretical understanding of the dynamics of erosion for a mix of particle sizes is incomplete. It is not clear, for example, exactly what results when the distribution of relative particle sizes is contradictory to the distribution of their relative transport capacities. In larger particles on stream bottoms, armoring will ultimately occur when smaller, more transportable particles are selectively removed, leaving behind an "armor" of large particles. For the smaller particle sizes found in the shallower flows and rapidly changing flow conditions characteristic of overland flow, however, there is considerably less understanding of the relations. Sufficient knowledge does exist, however, to use the following assumptions in the formulation of K2:

(1) If the largest particle size in a soil of mixed sizes is below its erosion threshold, the erosion of smaller sizes will be limited, since otherwise armoring will soon stop the erosion process.
(2) When erosive conditions exist for all particle sizes, particle erosion rates will be proportional to the relative occurrence of the particle sizes in the surface soil. The same is true of erosion by rain splash.
(3) Particle settling velocities, when concentrations exceed transportability, are independent of the concentration of other particle sizes.

Treatment of a mix of sizes is most critical for cases where the sediment characterizing the bed of the channels is significantly different from that of the upland slopes, and where impoundments exist in which there is significant opportunity for selective settling.

Numerical method for sediment transport

Equations (5.19)–(5.24) are solved numerically at each time-step used by the surface-water flow equations, and for each particle-size class. A four-point finite-difference scheme is again used; but iteration is not required since, given current and immediate past values for A and Q and previous values for C_s, the finite difference form of this equation is explicit, i.e.:

$$C_{sj+1}^{i+1} = f(C_{sj}^i, C_{sj+1}^i, C_{sj}^{i+1}). \qquad (5.28)$$

The value of C_m is found from Equation (5.24) using current hydraulic conditions.

Initial conditions for erosion

When runoff commences during a period when rainfall is creating splash erosion, the initial condition on the vector C_s should not be taken as zero. The initial sediment concentration at ponding, $C_s(t = t_p)$, can be found by simplifying Equation (5.19) for conditions at that time. Variation with respect to x vanishes, and hydraulic erosion is zero. Then:

$$\frac{\partial (AC_s)}{\partial t} = e(x, t) = c_f rq - C_s v_s, \qquad (5.29)$$

where $k(h)$ is assumed to be 1.0 since depth is zero. Since A is zero at time of ponding, and dA/dt is the rainfall excess rate (q), expanding the left-hand side of Equation (5.29) results in:

$$C_s(t = t_p) = \frac{c_f rq}{q + v_s}. \qquad (5.30)$$

The sediment concentration at the upper boundary of a single overland flow element, $C_s(0, t)$, is given by an expression identical to Equation (5.30), and a similar expression is used at the upper boundary of a channel.

Channel erosion and sediment transport

The general approach to sediment-transport simulation for channels is nearly the same as that for upland areas. The major difference in the equations is that splash erosion (e_s) is neglected in channel flow, and the term q_s becomes important in representing lateral inflows. Equations (5.19) and (5.23) are equally applicable to either channel or distributed surface flow, but the choice of transport-capacity relation may be different for the two flow conditions. For upland areas, q_s will be zero, whereas for channels it will be the important addition that comes with lateral inflow from surface elements. The close similarity of the treatment of the two types of elements allows the program to use the same algorithms for both types of elements.

The erosion computational scheme for any element uses the same time and space steps employed by the numerical solution of the surface-water flow equations. In that context, Equations (5.19) and (5.23) are solved for $C_s(x, t)$, starting at the first node below the upstream boundary, and from the upstream conditions for channel elements. If there is no inflow at the upper end of the channel, the transport capacity at the upper node is zero and any lateral input of sediment will be subject there to deposition. The upper boundary condition is then:

$$C_s(0, t) = \frac{q_s}{q_c + v_s W_B}, \qquad (5.31)$$

where W_B is the channel bottom width. $A(x, t)$ and $Q(x, t)$ are assumed to be known from the surface-water solution.

5.3   AGWA GIS INTERFACE

5.3.1   Background

AGWA was developed as a collaborative effort between the USDA-ARS, EPA-ORD, and UA under the following guidelines: (1) that its parameterization routines be simple, direct, transparent, and repeatable; (2) that it be compatible with commonly available GIS data layers; and (3) that it be useful for scenario development and assessment at multiple scales.

Over the past decade numerous significant advances have been made in the linkage of GIS and various research and application models (e.g., HEC-GeoHMS, USACE, 2003; AGNPS, Bingner and Theurer, 2001; BASINS, Lahlou *et al.*, 1998). These GIS-based systems have greatly enhanced the capacity for research scientists to develop and apply models due to the improved data management and rapid parameter-estimation tools that can be built into a GIS driver. As one of these GIS-based modelling tools, AGWA provides the functionality to conduct all phases of a watershed assessment for two widely used watershed hydrologic models: K2 and the Soil and Water Assessment Tool (SWAT; Arnold *et al.*, 1995). SWAT is a continuous-simulation model for use in large (river-basin scale) watersheds and in humid regions, where K2 cannot be applied with confidence. The AGWA tool provides an intuitive interface to these models for performing multi-scale modelling and change assessment in a variety of geographies. Data requirements include elevation, classified land cover, soils, and precipitation data, all of which are typically available at no cost over the Internet. Model input parameters are derived directly from these data using optimized look-up tables that are provided with the tool.

AGWA shares the same ArcView GIS framework as the US EPA Analytical Tools Interface for Landscape Assessment (ATtILA; Ebert and Wade, 2004), and Better Assessment Science Integrating Point and Nonpoint Sources (BASINS; Lahlou *et al.*, 1998), and can be used in concert with these tools to improve scientific understanding (Miller *et al.*, 2002). Watershed analyses may benefit from the integration of multiple model outputs as this approach facilitates comparative analyses and is particularly valuable for

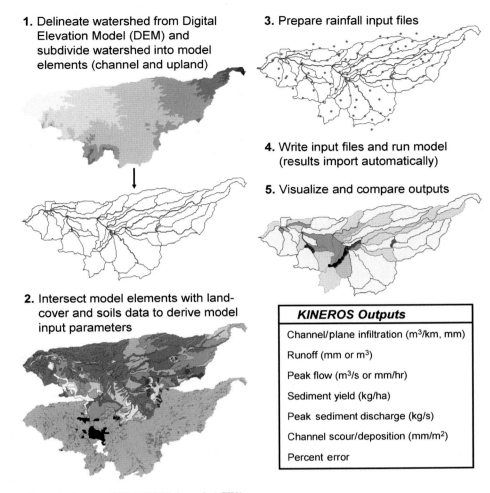

1. Delineate watershed from Digital Elevation Model (DEM) and subdivide watershed into model elements (channel and upland)

2. Intersect model elements with land-cover and soils data to derive model input parameters

3. Prepare rainfall input files

4. Write input files and run model (results import automatically)

5. Visualize and compare outputs

KINEROS Outputs
Channel/plane infiltration (m³/km, mm)
Runoff (mm or m³)
Peak flow (m³/s or mm/hr)
Sediment yield (kg/ha)
Peak sediment discharge (kg/s)
Channel scour/deposition (mm/m²)
Percent error

Figure 5.11 Sequence of steps in the use of KINEROS2 through AGWA

inter-disciplinary studies, scenario development, and alternative futures simulation work.

The following description of AGWA focuses on the K2 interface. Specifically, the interface design, processes, and ongoing research relating to the application of K2 are presented in detail. Miller *et al.* (2002) provide a more detailed description of AGWA-SWAT and its application in conjunction with K2 for multi-scale analyses. Hernandez *et al.* (2003) describe the integration of AGWA and ATtILA. Kepner *et al.* (2004) describe the use of AGWA for the analysis of alternative future land-use/cover scenarios, and the potential benefit to planning efforts. Burns *et al.* (2004) describe the development of a version of AGWA that was fully integrated into BASINS version 3.1.

5.3.2 Design

The conceptual design of AGWA is presented in Fig. 5.11. A fundamental assumption of AGWA is that the user has previously compiled the necessary GIS data layers, all of which are easily obtained in most countries. The AGWA extension for ArcView adds the "AGWA Tools" menu to the View window, and must be run from an active view. Pre-processing of the DEM to ensure hydrologic connectivity within the study area is required, and tools are provided in AGWA to aid in this task. Once the user has compiled all relevant GIS data and initiated an AGWA session, the program is designed to lead the user in a stepwise fashion through the transformation of GIS data into simulation results. The AGWA Tools menu is designed to reflect the order of tasks necessary to conduct a watershed assessment, which is broken out into five major steps: (1) watershed delineation and subdivision; (2) land-cover and soils parameterization; (3) preparation of rainfall input files; (4) writing of input files and running of model; and (5) visualization and comparison of results.

Following model execution, AGWA will automatically import the model results and add them to the polygon and stream map-attribute tables for display. A separate module controls the visualization of model results. Users can toggle among viewing various model outputs for both upland and channel elements, enabling the problem areas to be identified visually. If multiple land-cover scenes exist, they can be used to derive multiple parameter sets with which the models can be run for a given watershed. Model results can then be compared on either an absolute- or percent-change basis for each model element, and overlain with other digital data layers to help prioritize management activities.

5.3.2.1 WATERSHED DELINEATION AND SUBDIVISION

The most widely used method, and that which is used in AGWA, for the extraction of stream networks is to compute the accumulated area upslope of each pixel through a network of cell-to-cell drainage paths. This flow accumulation grid is subsequently pruned by eliminating all cells for which the accumulated flow area is less than a user-defined threshold drainage area, called the Channel, or Contributing Source Area (CSA). The watershed is then further subdivided into upland and channel elements as a function of the stream-network density. In this way, a user-defined CSA controls the spatial complexity of the watershed subdivision. This approach often results in a large number of spurious polygons and disconnected model elements. A suite of algorithms has been implemented in AGWA that refines the watershed elements by eliminating spurious elements and ensuring downstream connectivity.

Numerous options are available during the delineation and subdivision step that are designed to facilitate the application of K2 for real-world problems. Channel reaches and associated upland areas can be split at user-defined locations, such as flow gauges, to facilitate comparisons with observational data. Multiple watersheds can be delineated simultaneously to accommodate areas of concern not falling neatly within watershed boundaries, such as governmental jurisdictions. Ponds (releasing or non-releasing) and riparian buffer strips can also be incorporated during the subdivision process.

5.3.2.2 PARAMETER ESTIMATION

Each of the overland and channel elements delineated by AGWA is represented in K2 by a set of parameter values. These values are assumed to be uniform within a given element. There may be a large degree of spatial variability in the topographic, soil, and land-cover characteristics within the watershed, and AGWA uses an area-weighting scheme to determine an average value for each parameter within an overland flow model element abstracted to an overland flow plane (Miller *et al.*, 2002). GIS layers are intersected with the subdivided watershed, and a series of look-up tables and spatial analyses are used to estimate parameter values

for the unique combinations of land cover and soils. K2 requires a host of parameter values, and estimating their values can be a tedious task; AGWA rapidly provides estimates based on an extensive literature review and calibration efforts. This convenience does not obviate the need for calibrating K2 applications, and uncalibrated model results should thus be used only in qualitative assessments. Users are able to modify input parameters by adjusting values in the provided look-up tables prior to running the land-cover and soils parameterization, or by adjusting the parameters associated with each element following the parameterization.

Soil parameters for upland elements as required by K2 (such as percent rock, suction head, porosity, saturated hydraulic conductivity) are initially estimated from soil texture according to the soil data following Woolhiser *et al.* (1990) and Rawls *et al.* (1982). Saturated hydraulic conductivity is reduced following Bouwer (1966) to account for air entrapment. Further adjustments are made following Stone *et al.* (1992) as a function of estimated canopy cover, and following Bouwer and Rice (1984) as a function of the amount of rock fragments in the soil. Area and depth weighting procedures are used to derive a single representative soil profile for each upland element. Cover parameters, including interception, canopy cover, Manning's roughness, and % paved area are estimated following expert opinion and previously published look-up tables (Woolhiser *et al.*, 1990). Upland element slope is estimated as the average slope within the element, and width and length are a function of the element shape, which is assumed to be rectangular with element length equal to longest flow length. Stream-channel slope is estimated as the difference in elevation between the reach endpoints divided by the reach length. Channel width and depth are parameterized using regional hydraulic-geometry relationships, or regression equations relating contributing area to channel dimensions (e.g., Miller *et al.*, 1996). An editable database of published hydraulic-geometry relationships is provided with AGWA, or custom relationships can be specified by the user. AGWA is unable to derive soil-related parameters for channels from GIS data and therefore assumes that all channels have a sandy bed. These parameters can be easily edited through the stream-reach attribute table if necessary.

Digital soil maps for different countries or regions vary considerably in terms of the information they contain, and how that information is organized in their associated database files. Automated use of soil maps for model parameterization is heavily dependent on this information structure, and thus not just any soil map can be used with AGWA. As a result, procedures to use the United Nations Food and Agriculture Organization's (FAO) Digital Soil Map of the World were developed to maximize the geographic extent of its applicability. Despite the relatively low spatial resolution of the FAO soil maps, K2 results derived using them compare well with results derived from higher resolution soil maps in the United States (Levick *et al.*, 2004; Levick *et al.*, 2006).

5.3.2.3 RAINFALL INPUT

Uniform rainfall input files for K2 can be created in AGWA using gridded return-period rainfall maps, a database of geographically specific return-period rainfall depths provided with the tool, or using data entered by the user. Return-period rainfall depths are converted to hyetographs using USDA Soil Conservation Service (SCS) methodology and a type-II distribution (USDA-SCS, 1973). The hypothetical type-II distribution is suitable for deriving the time distribution of 24-hour rainfall for extreme events in many regions, but may result in overestimated peak flows, particularly when applied to shorter-duration events.

If return-period rainfall grids are available, then AGWA extracts the rainfall depth for the grid cell containing the centroid of the watershed for which the rainfall input file is being generated. The depth is then converted into a hyetograph for the specified return period using the SCS methodology described above. This process has been automated for convenient use with common datasets available in the United States, and can be easily modified to accommodate other formats.

If return-period rainfall maps are not available, or a specific depth and duration are desired, the provided design-storm database file can be easily edited to add new data. Data are entered in the form of a location, recurrence interval, duration, and rainfall depth in millimeters. The design-storm database further provides the option to incorporate an area-reduction factor, if known, which can be particularly convenient when working in regions characterized by convective thunderstorms.

In the event that gauge observations of rainfall depth are available, or a specific hyetograph is desired, then data may be entered manually by the user through the AGWA interface. User-defined storms are entered as time-depth pairs, thus providing the flexibility to define any hyetograph.

Version 1.5 of AGWA also incorporates the ability to create distributed rainfall input files for K2. This feature allows users to select the gauges, include gauges with zero depths for more accurate interpolation by K2, set a minimum depth that the gauges must exceed before creating a rainfall file, and generate files for any number of events between a given range of dates.

5.3.2.4 MODELLING

Once model element parameters have been assembled and a rainfall input file has been written, AGWA can write the K2 parameter file and run the model. When this option is selected the user is presented with the opportunity to enter parameter multipliers for the most sensitive channel and upland parameters. Multipliers, which default to 1.0, are entered as real numbers and can be used to manipulate parameters as they are written to the parameter file. This option is particularly useful during calibration and sensitivity exercises. Parameter files generated or modified outside of AGWA may also be run through the AGWA interface, provided that they correspond with an existing watershed configuration within the AGWA project.

Once parameter files are written or selected, K2 is called automatically and runs in a separate command window. When the simulation is complete, AGWA reads the output file and results for each model element are parsed back into an ArcView database file, which is stored in the AGWA project and available for viewing.

5.3.2.5 RESULTS VISUALIZATION AND COMPARISON

Simulation results can be selected and viewed using the AGWA visualization tool. When selected, results database files are joined with the polygon and stream map-attribute tables to associate output for each model element with its corresponding polygon or line on the map. Once selected from a list of available outputs, results are automatically divided into nine equal-interval classes and mapped with graduated colors to provide users with the ability to rapidly assess the spatial variability of results within the watershed.

Results may also be easily compared for multiple simulations in the same watershed by computing differences in terms of absolute values or percentages. Differences are written to a new database file that is treated as a separate simulation result and available for visualization. Common comparisons, or relative assessments, include results from simulations based on different land-use/cover conditions, which may represent historic observations or projected future conditions. This option makes it possible to rapidly evaluate the spatial patterns of hydrologic response to landscape change, and to target mitigation and restoration activities for maximum effect.

5.3.3 Scenario building

The use of AGWA as a strategic planning tool has been accommodated through the addition of a land-cover modification tool that allows users to manipulate land-cover grids to represent alternative future land-use/cover scenarios. Changes are carried out within polygons that can be interactively drawn on the screen or taken from imported shapefiles. A variety of types of change can be prescribed, including:

(1) Change entire area to a new land-cover type (e.g., to urban).
(2) Change one land-cover type to another within a user-defined area (e.g., to simulate unpaved road restoration, change barren to desert scrub).
(3) Create a random land-cover pattern (e.g., to represent a burn pattern, change to 64 % barren, 31 % desert scrub, and 5 % mesquite woodland). Patterns can be completely spatially

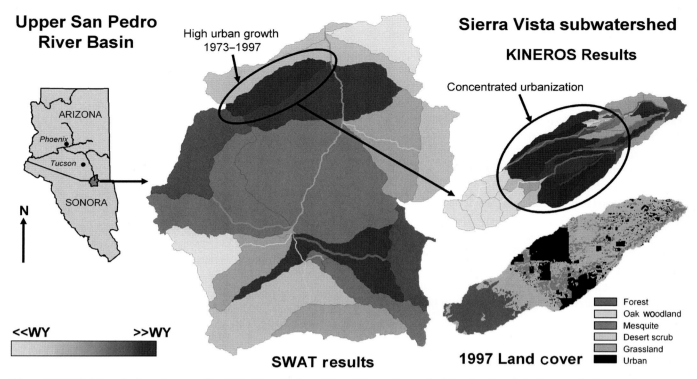

Figure 5.12 Model results from the upper San Pedro River Basin and Sierra Vista subwatershed showing the relative increase in simulated water yield as a result of urbanization between 1973 and 1997. Also demonstrated is the multi-scale assessment capability of AGWA; basin-scale effects observed with SWAT can be investigated at the small-watershed scale with KINEROS2

random or multifractal randomly distributed patchy, with patch size (degree of clustering) controlled by the user.

(4) Modify a land-cover grid based on a burn-severity map.

Using the land-cover modification tool in combination with the relative assessment functionality described in the previous section, it is possible to build a suite of alternative future scenarios and evaluate their relative merits in terms of impact to the local hydrology.

5.3.4 Example application: upper San Pedro River multi-scale assessment

Flowing north from Sonora, Mexico into southeastern Arizona, the San Pedro River Basin has a wide variety of topographic, hydrologic, cultural, and political characteristics. The basin is an exceptional example of desert biodiversity in the semi-arid southwestern United States, and a unique study area for addressing a range of scientific and management issues. It is also a region in socioeconomic transition as the previously dominant rural ranching economy is shifting to increasing areas of urban development. The area is a transition zone between the Chihuahuan and Sonoran

deserts and has a highly variable climate with significant biodiversity. The tested watershed is approximately 3150 km^2 and is dominated by desert shrub-steppe, riparian, grasslands, agriculture, oak and mesquite woodlands, and pine forests.

The AGWA tool was used to delineate the upper San Pedro above the USGS Charleston gauge, and prepare input parameter files for SWAT. The watershed was subdivided using the AGWA default CSA value of 2.5 % of the total watershed area, or approximately 79 km^2. Parameter files were built using both the 1973 and 1997 Landsat satellite classified land-cover scenes. SWAT was run for each of these using the same ten years of observed daily precipitation and temperature data for a single location. By using the same rainfall and temperature inputs, simulated changes in water yield are due solely to altered land cover within the watershed. The "differencing" feature in AGWA was used to compute the % change between the two simulation results and display it visually (Fig. 5.12). This analysis showed that a small watershed running through the developing city of Sierra Vista, shown in Fig. 5.12 as the "Sierra Vista subwatershed," underwent changes in its land cover that profoundly affected the hydrologic regime.

The Sierra Vista subwatershed (92 km^2) was modelled in greater detail using K2. It was also subdivided using a CSA value of 2.5 %, and run using both the 1973 and 1997 land-cover data. A uniform design storm representing the 10-year, 60-minute event (Osborn *et al.*, 1985) was used in both simulations. Since applying point estimates for design storms across larger areas tends to lead to the over-prediction of runoff due to the lack of spatial heterogeneity in input data, an area-reduction method developed by Osborn *et al.* (1980) is used in AGWA to reduce rainfall estimates for watersheds in the San Pedro Basin. Percent change in runoff between the two simulations was computed using the "differencing" tool in AGWA, and the results are presented (directly from AGWA) in Fig. 5.12. From this analysis it is clear that the hydrologic response of the region of concentrated urban growth is adequately represented. Increasing impervious area associated with urban growth has resulted in large increases in runoff from those areas where urbanization is highest.

This type of relative-change assessment is considered to be the most effective use of the AGWA tool without calibrating its component models for a particular site. Without calibration, absolute values of model output parameters should not be considered accurate, nor should the magnitude of computed changes. In a relative sense, however, AGWA can still be useful for inexpensively identifying locations in ungauged watersheds that are particularly vulnerable to degradation, and where restoration activities may therefore be most effective. The ability to use a second model to zoom in on sensitive areas provides a further means of focusing restoration efforts, or preventative measures if the tool is being used to assess potential future scenarios.

5.4   RESEARCH AND DEVELOPMENT

5.4.1   KINEROS2

K2 is a Fortran 77 code designed to handle an unlimited number of model elements (planes, channels, etc.) while keeping the compiled program size under the 640 KB limit imposed by the original MS-DOS operating system. The memory-efficient design features of K2 served it well during the early period of personal computing. At the present time, however, there is tremendous processing power and huge memory resources available on personal computers, both in hardware and through the use of virtual memory strategies like page file swapping. Therefore the hardware and operating system issues that K2 was designed to address no longer exist. Fortran itself has also advanced to a new standard, Fortran 90/95. Fortran 90/95 provides dynamic memory allocation, a proprietary pointer mechanism and modules that encapsulate data structures and procedures, allowing a rudimentary object-oriented programming approach. Also, although K2 is composed of well-defined components, those components were designed to be parts of a

whole and not to function independently. This monolithic nature of K2 has led to a number of modified versions, each of which must be maintained as a separate program.

To overcome the design limitations of K2, and to take advantage of features offered by Fortran 90/95, the K2 code was deconstructed and rebuilt into a library of Fortran 90/95 modules, with each module implementing a single process model, or object (Goodrich *et al.*, 2006). Two major goals of this restructuring are to make K2 technology more readily available to developers of new programs that could benefit from its capabilities, and to make versions of K2 that have been modified for specialized applications easier to maintain. In the restructured version of K2 each object is controlled through an interface, or set of procedures, that are designed to simplify use of the object by non-Fortran programs. This is desirable in that none of the popular and full-featured graphical user interface development products are based on Fortran.

The new structure makes it possible to iterate over all elements at each time-step, rather than each element over all time-steps, as is done by K2. Examples would be open-ended simulations, such as real-time operation, or to graphically display the spatial distribution of simulated quantities, such as runoff from each element, at each time-step during a simulation. In addition to the core process models, there are utility modules to conveniently support backward-compatibility, such as one to extract parameters from a K2 input file. Compatibility between future versions of the module library is also ensured by not allowing existing procedures to be removed, or their names or argument lists to change, although they can change internally. Additional procedures that support extensions to a module's capabilities can be added in the future as long as suitable defaults can allow existing programs to use the module without calling the procedures.

5.4.2   AGWA

A number of ongoing research projects are designed to develop and evaluate strategies for improving the accuracy and usability of K2 through the AGWA interface. These will ultimately be implemented as new tools that will be available to AGWA users, so they are summarized here to provide the reader with an idea of how AGWA will be enhanced in the near future.

Improvements to the accuracy of K2 simulations developed through the AGWA interface are focused on improving its ability to utilize remotely sensed data, including new sources that are becoming increasingly available. One such project is evaluating the potential to improve the watershed subdivision procedure by utilizing additional information available on topographic, land-cover, and soil maps. The goal of this effort is to improve the automated recognition of hydrologic response units in terms of slope, cover, and soil type so that they may be split out during the subdivision process and thus minimize intra-element parameter variability.

Radar rainfall data is another source of remotely sensed data that is becoming more popular as a source of data for hydrologic models (e.g., Morin *et al.*, 2003). A project currently underway is evaluating the potential to utilize this data in real time for the purpose of predicting flash flooding in arid regions. A customized version of AGWA has been developed to read in level II NEXRAD 1 km × 1° radar images at five-minute intervals, and process that information for distributed input to the restructured version of K2, which can be run one time-step at a time.

Airborne Light Detection and Ranging (LIDAR) data is another type of remotely sensed data that holds a large potential benefit to GIS-based hydrologic modelling. It provides high-resolution (∼1 m) topographic information that can improve channel characterization. AGWA currently uses simple hydraulic-geometry relationships to estimate channel dimensions because they cannot be resolved from typical DEM data (Miller *et al.*, 2004). With LIDAR data, however, it is possible to derive detailed channel morphologic information, and a tool is being developed to extract it for the purpose of reach-based characterization (Semmens *et al.*, 2006) as needed by K2 and numerous other models.

Other planned improvements to the AGWA interface focus on providing users with tools to support environmental management and planning. Scenario development is a key component of this work, and is being approached from three different perspectives: climate, land-cover/use, and management activities. A climatic-scenario generator will permit the development of prescribed climate changes and evaluation of their associated impacts given current or anticipated future management conditions. A new modelling tool will permit the simulation of large-scale changes in land-cover/use, in addition to the prescribed changes already possible through the land-cover modification tool, for the evaluation of different spatial patterns of potential landscape change. A best management practice (BMP) tool will allow users to modify land-cover surfaces based on defined plant transitions and the management practices driving them (Scott, 2005). Finally, the functionality to parameterize the K2 urban element feature is presently under development and will permit the rapid evaluation of alternative impervious-surface management strategies.

5.4.2.1 THE NEXT GENERATION OF AGWA

The final version of AGWA for ArcView 3.X, version 1.5, was released in early 2006. This software will continue to be maintained, but research and development will be focused on two new versions of AGWA: one for ArcGIS 9.X (AGWA 2.0), and one for the Internet (DotAGWA). AGWA 2.0 and DotAGWA, due for Beta release in 2006 and 2007, respectively, will incorporate the same functionality as AGWA 1.5, but are designed to meet several additional criteria: (1) maximize the capacity to incorporate additional models, including non-hydrologic models, (2) facilitate the interaction between observed and modelled information at multiple scales, and (3) maximize potential user audiences.

AGWA 2.0 will be distributed as a custom tool for ArcGIS 9.X that can be loaded to ArcMap. It will include greater flexibility to incorporate user-provided/defined information, additional tools for scenario development and the analysis and visualization of model results, and improved watershed subdivision and parameterization functionality to enhance the performance and application of K2. AGWA 2.0 will provide the maximum flexibility to work with input provided by the user and to manipulate the parameters and settings of K2 simulations, thus facilitating model calibration.

DotAGWA is designed to assist in the decision-making processes by making AGWA functionality available to a much larger audience, namely those without access to proprietary GIS software and/or the GIS skills needed to assemble necessary input datasets (Cate *et al.*, 2005; Cate *et al.*, 2006). The DotAGWA design includes features that will help users share and visualize data by providing access to the application through an Internet browser interface. Different stakeholder groups will be able to interact with the application to help facilitate the communication and decision-making processes. Users will be able to define management scenarios, attach models to a plan, and have the application parameterize and run the models for the defined management plan. Multiple file-format options (e.g., text, XML, and HTML) will be available for exporting simulation outputs.

5.4.3 AGWA-KINEROS

Rather than running K2 as an external program, future versions of AGWA 2.0 and DotAGWA will be able to utilize the restructured K2 object library to enhance interaction between model and GIS interface. This will open up many possibilities by giving AGWA access to complete information about every element at each time step. For example, AGWA could animate the spatial development of runoff, infiltration or sediment production during a simulation. The user could have control over the animation speed, pause progress, change which quantities are displayed, or terminate the simulation at any time. Breakpoints could be set to pause the simulation at predetermined times, and it would be possible to "rewind" the simulation back to a previous breakpoint. Additional windows could monitor the longitudinal profile of the water surface in channels, as well as sediment concentration and bed deposition and scour. Using the object library would also make it easier to expand the sources of input data, such as radar rainfall and LIDAR topographic data.

5.5 REFERENCES

Arnold, J. G., J. R. Williams, and D. A. Maidment. (1995). A continuous time water and sediment routing model for large basins. *J. Hydr. Eng.*, **121 (2)**, 171–83.

Bennett, J. P. (1974). Concepts of mathematical modeling of sediment yield. *Water Resour. Res.*, **10 (3)**, 485–92.

Bingner, R. L. and Theurer, F. D. (2001). AGNPS 98: A suite of water quality models for watershed use. *Proceedings of the 7th Federal Interagency Sedimentation Conference*, Reno, NV, March 25–29, 2001, VII–1–VII–8.

Bouwer, H. and Rice, R. C. (1984). Hydraulic properties of stony vadose zones. *Groundwater*, **22 (6)**, 696–705.

Bouwer, H. (1966). Rapid field measurement of air entry and hydraulic conductivity as significant parameters in flow systems analysis. *Water Resour. Res.*, **2**, 729–38.

Burns, I. S., Semmens, D. J., Scott, S. N., Levick, L. R., Goodrich, D. C., and Kepner, W. G. (2004). The Automated Geospatial Watershed Assessment (AGWA) tool – enhanced capabilities of just released version 1.4. *Proceedings of the Arizona Hydrological Society 17th Annual Symposium*, Sept. 5–18, Tucson, AZ.

Cate, A. J. Jr., Semmens, D. J., Burns, I. S., Goodrich, D. C., and Kepner, W. G. (2005). *AGWA Design Documentation: Migrating to ArcGIS and the Internet*. US EPA Report EPA/600/R-05/056, USDA-ARS Report ARS/181027.

Cate, A. Jr., Goodrich, D. C., and Guertin, D. P. (2006). Integrating hydrologic models and spatial data in a distributed Internet application. *Proceedings of the Third Federal Interagency Hydrologic Modeling Conference*, April 3–6, 2006, Reno, NV (http://www.tucson.ars.ag.gov/unit/publications/pdffiles/1801.pdf).

Corradini, C., Melone, F., and Smith, R. E. (2000). Modeling local infiltration for a two-layered soil under complex rainfall patterns. *J. Hydrol.*, **237 (1–2)**, 58–73.

Ebert, D. W. and Wade, T. G. (2004). *Analytical Tools Interface for Landscape Assessments (ATtILA) Version 2004 User Manual*. US EPA Report EPA/600/R-04/083.

Engelund, F. and Hansen, E. (1967). *A Monograph on Sediment Transport in Alluvial Streams*. Copenhagen: Teknisk Forlag.

ESRI (2001). *ArcView Version 3.2a Software and User Manual*. Redlands: Environmental Systems Research Institute.

Fair, G. M. and Geyer J. C. (1954). *Water Supply and Wastewater Disposal*. New York: Wiley.

Goodrich, D. C., Unkrich, C. L., Smith, R. E., and Woolhiser, D. A. (2002). KINEROS2 – a distributed kinematic runoff and erosion model. *Proceedings of the Second Federal Interagency Hydrologic Modeling Conference*, July 29-Aug. 1, Las Vegas, NV.

Goodrich, D. C., Unkrich, C. L., Smith, R. E., and Woolhiser, D. A. (2006). KINEROS2 – new features and capabilities. *Proceedings of the Third Federal Interagency Hydrologic Modeling Conference*, April 3–6, 2006, Reno, NV (http://www.tucson.ars.ag.gov/unit/publications/pdf files/1799.pdf).

Green, W. H. and Ampt, G. (1911). Studies of soil physics. Part I: The flow of air and water through soils. *J. Agri. Sci.*, **4**, 1–24.

Hernandez, M., Kepner, W. G., Semmens, D. J., *et al.* (2003). Integrating a landscape/hydrologic analysis for watershed assessment. *Proceedings of the First Interagency Conference on Research in the Watersheds*, Benson, AZ, Oct 27–30. USDA, Agricultural Research Service, 461–466.

Kepner, W. G., Semmens, D. J., Basset, S. D., Mouat, D. A., and Goodrich, D. C. (2004). Scenario analysis for the San Pedro River, analyzing hydrological consequences for a future environment. *Environ. Mod. Assess.*, **94**, 115–27.

Lahlou, M., Shoemaker, L., Choudry, S., *et al.* (1998). *Better Assessment Science Integrating Point and Nonpoint Sources: BASINS 2.0 User's Manual*. US EPA Report EPA-823-B-98–006.

Levick, L. R., Semmens, D. J., Guertin, D. P., *et al.* (2004). Adding global soils data to the Automated Geospatial Watershed Assessment tool (AGWA). *Proceedings of the Second International Symposium on Transboundary Waters Management*, Nov 16–19, Tucson, AZ.

Levick, L. R., Guertin, D. P., Scott, S. N., Semmens, D. J., and Goodrich, D. C. (2006). Automated Geospatial Watershed Assessment tool (AGWA): uncertainty analysis of common input data. *Proceedings of the Third Federal Interagency Hydrologic Modeling Conference*, April 3–6, Reno, NV. (http://www.tucson.ars.ag.gov/unit/publications/pdffiles/1797.pdf)

Meyer, L. D. and Wischmeier, W. H. (1969). Mathematical simulation of the process of soil erosion by water. *Trans. Am. Soc. Agri. Eng.*, **12 (6)**, 754–62.

Miller, S. N., Guertin, D. P., and Goodrich, D. C. (1996). Linking GIS and geomorphology field research at Walnut Gulch. *Proceedings of the American Water Resources Association 32nd Annual Conference and Symposium: "GIS and Water Resources"*, Sept. 22–26, Ft. Lauderdale, FL, 327–35.

Miller, S. N., Semmens, D. J., Miller, R. C., *et al.* (2002). GIS-based hydrologic modeling: the Automated Geospatial Watershed Assessment tool. *Proceedings of the Second Federal Interagency Hydrologic Modeling Conference*, July 29–Aug 1, Las Vegas, NV.

Miller, S. N., Shrestha, S. R., and Semmens, D. J. (2004). Semi-automated extraction and validation of channel morphology from LIDAR and IFSAR terrain data. *Proceedings of the 2004 American Society for Photogrammetry and Remote Sensing Annual Conference*, May 23–28, Denver, CO.

Morin, E., Krajewski, W. F., Goodrich, D. C., Gao, X., and Sorooshian, S. (2003). Estimating rainfall intensities from weather radar data: the scale-dependency problem. *J. Hydrometeorol.*, **4**, 782–97.

Morris, E. M. and Woolhiser, D. A. (1980). Unsteady one-dimensional flow over a plane: partial equilibrium and recession hydrographs. *Water Resour. Res.*, **16 (2)**, 355–60.

Osborn, H. B., Lane, L. J., and Myers, V. A. (1980). Rainfall/watershed relationships for southwestern thunderstorms. *Trans. Am. Soc. Civ. Eng.*, **23 (1)**, 82–91.

Osborn, H. B., Unkrich, C. L., and Frykman, L. (1985). Problems of simplification in hydrologic modeling. *Hydrology and Water Resources in Arizona and the Southwest, Arizona-Nevada Academy of Science*, **15**, 7–20.

Parlange, J.-Y., Lisle, I., Braddock, R. D., and Smith, R. E. (1982). The three-parameter infiltration equation. *Soil Sci.*, **133 (6)**, 337–41.

Rawls, W. J., Brakensiek, D. L., and Saxton, K. E. (1982). Estimation of soil water properties. *Trans. Am. Soc. Agri. Eng.*, **25 (5)**, 1316–20, 1328.

Reeves, M. and Miller, E. E. (1975). Estimating infiltration for erratic rainfall. *Water Resour. Res.*, **11 (1)**, 102–10.

Rovey, E. W. (1974). A kinematic model for upland watersheds. Unpublished M.Sc. thesis, Colorado State University, Fort Collins.

Rovey, E. W., Woolhiser, D. A., and Smith, R. E. (1977). *A Distributed Kinematic Model for Upland Watersheds*. Hydrology Paper No. 93, Colorado State University.

Scott, S. (2005). Implementing best management practices in hydrologic modeling using KINEROS2 and the Automated Geospatial Watershed Assessment (AGWA) tool. Unpublished M.Sc. thesis, University of Arizona, Tucson.

Semmens, D. J., Miller, S. N. and Goodrich, D. C. (2006). Towards an automated tool for channel-network characterization, modeling and assessment. *Proceedings of the Third Federal Interagency Hydrologic Modeling Conference*, April 3–6, Reno, NV. (http://www.tucson.ars.ag.gov/unit/publications/pdffiles/1800.pdf)

Smith, R. E. (1990). Analysis of infiltration through a two-layer soil profile. *Soil Sci. Soc. Am. J.*, **54 (5)**, 1219–27.

Smith, R. E. and Goodrich, D. C. (2000). Model for rainfall excess patterns on randomly heterogeneous areas. *J. Hydrol. Eng.*, **5 (4)**, 355–62.

Smith, R. E. and Parlange, J.-Y. (1978). A parameter-efficient hydrologic infiltration model. *Water Resour. Res.*, **14 (3)**, 533–8.

Smith, R. E., Corradini, C., and Melone, F. (1993). Modeling infiltration for multistorm runoff events. *Water Resour. Res.*, **29 (1)**, 133–44.

Smith, R. E., Goodrich, D. C., and Quinton, J. N. (1995b). Dynamic, distributed simulation of watershed erosion: the KINEROS2 and EUROSEM models. *J. Soil Water Conserv.*, **50(5)**, 517–20.

Smith, R. E., Goodrich, D. C., Woolhiser, D. A., and Unkrich, C. L. (1995a). KINEROS – a kinematic runoff and erosion model. In *Computer Models of Watershed Hydrology*, ed. V. J. Singh, Highlands Ranch, CO: Water Resources Publications, 697–732.

Smith, R. E., Smettem, K. R. J., Broadbridge, P., and Woolhiser, D. A. (2002). *Infiltration Theory for Hydrologic Applications*. Water Resources Monograph Series, Vol. 15. Washington, DC: American Geophysical Union.

Stone, J. J., Lane, L. J., and Shirley, E. D. (1992). Infiltration and runoff simulation on a plane. *Trans. Am. Soc. Agri. Eng.*, **35 (1)**, 161–70.

Unkrich, C. L. and Osborn, H. B. (1987). Apparent abstraction rates in ephemeral stream channels. *Hydrology and Water Resources in Arizona and the Southwest, Arizona-Nevada Academy of Science*, **17**, 34–41.

USACE (2003). *Geospatial Hydrologic Modeling Extension: HEC-GeoHMS User's Manual.* US Army Corps of Engineers, Hydrologic Engineering Center, Report CPD-77.

USDA-SCS. (1973). *A Method for Estimating Volume and Rate of Runoff in Small Watersheds.* SCS-TP-149. Washington, DC: US Department of Agriculture, Soil Conservation Service.

Wilgoose, G. and Kuczera, G. (1995). Estimation of subgrid scale kinematic wave parameters for hillslopes. In *Scale Issues in Hydrologic Modeling*, ed. J. D. Kalma and M. Sivapalan, Chichester: John Wiley & Sons, pp. 227–240.

Woolhiser, D. A. and Liggett, J. A. (1967). Unsteady, one-dimensional flow over a plane – the rising hydrograph. *Water Resour. Res.*, **3 (3)**, 753–71.

Woolhiser, D. A., Hanson, C. L., and Kuhlman, A. R. (1970). Overland flow on rangeland watersheds. *J. Hydrol. (New Zealand)*, **9 (2)**, 336–56.

Woolhiser, D. A., Smith, R. E., and Goodrich, D. C. (1990). *KINEROS, A Kinematic Runoff and Erosion Model: Documentation and User Manual.* US Department of Agriculture, Agricultural Research Service, ARS-77.

Woolhiser, D. A., Smith, R. E., and Giraldez, J.-V. (1997). Effects of spatial variability of saturated hydraulic conductivity on Hortonian overland flow. *Water Resour. Res.*, **32 (3)**, 671–8.

6 Ephemeral flow and sediment delivery modelling in the Indian arid zone

K. D. Sharma

6.1 STUDY AREA: THE LUNI BASIN

The Luni and its tributaries form the only integrated drainage system in north-west arid India. The Luni originates in the Aravalli hill ranges near Ajmer (20.5°N, 74.7°E) and after an initial west-south-westerly course, flows south-west until it discharges into the Rann of Kachchh (Fig. 6.1). The area of the Luni drainage basin is 34 866 km^2 and elevations range from 886 m at the source to 10 m at the outlet. In the upland region the mean depth, width, and gradient of the flow channels are 1.2 m, 158 m, and 0.00245, respectively, whereas in the channel phase these are 3.6 m, 1958 m and 0.0012, respectively. The drainage basin areas vary from 104 to 950 km^2 in the upland region and 1449 to 5492 km^2 in the downstream valley. The eastern part of the drainage basin is a hilly and rocky piedmont, underlaid by igneous and metamorphic rocks of Precambrian and Paleozoic age. Of the drainage basin, 52 % is rugged mountainous terrain with shallow soils and minor amounts of unconsolidated alluvium. The western part of the drainage basin consists of Pleistocene alluvium and Holocene sand ranging from 1 to 40 m in depth.

Annual precipitation in the Luni drainage basin ranges between 600 mm in the south-east and 300 mm in the north-west. The rainfall season is relatively short, starting in June and ending in September with 80 % of rainfall occurring during July and August. The number of rainy days is low (about 14). Typical of the desert climate, the rainfall is characterized by a rapid onset and short duration. It is infrequent, localized, and variable within the drainage basin. Annual average pan evaporation is 2640 mm – five times the precipitation – so that the streamflow is non-existent for much of the year.

Study of the spatial variation in stream discharge was based on 34 gauging stations, which are located on various channel sections in the drainage basin (Figure 6.1). Each channel reach was gauged at a minimum of two stations, one in the upland region and the other in the down-channel section. Hourly stage heights were observed at each station during flows and discharge was calc-ulated, taking the values of hydraulic roughness from Vangani and Kalla (1985). The first water samples for sediment concentration were collected when flow began, with subsequent samples collected at times when there were significant changes in the flow discharge. The measurements of flow rate and sediment concentration allowed computation of sediment delivery rates for each flow event. These data were collected for a period of nine years during 1979–87.

6.2 EPHEMERAL FLOW MODELLING

The ephemeral flow in the Luni and its tributaries occurs sporadically during short, isolated flow periods separated by longer periods of low or zero flow; sustained flow is rare and baseflow is essentially absent. Peak flow rates occur within a few hours of the start of a rise, partly due to the nature of arid zone rainfall, and sometimes from the steepness of the channels draining the well-defined runoff-generating zones. Frequently, peak flow rates are reached almost instantaneously because the ephemeral flood wave forms a steep wave front, or the "wall of water" of folklore, in its travel downstream (Jones, 1981; Pilgrim *et al.*, 1988; Sharma and Murthy, 1996). Two mechanisms contribute to the formation of the steep wave front. First, the rate of infiltration into the permeable dry streambed is highest at the wave front and decreases in the upstream direction, with the effect that the leading edge of the wave steepens as it moves downstream. Second, the deeper portion of the flood wave near the peak travels faster than the leading edge of the wave, with the result that the wave peak approaches the front until the peak and front almost coincide and a shock front is formed.

For a simple individual flow event generated by discrete storms, the rapid rise to peak discharge (almost instantaneous) is followed by the recession portion of the hydrograph. The duration of recession is generally much longer than the time required to reach peak flow. The resulting flood wave shape is such that almost the entire

Hydrological Modeling in Arid and Semi-Arid Areas, ed. Howard Wheater, Soroosh Sorooshian, and K. D. Sharma. Published by Cambridge University Press.
© Cambridge University Press 2008.

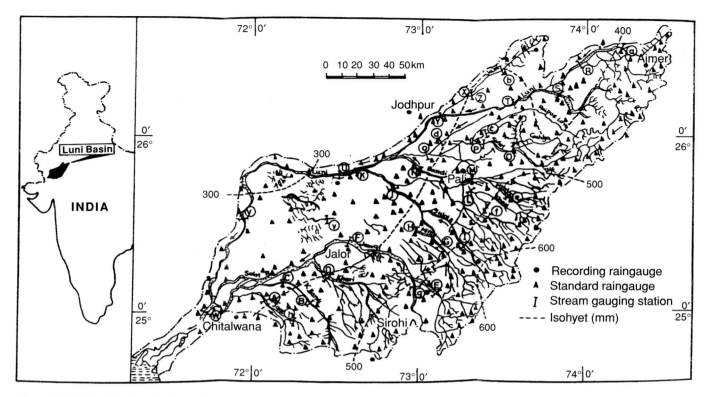

Figure 6.1 The Luni drainage basin, north-west India

hydrograph consists of the recession curve (Fig. 6.2), which is faster than the exponential decay, i.e., curvilinear decay, thereby highlighting the interaction of alternative drainage basin processes in the development of the recession flow in ephemeral channels. Wu (1972) and Yakowitz (1973) have also reported the curvilinear decay of the ephemeral stream recession curves.

6.2.1   Ephemeral flow modelling in the upland phase

Choosing a series of peak flow values and attaching to these properly shaped recession curves may simulate the outflow hydrograph in arid upland basins. The peak discharge may be estimated by: (1) using a peak-flow value from an observed recession curve, (2) choosing a peak-flow value from a probability distribution function of peak discharge, or (3) by regression analysis relating the peak discharge and time-to-peak with effective rainfall and a few selected basin parameters.

We hypothesize that at some fixed reference point on the stream draining the upland basin, just subsequent to the peak discharge, some portion of the upstream channel network contains the flood. Floodwater leaves the upstream basin either by draining past the reference point or by infiltration at the reference point. Thus, a single leaky reservoir substitutes the upland basin at the time of peak discharge at the point of reference to simulate the behavior

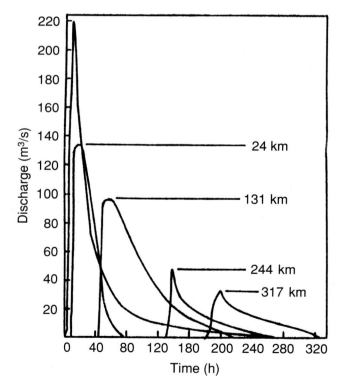

Figure 6.2 Discharge hydrographs at stations on River Luni during July 19– August 2, 1981

of the ephemeral channel network. The reservoir loss rate may be set equal to the estimated drainage-basin infiltration rate, and the reservoir configuration and outflow rate may be designed to reflect the topography and possible physical mechanism operating during the flow recession. The use of a reservoir as a flow source is an established practice in hydrology; it has been used to model the rainfall-runoff relations (Nash, 1957), flow recession in irrigation (Wu, 1972) and in flood routing (Chow et al., 1988).

Unsteady open-channel flow in channels where infiltration loss takes place is described by a continuity equation (Henderson, 1966):

$$\delta Q/\delta X + \delta A/\delta t = -i(X, t), \quad (6.1)$$

and a momentum equation (Ball, 1987):

$$S_f = S_c - \delta y/\delta X - u\delta u/g\delta X - \delta u/g\delta t - i(X, t)u/gA$$
$$= u^2 n^2/R_h^{4/3}, \quad (6.2)$$

where $Q(m^3/s)$ is discharge, $A(m^2)$ is cross-sectional area of flow, $i(m^3s^{-1}m^{-1})$ is local outflow due to infiltration, $y(m)$ is flow depth, S_f is friction slope, S_c is channel bed slope, u (m/s) is local mean flow velocity, g(m/s) is acceleration due to gravity, n is Manning's roughness coefficient, R_h(m) is hydraulic radius, X(m) is distance along the channel and t(s) is time. The order of magnitude of various terms in the momentum Equation (6.2), as applicable to the arid upland drainage basins, indicated that the terms $\delta y/\delta X$, $u\delta u/g\delta X$, $\delta u/g\delta t$, and $i(X, t)u/gA$ are 100 or more times less than the gravity force due to S_c and friction force due to S_f and hence are negligible. This leaves the equation of motion as:

$$S_f = S_c = u^2 n^2/R_h^{4/3}, \quad (6.3)$$

which is a special case of non-linear kinematic approximation to the momentum equation (Singh, 1976):

$$Q = \alpha h^m \quad (6.4)$$

where h(m) is stage, and α and m are empirical constants.

Combining the continuity Equation (6.1) with discharge-stage (Equation 6.4) and storage-stage ($S = \beta h^p$; where $S(m^3)$ is storage, and β and p are empirical constants) relationships, Peebles et al. (1981) obtained a general form of the equation for leaky reservoir discharge as:

$$-c^*(Q, t) - Q(t) = (\beta/\alpha^{p/m})d[Q(t)]^{p/m}/dt, \quad (6.5)$$

where $c^*(Q, t)(m^3s^{-1})$ is reservoir loss rate.

For a rectangular cross-section, i.e., $S \propto h^2$, Peebles et al. (1981) derived a variable loss rate (VLR2) model for the dis-

charging reservoir as:

$$dQ/dt = -mc'\alpha^{1/m}Q^{(1-1/m)} - (m\alpha^{2/m}/Wa^o)Q^{2(1-1/m)}, \quad (6.6)$$

where $c'(m^3s^{-1}m^{-2})$ is loss rate per unit area, W(m) is reservoir width and a^o is a reservoir shape factor.

An alternative version of the variable loss rate model can also be derived assuming a triangular cross-section, i.e., $S \propto h^3$, and named as VLR3 model (Sharma and Murthy, 1996). Storage of a triangular cross-section is given by:

$$S = Wa^o h^2/6. \quad (6.7)$$

Here p in the $S = \beta h^p$ relationship, is equal to three. Combining Equation (6.7) with the discharge-stage relation (Equation 6.4), and substituting for h in the total-loss rate Equation (6.5), the governing differential equation for the VLR3 model becomes:

$$dQ/dt = -mc'\alpha^{1/m}Q^{(1-1/m)} - (m\alpha^{3/m}/Wa^o)Q^{(2-3/m)}. \quad (6.8)$$

Solutions to Equations (6.6) and (6.8) were obtained numerically.

6.2.2 Ephemeral flow modelling in the channel phase

A dry ephemeral channel reach may be conceptualized as a reservoir such that storage is a function of discharge. Inflow to this channel reach begins and storage builds up so that the peak outflow coincides with the maximum storage, i.e., with the outflow and storage peak, concurrently. For an ephemeral channel reach, input $I(t)$, output $Q(t)$, transmission loss $Q_L(t)$ (all expressed in m^3s^{-1}) and storage $S(m^3)$ are related by a spatially lumped form of the continuity equation:

$$dS/dt = I(t) - Q(t) - Q_L(t). \quad (6.9)$$

If the inflow hydrograph $I(t)$ is known, Equation (6.9) cannot be solved directly to obtain the outflow hydrograph $Q(t)$, because $Q(t)$, $Q_L(t)$ and S are unknown. Two other equations relating S to $I(t)$, $Q(t)$ and $Q_L(t)$ are required to obtain a unique solution for Equation (6.9). However, for the time being, let $Q_L(t)$ be assumed known; thus, in this case, one additional equation would be needed. A storage function S may be written as an arbitrary function F of I, Q, and Q_L, and their time derivatives as:

$$S = F(I, dI/dt, d^2I/dt^2, \ldots, Q, dQ/dt, d^2Q/dt^2, \ldots,$$
$$Q_L, dQ_L/dt, d^2Q_L/dt^2, \ldots). \quad (6.10)$$

The S is only a temporary storage and is equal to the total volume of water in transit to the channel outlet. The relationship in Equation (6.10) exists for long narrow reservoirs or open channels where S is a function of both the inflow and outflow. A finite difference solution method is applied to Equations (6.9) and (6.10). The time horizon is divided into finite intervals, and continuity Equation (6.9) is solved recursively from one time point to the next, using

Table 6.1 *Physical characteristics of representative upland drainage basins*

Drainage basin	Area (km^2)	Stream order	Stream network	Geomorphology	Lithology
Alniawas (a^a)	950	6	Organized	Shallow alluvium	Schist–gneiss
Pipar (b^a)	631	5	Organized	Shallow alluvium	Limestone
Sanderao (G^a)	597	6	Discontinuous	Deep alluvium	Gneiss
Bhuti (H^a)	974	6	Discontinuous	Deep alluvium	Gneiss
Posalia (g^a)	286	6	Organized	Shallow alluvium	Granite
Ramnia (v^a)	130	5	Discontinuous	Blown sand	Granite

^{a} Refers to Fig. 6.1

the storage function (Equation 6.10) to account for the value of storage at each time period (Sharma and Murthy, 1995).

6.2.3 Estimation of transmission losses

The most uncertain parameter for the hydrologic routing of flow in ephemeral channels is the time-distributed volumetric rate of transmission loss, $Q_L(t)$. An empirical method to estimate time-dependent $Q_L(t)$ is as follows.

From the measured peak-inflow and peak-outflow discharges, the peak-transmission loss rate in the channel segment Δx(m), $Q_{Lp\Delta x}$(m^{3}s^{-1}) may be estimated from:

$$Q_{Lp\Delta x} = Q_u[1 - (Q_d/Q_u)^{\Delta x/X}], \qquad (6.11)$$

where Q_u(m^{3}s^{-1}) is peak inflow and Q_d(m^{3}s^{-1}) is peak outflow from a channel reach of length X(m). The concept of non-linearity in peak flow reduction between two consecutive gauging stations is inherent in Equation (6.11). Equation (6.11) is based on the following assumptions:

(1) $Q_L = F(Q, t)$.
(2) Peak-transmission loss rate Q_{Lp} occurs when the stream discharges at its peak, because an increase in the wetted parameter with increasing discharge is an important variable in reckoning the transmission losses.
(3) The distance Δx is small, so Q_u and Q_d occur at nearly the same time.

The estimates of Q_{Lp} can be made for any channel segment by calculating the peak loss rate at the beginning of the segment as the peak inflow minus the Q_{Lp} in the previous segment.

From the observed duration of inflow and outflow, the duration of flow in each channel segment may be estimated as:

$$T_{xi} = T_I + [|T_I - T_O|/X](x_i), \qquad (6.12)$$

where T_{xi}(s) is duration of flow in the xith segment of the channel reach, T_I(s) is duration of inflow and T_O(s) is duration of outflow. Similarly, the time to peak flow in each channel segment may also

be estimated from:

$$T_{xip} = T_{Ip} + [|T_{Ip} - T_{Op}|/X](x_i), \qquad (6.13)$$

where T_{xip}(s) is time to peak flow in the xith segment of the channel reach, T_{Ip}(s) is time to peak flow at the inlet section and T_{Op}(s) is time to peak flow at the outlet section.

Using Equations (6.11) to (6.13), a triangular transmission loss rate hydrograph can be drawn for each segment of the channel reach. These hydrographs are staggered by the observed advance rate of flow. The loss rate hydrograph for the entire channel reach may be prepared by summing up the ordinates of Q_L at a given instance from each segment of the channel reach.

6.2.4 Results and comments

Fifty-one ephemeral flow hydrographs from six representative upland drainage basins (Table 6.1) recorded between June 26, 1980 and June 13, 1987 were selected for simulation. The majority of flows in these drainage basins lasted from 4 to 20 h; none among the group exceeded 50 h. Records are continuous recordings; in order to compare with model curves, discrete values of discharge were picked at half-hourly intervals for the duration of flow. For the purpose of simulation, recession curves were defined as: (1) decay from peak flow up to a point where the stream stage does not change or increase due to the arrival of a second peak, or (2) to a flow depth at 5 % of the peak discharge.

6.2.4.1 MODELLING THE RISING LIMB OF A HYDROGRAPH IN THE UPLAND PHASE

For a given drainage basin the peak flow and time-to-peak are a function of the effective rainfall (Holder, 1985; Sharma and Murthy, 1996). Therefore, graphical relationships among the peak discharge-effective rainfall (the part of precipitation that produces runoff (Langbein and Iseri, 1960), basin area (Fig. 6.3), and time to peak-effective rainfall-basin area (Fig. 6.4) were developed for the studied drainage basins. These two relationships were used to predict the rising limb of the ephemeral flow hydrographs.

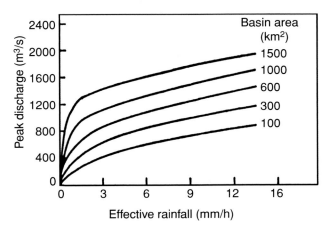

Figure 6.3 Peak discharge as a function of effective rainfall for arid upland drainage basins

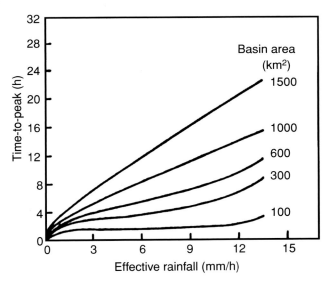

Figure 6.4 Time-to-peak as a function of effective rainfall for arid upland drainage basins

6.2.4.2 MODELLING THE RECESSION FLOW IN THE UPLAND PHASE

The model parameters c' and a^o (Equation 6.6) were fitted to observed data by optimization. Comparison of model and observed curves was carried out by defining an objective function:

$$I(c', a^o) = [1/(N-1) \sum_{i=2}^{N} (X_i - Y_i)^2]^{1/2}, \qquad (6.14)$$

where X_i and Y_i are the ith points on the observed and simulated recession curves and N is number of points on the curves. About three to ten flow events occur annually in the studied upland drainage basins with flow duration of a few hours to two days.

Results from the analysis of recessions recorded in a representative discontinuous upland drainage basin are presented. The main stream emanating from the hills at Sanderao (G in Fig. 6.1) disappears into deep alluvium/wind deposited sand after flowing a certain distance in the foothills and flood plains. Due to high transmission losses the stream fails to reach its ultimate outlet and cannot join the trunk stream. In deserts such "disorganized" stream networks are common (Mabbutt, 1977). Observed stage records at the gauging station were converted to discharge using the relation:

$$Q = 199.20 \, h^{2.22}. \qquad (6.15)$$

Nine recession curves were simulated at Sanderao. It was observed that the minimum trend of the objective function for four events with peak discharges 420.0, 144.0, 143.0, and 114.7 m³/s indicates that the VLR2 model was more sensitive to the loss rate changes for these events (Fig. 6.5). Incidentally, these were either the first flow events of the season or there was a gap of 19 to 27 days of no flow which dried up the streambed and therefore, the loss rate was comparatively higher. High infiltration losses were also reported in the dry streambed of Saudi Arabian discontinuous drainage systems (Walters, 1989). The VLR2 model gave a better fit for all the nine simulated recession curves (Table 6.2), even at a negligible loss rate for two events. A high peak flow of 160 m³/s on July 9, 1981 was preceded by a flow on July 7, 1981, only two days earlier (due to silting of the gauge well, this flow could not be recorded). A low peak flow of 5 m³/s on July 15, 1982 was preceded by a flow on July 14, 1982, only a day earlier (Table 6.2). The best loss rate of zero might reflect a real condition in the channel, i.e., a high antecedent moisture condition causing streambed infiltration to be negligible. Two identical flow peaks of 140 m³/s were recorded on July 1, 1983 and August 16, 1983. Since the former was the first flow event of the season, it gave a better fit for the VLR2 model with a higher loss rate of 13.0×10^{-6} m/s than did the latter, which was preceded by a flow event on August 14, 1983, and therefore, had a lower loss rate of 7.2×10^{-6} m/s. However, the reservoir storage was more for the latter than the former ($a^o = 0.48 \times 10^5$ versus 0.19×10^5) since the recession period was longer, 10 h versus 5 h, respectively. Therefore, a^o seems to be a function of the flow duration; the longer the duration of flow, the greater the reservoir storage. The results of the search for best fit are depicted in Fig. 6.6 where VLR2 and VLR3 model curves are plotted against the observed curves.

Four flow events simulated at Sanderao were also observed at Bhuti (H in Fig. 6.1), a downstream gauging station at a distance of 22 km, before the stream disappears. The derived discharge-stage relationship at Bhuti was:

$$Q = 45.31 h^{2.61}. \qquad (6.16)$$

This rating differs from that at Sanderao (Equation 6.15) in the sense that it has a lower value of multiplier, therefore reflecting

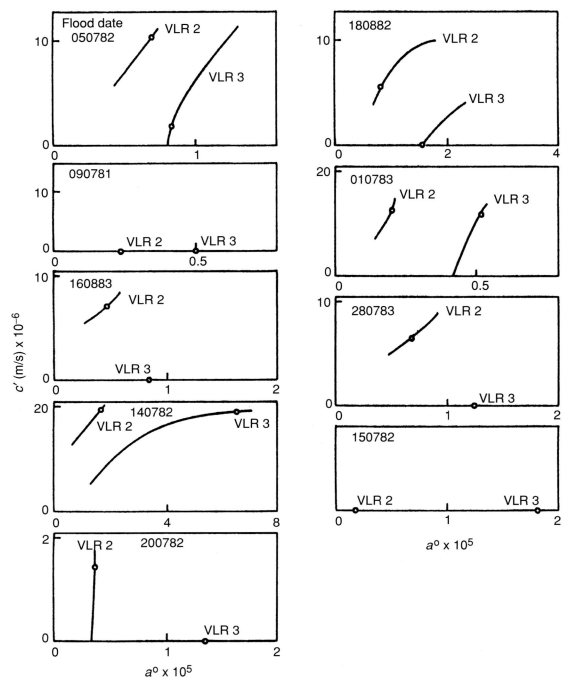

Figure 6.5 Objective function maps for nine recession curves at Sanderao. ____ trend of minimum, **o** objective function minimum

lower discharge at Bhuti for the same stage since the stream cross-section narrows between these two gauging stations. The VLR2 model gave a better fit for all the four events of 168.2, 107.5, 79.1, and 26.0 m³/s peak discharges. The reason for choosing the same flow events at Bhuti was to examine the behavior of the best-fit parameters as the flood moves downstream.

Figure 6.7 shows that a^o increases for all the events as the flood moves downstream. Increase in a^o means increase in the reservoir length for a fixed peak stage at the point of discharge. In general, the recession curves at Bhuti had a longer duration than at Sanderao (average 13 h versus 9 h) for the same peak discharge. Also, as the flood moves downstream through the stream network

Table 6.2 *Values of* c′, a° *and objective function* (I) *for the best fit model recession curves at Sanderao (G in Fig. 6.1)*

Date	Peak discharge (m³/s)	VLR2			VLR3		
		$c'(m/s) \times 10^{-6}$	$a^{o} \times 10^{5}$	I	$c'(m/s) \times 10^{-6}$	$a^{o} \times 10^{5}$	I
05.07.82	420.0	10.6	0.70	12.54	2.0	0.84	25.85
18.08.82	204.0	5.6	0.83	8.11	0	1.55	17.71
09.07.81	160.0	0	0.24	9.45	0	0.50	17.10
01.07.83	144.0	13.0	0.19	2.69	12.0	0.52	2.80
16.08.83	144.0	7.2	0.48	4.78	0	0.85	4.88
28.07.83	114.7	6.6	0.68	3.77	0	1.23	4.90
14.07.82	53.7	19.6	1.58	3.24	19.0	6.50	3.55
20.07.82	32.2	1.4	0.34	2.0	0	1.35	2.99
15.07.82	5.0	0	0.17	0.13	0	1.80	0.30

the best loss rate increases (Fig. 6.7). Thus, the simulation of recession at successive gauging stations as the flood moves downstream in a discontinuous drainage system indicates the physical processes of infiltration responsible for the disappearance of runoff in the thick sediment deposits.

There appeared to be a group of best reservoir shapes at Sanderao and a group at Bhuti (Fig. 6.7). This suggested that a characteristic shape might exist for all recession curves at the point of reservoir discharge. For the VLR2 model at the Sanderao gauging station the best shape is near $a^{o} = 0.80 \times 10^{5}$, and at the Bhuti gauging station the best shape is near $a^{o} = 2.68 \times 10^{5}$. Similarly, for the VLR3 model the best shapes are $a^{o} = 2.27 \times 10^{5}$ and 4.75×10^{5} at the Sanderao and at the Bhuti gauging stations, respectively.

The regression-type relationships from observed records on the gauged drainage basins are the simplest hydrological models. This type of analysis requires less information about the physical features of the drainage basin, yet it is transferable to the ungauged basins in the region (Holder, 1985). The major emphasis of this type of study is on the known causal relationships so that the model parameters have some physical interpretation as well as a statistical one. Lane (1983) reported a significant relationship for peak discharge and time-to-peak with effective rainfall and the drainage-basin area. In the present study the graphical relationship among these parameters (Figs. 6.3 and 6.4) was also found significant and, therefore, used for prediction of the rising limb of the flow hydrograph.

Generally, the best fit for ephemeral flow recession curves obtained through the search for optimization was a close fit to the observed curves. However, the convex and concave portions of the observed recession curves did not show a close fit. The shape of the objective function surface and trend of the minimum indicated the sensitivity of the model to changes in the c' and a^{o} values. The trend of the objective function for all the modelled

recessions (except for the first flow event of the season) indicated the greater sensitivity of VLR2 model to initial-storage or reservoir-length changes than to the loss-rate changes. In terms of these arid upland drainage basins, this result suggested that for the higher peak discharges, the shape of the observed recession curve depended primarily on the volume of water stored upstream, as high peak discharges were associated with torrential rainfall, which produced larger volumes of runoff. The effect of infiltration rate changes would tend to be overshadowed by this large volume of water stored at the peak stage. However, the reversal of slope of the objective function for the first flow event of the season suggested that the VLR2 model was more sensitive to the loss-rate changes than to the reservoir-length changes due to high infiltration losses in the initially dry streambed.

Comparison of the objective function for all the flow events showed the trend of changes in the direction of increasing a^{o}, as the model changed from VLR2 to VLR3. Increased values of a^{o} compensated for the decrease in initial storage. Initial storage, i.e., storage at peak discharge, decreased from VLR2 ($S = WhL/2$) to the VLR3 ($S = WhL/6$). Further, the best loss rate for VLR2 was higher than for VLR3. For the recession curves of both the models to be alike it is necessary that the high initial storage be compensated by a high loss rate. Therefore, the trend of the loss rate decreased from VLR2 to VLR3. Of the features shown by the plots of objective function, the most interesting was that the group of minima fell in a narrow range on the a^{o} axis (Fig. 6.7). This result suggested that some characteristic shape might exist that could be applied to that drainage basin to synthesize the recession flows. Physically, this implied that the flood wave, after having traveled some distance down a dry, infiltrating stream channel, developed a shape, which was determined by the channel properties with the result that the shape of the input was obliterated.

The concept of a leaky reservoir model to simulate the outflow hydrograph from arid upland drainage basins has a few weak

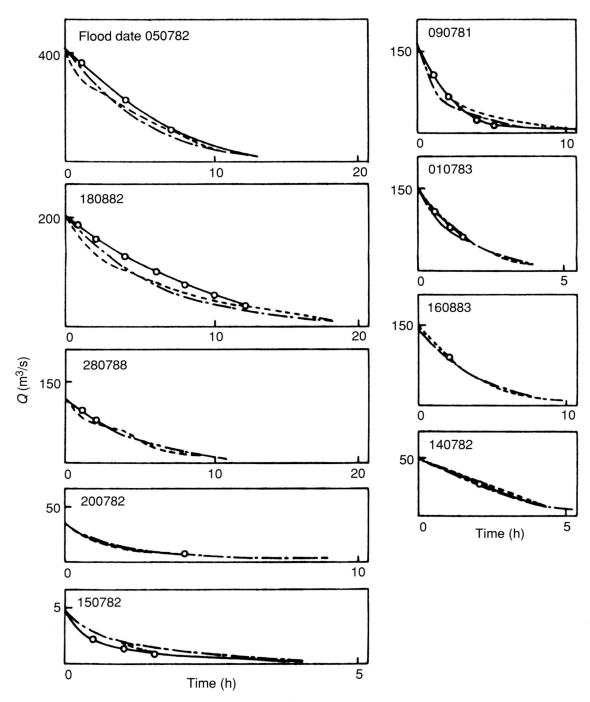

Figure 6.6 Best model and observed recession at Sanderao _____ observed, _____ VLR2, o–o VLR3.

points. The choice of a model rating curve, $Q = \alpha h^{m}$, implies that the discharge-stage relation is unique and does not depend on time. Both these conditions are probably never met in the arid regions. In fact, a unique discharge-stage relation might never exist for channels with movable beds. The beds of ephemeral channels often consist of unconsolidated sediments. There is sig-

nificant scour and fill of the cross-section during a flow event and the channel morphology can change significantly between flows. Further, the choice of only two parameters, c' and a^{o}, was arbitrary. It was based on the observation that c' and a^{o} were the least known of the variables that must be estimated to model the recession.

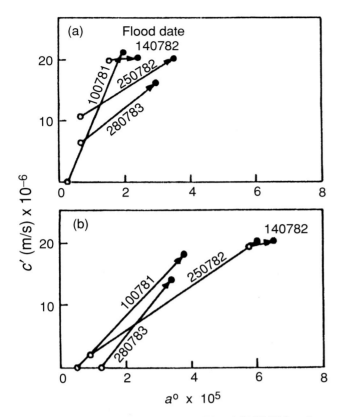

Figure 6.7 Downstream changes in (a) VLR2 and (b) VLR3 best fit parameters. o Sanderao ● Bhuti

6.2.4.3 FLOW ROUTING IN EPHEMERAL CHANNELS

Four representative channel reaches were selected within the Luni Basin to model the ephemeral flow in the channel phase. These reaches represent the entire range of channel characteristics and dominant lithologies encountered in India's arid zones.

The form of storage function to be employed in flow routing depends on the nature of the system being analyzed (Chow *et al.*, 1988). In the hydrologic routing of flow by the level-pool method, storage is a non-linear function of Q only, i.e.,

$$S = F(Q), \qquad (6.17)$$

and the function $F(Q)$ is determined by relating the channel storage and outflow to the stage. In the ephemeral channels, because the retarding effect due to backwater is negligible (Jones, 1981), an invariable storage–discharge relationship has been worked out for the test reaches (Table 6.3). Each $S = F(Q)$ relationship for the test reaches has two segments; the lower segment is a straight line represented by:

$$S = (dy/dx)Q, \qquad (6.18)$$

thereby indicating the within-bank flow discharge. The upper segment is a curvilinear relationship, showing the over-bank discharge.

The flow-routing technique was evaluated through 37 flow events recorded in the test reaches during the study period. The values of coefficient of determination, R^2, for the hourly discharges between the observed and calculated hydrographs show a good fit (Table 6.4). In addition, the outflow hydrographs generated by the flow-routing technique gave a better fit than the triangular approximation by Lane's (1982) lumped regression model. As an example, three hydrographs covering the range of flow, from the lowest to the highest, for the Sanderao–Bhuti channel reach are depicted in Fig. 6.8 for visual interpretation.

6.3 MODELLING SEDIMENT DELIVERY

Arid regions have a potential for generating and transporting large quantities of sediment (Schick 1970) mainly due to the torrential

Table 6.3 *Storage [S(m^3)] versus discharge [Q(m^3/s)] relationships in test channel reaches*

Channel reach	Reach identification[a]	S versus Q	R^2 [b]
Alniawas–Jasnagar	a–R	$S = 1\,000Q; Q < 20$	0.93
		$S = 200\,000 + (Q - 20)Q^{2.91}; Q > 20$	0.92
Pipar–Binawas	b–x	$S = 90\,000Q; Q < 18$	0.88
		$S = 1\,620\,000 + (Q - 18)Q^{2.47}; Q18$	0.91
Sanderao–Bhuti	G–H	$S = 9\,755Q; Q > 20$	0.96
		$S = 19\,5100 + (Q - 20)Q^{2.28}; Q > 20$	0.93
Posalia–Nawakhera	g–F	$S = 10\,000Q; Q > 3$	0.95
		$S = 30\,000 + (Q - 3)Q^{3.51}; Q > 33$	0.88

[a] Refers to Fig. 6.1

[b] R^2 is significant at 0.01 level of probability

Table 6.4 *Comparison of flow routing techniques in ephemeral channels*

	Counts of events within limits	
R^2 limits	Flow routing	Regression analysis[a]
0.98–1.00	9	None
0.95–0.98	9	6
0.90–0.95	18	6
0.80–0.90	1	19
< 0.80	None	6
Total events	$37(0.82^b)$	$37(0.95^b)$

[a] Sharma *et al.* (1984)
[b] Average R^2; $P > 0.01$

rainfall (Bell, 1979), weathering (Goudie and Wilkinson, 1977), almost total lack of natural protection against soil erosion due to sparse vegetative cover (Magfed, 1986; Pilgrim *et al.*, 1988), aeolian surficial deposits providing readily available material to be eroded by runoff (Jones, 1981) and biotic interference (FAO, 1973). A compilation of river-suspended sediment yields for moderate size drainage basins suggests that arid basins produce 36 times more sediment than humid temperate basins and 21 times more than the humid tropical equivalents (Reid and Frostick, 1987). However, significant sediment delivery is limited to the major flood flows – a characteristic of the arid regions (Chang and Stow, 1988).

6.3.1   Sediment delivery in upland phase

Models of sediment delivery by water in uplands may dynamically route the sediment by solving the continuity equation for sediment transport (Bennett, 1974). The solution of this equation is generally accomplished using numerical methods in association with a water-balance model, which provides the required hydrologic inputs. These dynamic models provide estimates of total soil loss from a drainage basin and predict the sediment delivery by considering the processes of soil detachment, transport, and deposition. However, in data-deficit arid regions, a closed-form solution to the governing differential equation under steady-state conditions is preferred. The closed-form solution reduces computational time and alleviates instabilities associated with numerical solutions (Chow *et al.*, 1988). Sediment movement downslope obeys the principle of continuity of mass expressed by Nearing *et al.* (1989):

$$\partial q_s/\partial X = D_f + D_i, \qquad (6.19)$$

where q_s(kg/(s/m)) is sediment transport rate per unit width, X(m) is downslope distance, D_f (kg /(s/m)) is net flow detachment rate and D_i(kg/s/m)) is net rainfall detachment rate. The assumption

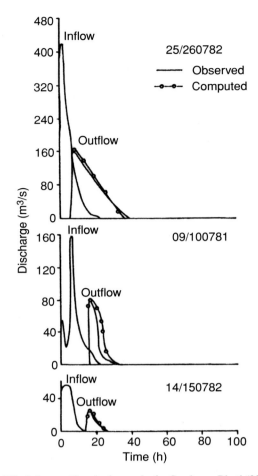

Figure 6.8 Inflow-outflow hydrographs for Sanderao–Bhuti (22 km) channel reach

of quasi-steady state allows Equation (6.19) to be written without an explicit time parameter. If D_i is assumed to be small (Lu *et al.*, 1989), Equation (6.19) can be written as:

$$\partial q_s/\partial X = D_f. \qquad (6.20)$$

The net flow detachment rate, D_f, is positive for detachment and negative for deposition. In the arid regions, since the initial potential sediment load is greater than the sediment transport capacity (Foster, 1982; Jones, 1981) detachment is assumed to occur at a rate:

$$D_f = G(T_c - q_s). \qquad (6.21)$$

This relationship is a diffusion-type equation (Foster and Meyer, 1972a), where G/(m) is a first-order reaction coefficient – equivalent to V_f/q, where V_f (m/s) is effective fall velocity for the sediment and q (m³/(s/m)) is flow discharge per unit of width – and T_c (kg/(s/m)) is sediment transport capacity.

The depth of flow is estimated by Manning's equation as:

$$h = \left(qnS_c^{-0.5}\right)^{0.6}, \qquad (6.22)$$

where h (m) is overland flow depth, n is Manning's roughness coefficient and is equal to 0.046 (for moderate vegetative cover and rough surface/depressions of 10 to 15 cm depth, a moderate value, Foster, 1982) and S_c is mean bed slope. Although the Darcy–Weisbach equation with a varying friction factor for laminar flow might be more accurate for the calculation of depth in some cases, most users are better acquainted with estimating Manning's n. The error in estimating a value for n is probably greater than the error in using the Manning's equation for the laminar flow. The effective duration of runoff (t_r, s) can be calculated as:

$$t_r = V_{up}/P_r, \qquad (6.23)$$

where V_{up} (m^3) is total runoff volume and P_r(m^3/s) is peak runoff rate. Shear stress acting on the soil, τ_s(kg/(m/s)), is calculated using the equation:

$$\tau_s = \gamma h S_c, \qquad (6.24)$$

where γ(kg/(m/s)) is the specific weight of water.

Several generalized formulae have been developed for computing the sediment transport capacity, T_c. Many of these equations were developed for streams, and were later applied to shallow overland and channel flows. However, Alonso et al. (1981) after evaluating nine sediment transport equations, concluded that the Yalin equation (Yalin, 1963) provided reliable estimates of transport capacity for shallow overland flow. Foster and Meyer (1972b) also concluded that the Yalin equation was the most appropriate for the shallow flows associated with soil erosion.

The Yalin equation is defined as:

$$T_c/(SG)d\delta_w^{0.5}\tau_s^{0.5} = 0.635\delta\{1 - (1/\beta)\ln(1+\beta)\}, \qquad (6.25)$$

with β, δ, and Y expressed as:

$$\beta = 2.45(SG)^{-0.4}(Y_{cr})^{0.5}\delta, \qquad (6.26)$$

$$\delta = (Y/Y_{cr}) - 1 \quad (\text{when } Y < Y_{cr}, \delta = 0), \qquad (6.27)$$

$$Y = (\tau_s/\delta_w)/(SG - 1)gd, \qquad (6.28)$$

where SG is particle specific gravity (2.65 for fine sand and silt), d (m) is particle diameter, δ_w(kg/m^3) is mass density of water, Y is dimensionless shear stress, Y_{cr} is dimensionless critical shear stress read from the Shield's diagram, revised as per Abrahams et al. (1988) for the overland flow on desert hill slopes, g(m/s^{-2}) is acceleration of gravity and β and δ are dimensionless parameters as defined by Equations (6.26) and (6.27), respectively. The modified Yalin equation, which considers a mixture of particles of varying size and density (Foster, 1982), was used. Combining Equations (6.20) and (6.21), the sediment delivery model was written as:

$$\partial q_s/\partial X + G(q_s - T_c) = 0. \qquad (6.29)$$

The closed-form solution of Equation (6.29) is:

$$\ln(T_c - q_s) = -GX + \ln(C) \qquad (6.30)$$

where C(kg/s/m)) is a constant of integration and is equal to $(T_c - q_s)$ at $X = 0$. Thus, C is the difference between sediment transport capacity and the actual sediment transport at the point of initiation of runoff within the drainage basin.

6.3.2 Sediment delivery in the channel phase

The channel storage of sediment in arid ephemeral channels has greater effect on sediment transport and, therefore, sediment supply has to be taken into account for sediment transport modelling in arid environments (Hadley, 1977; Reid and Frostick, 1987). In fact, as the flood flows traverse coarse, unsaturated sediments in the ephemeral channels, the sediment transport capacity reduced progressively by transmission losses of the flood flows resulting in deposition of sediment (Sharma and Murthy, 1996). Due to infrequent flow events and inaccessibility in arid regions, a simple practical model, based on a limited number of parameters, needs consideration.

The sediment graph can be predicted by convolution of an Instantaneous Unit Sediment Graph (IUSG) with actual sediment data that describes the changing concentration of sediment supply. The IUSG describes the amount of sediment supplied by an instantaneous burst of runoff, i.e., that producing one unit of sediment. The conceptual model was formulated as a routing of the sediment supply through a cascade of identical linear reservoirs to represent an IUSG.

A spatially lumped continuity equation and a linear storage law can represent the sediment dynamics. For a time interval Δt(s), those relationships can be written as:

$$I_s(t) = Q_s(t) + dS_s(t)/dt \qquad (6.31)$$

and

$$S_s = K_s Q_s, \qquad (6.32)$$

where $I_s(t)$(kg/s) is sediment input to a reservoir, $Q_s(t)$ (kg/s) is sediment discharge from the reservoir, $S_s(t)$ (kg) is sediment storage within the reservoir, K_s (s) is a sediment storage coefficient and t (s) is time since the beginning of sediment discharge. The model assumes the concept of linearity in the storage/outflow relationship as inherent in Equation (6.32). For an instantaneous sediment inflow of amount V_s (kg), the outflow from the first linear reservoir may be expressed as:

$$Q_s(t) = (V_s/K_s)\exp(-t/K_s), \qquad (6.33)$$

with

$$V_s = A_b E_s \qquad (6.34)$$

where A_b (km^2) is drainage basin area and E_s (kg/km) is supplied sediment. By successively routing through n identical linear reservoirs, the sediment outflow from the nth reservoir is:

$$Q_s(t) = [A_b E_s/K_s](ns-1)](t/K_s)^{(ns-1)}\exp(-t/K_s), \quad (6.35)$$

where ns is a dimensionless shape parameter. Differentiating Equation (6.35) with respect to time and using the condition $dQ_s(t)/dt = 0$ at $t = t_p$ (where t_p (s) is time to peak of sediment discharge):

$$t_p = (ns-1)K_s. \quad (6.36)$$

On substitution of the value of K_s in Equation (6.35), the sediment impulse response becomes:

$$U_s(0,t) = [Q_s(t)/A_b E_s]$$
$$= [(ns-1)^{ns}/t_p](ns-1)][(t/t_p)\exp(-t/t_p)]^{(ns-1)}, \quad (6.37)$$

where $U_s(0,t)/$(s) is the IUSG ordinate at time t. Equation (6.37) is defined by the values of two parameters, ns and t_p, which may be used for determining the shape of the IUSG.

The linkage of sediment delivery between the upland and channel phases is based on the transmission losses of runoff and is obtained through a regression model:

$$V_s = a + bV_i + c[V_{up}(X, W) - V(X, W)], \quad (6.38)$$

where V_i (kg) is the inflow sediment calculated from the area integration of the upland sediment graph (Equation 6.30), V_{up} (m^3) is inflow runoff volume, V (m^3) is outflow runoff volume – both in a channel reach of length X (m) and average width W (m) – and a, b, and c are relationship parameters.

6.3.3 Model calibration

Absolute values of sediment concentrations in the upland basins receiving runoff from different terrains were as follows:

Limestone	0.2 to 13.0 g/l
Phyllite, schist and shale/slate	0.4 to 29.0 g/l
Gneiss and granite	0.2 to 18.0 g/l
Rhyolite and sandstone	5.7 to 28.9 g/l

Downstream in the channel phase the sediment concentrations rose sharply to between 1.0 and 453.6 g/l. Nearly 90 % of the suspended sediments by weight have particle sizes ranging between 0.002 and 0.2 mm. Increases in the sediment concentrations in the channel phase were attributed to the cessation of smaller flows as a result of high transmission losses in the channel alluvium, leaving a large amount of loose material to be picked up by subsequent flows of greater magnitude (Hadley, 1977; Sharma et al., 1984). This is contrary to the corresponding process in humid regions, where the sediment concentration is further diluted with downstream increase in the discharge.

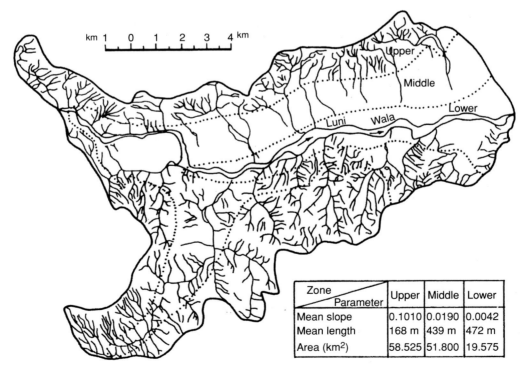

Zone / Parameter	Upper	Middle	Lower
Mean slope	0.1010	0.0190	0.0042
Mean length	168 m	439 m	472 m
Area (km^2)	58.525	51.800	19.575

Figure 6.9 Segmentation of Ramnia drainage basin into small uniform zones to account for the complexity

Table 6.5 *Summary of statistical analysis of three calibration methods for the sediment delivery model*

Hydrograph stage	Calibration method	Sum of squares	Root-mean-squared difference	Maximum deviation	Number of observations
Rising	Reference slope	3.46	0.20	6.1	84
	Dual slope	4.73	0.24	6.4	
	Average shear	5.75	0.26	15.0	
Peak	Reference slope	114.51	1.33	25.5	65
	Dual slope	41.68	0.80	6.4	
	Average shear	43.95	0.82	6.7	
Recession	Reference slope	3.73	0.23	31.2	70
	Dual slope	1.21	0.13	4.5	
	Average shear	1.03	0.12	3.9	

The outlet of a drainage basin may control the amount of sediment leaving the basin. Therefore, the conditions at the outlet of a drainage basin can be used to calibrate a sediment-delivery model with the expectation that the model would give the best estimates of drainage-basin sediment yield using parameters based on the characteristics of the outlet (Finkner *et al.*, 1989).

The first calibration option determined τ_s based on the sediment discharge at the drainage-basin outlet and the reference slope (mean slope of the drainage basin), S_0. The dual-slope method considered the average of two τ_s values – the first value of τ_s based on the reference slope while the second τ_s value was based on the actual slope at the outlet. The third option for calibrating the sediment-delivery model, average shear method, would take into account the combination of slope and discharge along the flow path. The average shear stress could be calculated as:

$$\tau_s = \int \tau_s(X)\mathrm{d}x/L, \qquad (6.39)$$

where L (m) is the length of flow path.

The sediment delivery model was tested for ten arid upland drainage basins forming a part of the Luni Basin in the Indian arid zone (Sharma *et al.*, 1993) and with areas ranging between 104 and 1520 km^2. To account for the basin complexity, each drainage basin was segmented into three zones, upper, middle and lower, based on the degree of steepness and stream order (Sharma, 1992). One such segmented basin is depicted in Fig. 6.9 as an example. The characteristics of these segments were used to calculate the values of the calibration options. The values of coefficients G and C, determined by the least-squares technique at each stage of the flow hydrograph namely rising, peak and recession, varied between 0.0022 and 0.0072/m, and 0.66 and 89.23/kg (s/m), respectively. For G values greater than unity most of the resulting sediment concentrations were negligible (Laguna and Giraldez, 1993). Singh and Regl (1983) suggested a reasonable value of G as 0.0030/m.

6.3.4 Model validation

The reference slope, dual slope, and average shear methods of calibrating the coefficients G and C were evaluated through independent runoff events at each stage of the flow hydrograph. At the rising stage of the hydrograph, the root-mean squared difference was consistently the lowest with the reference slope method. This is because at the rising stage the desert streams convey the highest sediment concentrations, which is attributed to the existence of a thin loose-surface layer produced by weathering, drying, and biotic interference within the drainage basin during the dry season (Sharma *et al.*, 1984). The splash erosion process may provide an additional amount of material during the time interval between the initiation of rainfall and that of runoff. The presence of this abundant loose erodible material within the drainage basin rendered only the reference slope as a controlling factor for the sediment delivery, i.e., the average conditions within the drainage basin affected the sediment transport rate at the rising stage of the hydrograph.

At peak flow the root-mean-squared difference was the lowest with the dual-slope method. This is because at peak flow the flow conditions within the drainage basin were at equilibrium, i.e., $\mathrm{d}Q/\mathrm{d}t \to 0$; where Q (m^3/s) is discharge and t (s) is time; and the flatter slopes at the drainage basin outlet resulted in the deposition of a significant proportion of the sediment eroded from the upstream area before it left the drainage basin. Finkner *et al.* (1989) also found the best agreement using the dual slope method of calibrating the sediment transport models.

During the recession stage of the hydrograph, the receding flow deposited the sediments on its flow path throughout the drainage basin. The flow velocity dropped below the critical value, thereby resulting in a rapid decrease in the sediment concentration towards the end of flow. The average shear stress showed the least root-mean-squared difference (Table 6.5) since it represented not only the slope of each segment within the drainage basin, but also

Table 6.6 *Time and shape parameters of test flood hydrographs*

| Drainage basin | Date | Time parameters | | Shape parameter, n_s |
		Time to peak, t_p (hours)	Storage coefficient, K_s (hours)	
Jasnagar (a[a])	July 10, 1098	1.00	32	1.03
	July 24, 1982	2.00	23	1.09
	August, 22 1983	7.00	6	2.17
Binawas (z[a])	July 20, 1981	2.00	4	1.50
	July 26, 1983	2.00	12	1.17
	July 28, 1986	0.50	20	1.03
Bhuti (H[a])	July 10, 1981	1.80	10	1.18
	July 14, 1982	0.83	17	1.05
	July 25, 1982	4.50	6	1.75
Nawakhera (F[a])	July 18, 1984	2.00	19	1.11
	August 12, 1984	1.83	16	1.11
	July 15, 1987	2.00	13	1.15

[a] Refers to Figure 6.1

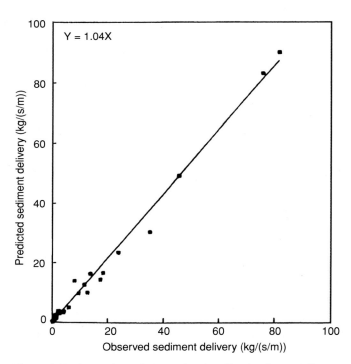

Figure 6.10 Comparison of observed and predicted sediment delivery rates from arid upland drainage basins

Table 6.7 *Parameters of the sediment-mobilization model (Equation 6.38)*

Stream channel	A	b	c	$R^{2\,a}$	Number of events
Luni	1119	0.61	−0.0001	0.99	9
Jojri	259	0.10	−20.5840	0.98	5
Mitri[b]	−4998	1.29	−0.0011	0.96	8
Jawai	4294	−0.03	0.0001	0.95	15

[a] Significance level ($P < 0.01$)

[b] Disorganized stream

rates are less than 10 % (ranges between 3.9 and 6.4 %) for the best-fit calibration methods. Thus, a deposition-based sediment-delivery model is a suitable approximation to sediment-transport rates in the arid-zone drainage basins.

To verify and validate the IUSG concept, the model parameter t_p was taken from an S-curve relationship between the peak flow and the time-to-peak flow at the channel outlet and K_s was approximated as the travel time of runoff crests in the trunk stream (Sharma *et al.*, 1994). These parameters are given in Table 6.6 for the selected test reaches. Table 6.7 gives the parameter values of the sediment-mobilization model (Equation 6.38), thereby predicting the sediment yield at the channel outlet.

Figure 6.11 compares the observed and predicted sediment graphs at the channel outlet. The sediment graphs generated by the IUSG technique have a better fit with the observed sediment

the combination of slope and discharge along the length of each segment, i.e., flow path and their cumulative effect at the outlet.

A comparison of observed and predicted sediment-delivery rates (Fig. 6.10) shows a good agreement. Further, the maximum deviation between the observed and predicted sediment-delivery

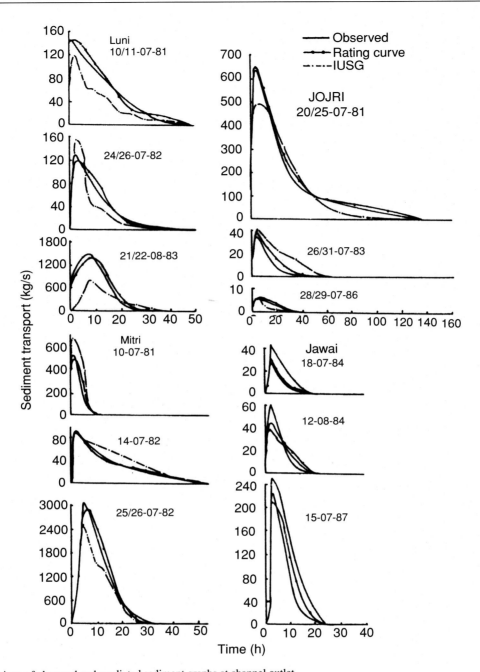

Figure 6.11 Comparison of observed and predicted sediment graphs at channel outlet

graphs than the sediment graphs generated using the sediment rating curves in all instances. Sediment discharge varies markedly between the rising and recession stages for the same discharge; this could not be taken into account by the sediment rating curves. This also implies that in ephemeral streams sediment transport is dependent on the presence of erodible material in the channel bed. Therefore, in arid regions the bed sediment supply has to be taken into account in sediment transport studies.

6.4 CONCLUSIONS

A combination of regression analysis for the rising limb of the hydrograph and a differential equation for the recession limb simulates acceptable hydrographs in arid upland drainage basins where runoff has a direct relationship with rainfall. The regression model involves effective rainfall and drainage-basin area, and predicts the peak discharge and time-to-peak with a high degree of accuracy.

The differential equation models the recession limb through an analog of the upland region consisting of a single discharging leaky reservoir. This model depends on two parameters, loss rate and reservoir shape, whose optimized values are reasonable from the physical point of view. The best reservoir shape lies in a narrow range thereby indicating a characteristic shape for all the recession curves, i.e., the control of the outlet properties rather than the shape of the input. In the channel phase, the continuity equation of flow was modified by including a non-linear volumetric transmission-loss rate term for routing of flow in ephemeral channels. The resulting outflow hydrographs at the channel outlet have a better fit with the observed data compared to the lumped regression outputs. The study is useful in estimating the water yield and a surface water balance incorporating the transmission losses, and in understanding of runoff-generation processes in arid regions.

Further, a closed-form solution of the continuity equation of sediment transport is preferred in arid-zone upland drainage basins, where initial potential sediment load is greater than the sediment transport capacity of overland flow and large amounts of sediment delivery are of concern. A model of this kind based on the Manning's turbulent flow equation and the Yalin sediment-transport capacity equation predicted the sediment delivery rates at the rising, peak, and recession stages of the hydrograph within $\pm 10\%$ accuracy. The dynamic forces active at each hydrograph stage can be accounted through various calibration methods involving slope and flow length at the upland drainage basin outlet. In case of channel phase, a lumped model using the inflow sediment volume and channel transmission loss in combination with the instantaneous unit sediment graph technique approximates the observed sediment graphs at the channel outlet. This simple approach allows a reasonable accuracy of prediction and satisfies practical requirements in arid regions. Our study shows that the sediment transport in arid ephemeral channels is hydraulically controlled and depends on the supply of erodible material.

6.5   REFERENCES

Abrahams, A. D., Luk, S. H., and Parsons, A. J. (1988). Threshold relations for the transport of sediment by overland flow on desert hillslopes. *Earth Surf. Proc. Landforms*, **13**, 407–19.

Alonso, C. V., Neibling, W. H., and Foster, G. R. (1981). Estimating sediment transport capacity in watershed modeling. *Trans. Am. Soc. Agri. Eng.*, **24**, 1211–20, 1226.

Ball, J. E. (1987). Importance of dynamic wave components for pipe flows. In *Proceedings of 22nd IAHR Congress and 4th International Conference on Urban Storm Drainage*. Lausanne: International Association of Hydraulic Research. 162–167.

Bell, F. C. (1979). Precipitation. In *Arid Land Ecosystem*, ed. D. W. Goodall and R. A. Perry. London: Cambridge University Press, 372–92.

Bennett, J. P. (1974). Concepts of mathematical modeling of sediment yield. *Water Resour. Res.*, **10**, 485–92.

Chang, H. H. and Stow, D. A. (1988). Sediment delivery in a semiarid coastal stream. *J. Hydrol.*, **99**, 201–14.

Chow, V. T., Maidment, D. R., and Mays, L. W. (1988). *Applied Hydrology*. New York: McGraw Hill.

FAO (1973). *Mans' Influence on the Hydrological Cycle*. Rome: FAO.

Finkner, S. C., Nearing, M. A., Foster, G. R., and Gilley, J. E. (1989). A simplified equation for modeling sediment transport capacity. *Trans. Am. Soc. Agri. Eng.*, **32**, 1545–50.

Foster, G. R. (1982). Modeling the erosion process. In *Hydrologic Modeling of Small Watersheds*, ed. C. T. Haan, H. P. Johnson, and D. L. Brakensiek. St. Joseph: American Society of Agricultural Engineers, 297–308.

Foster, G. R. and Meyer, L. D. (1972a). Transport of soil particles by shallow flow. *Trans. Am. Soc. Agri. Eng.*, **15**, 99–102.

Foster, G. R. and Meyer, L. D. (1972b). A closed-form soil erosion equation for upland area. In *Sedimentation*, ed. H. W. Shen. Fort Collins: Colorado State University, Chapter 12.

Goudie, A. and Wilkinson, J. (1977). *The Warm Desert Environment*. Cambridge: Cambridge University Press.

Hadley, R. F. (1977). Some concepts of erosional processes and sediment yield in a semi-arid environment. In *Erosion: Research Techniques, Erodibility and Sediment Delivery*, ed. T. J. Troy. Norwich: Geo Books, 73–81.

Henderson, F. M. (1966). *Open Channel Flow*. New York: Macmillan Publishers.

Holder, R. L. (1985). *Multiple Regression in Hydrology*. Wallingford: Institute of Hydrology.

Jones, K. R. (1981). *Arid Zone Hydrology*. Rome: FAO.

Laguna, A. and Giraldez, J. V. (1993). The description of soil erosion through a kinematic wave model. *J. Hydrol.*, **145**, 65–82.

Lane, L. J. (1982). Distributed model for small semiarid watersheds. *J. Hydr. Eng. ASCE*, **108**, 1114–31.

Lane, L. J. (1983). Transmission losses. In *National Engineering Handbook IV. Hydrology*. Washington DC: Soil Conservation Service, US Department of Agriculture.

Langbein, W. B. and Iseri, K. T. (1960). General introduction and hydrologic definitions. Manual of Hydrology, Part I: General surface water techniques. *US Geological Survey Water Supply Paper 154*. Washington DC: US Geological Survey.

Lu, J. Y., Cassol, E. A., and Moldenhauer, W. C. (1989). Sediment transport relationships for sand and silt loam soils. *Trans. Am. Soc. Agri. Eng.*, **32**, 1923–31.

Mabbutt, J. A. (1977). *Desert Landforms*. Canberra: Australian National University.

Magfed, Y. A. (1986). *Assessment of Water Resources in Arid and Semiarid Regions*. Nairobi: UNEP.

Nash, J. E. (1957). The form of instantaneous unit hydrograph. *Hydrolog. Sci. Bull.*, **3**, 114–21.

Nearing, M. A., Foster, G. R., Lane, L. J., and Finkner, S. C. (1989). A process-based soil erosion model for USDA-water erosion prediction project technology. *Trans. Am. Soc. Agri. Eng.*, **32**, 1587–93.

Peebles, R. W., Smith, R. E., and Yakowitz, S. J. (1981). A leaky reservoir model for ephemeral flow recession. *Water Res. Res.*, **17**, 628–36.

Pilgrim, D. H., Chapman, T. C., and Doran, D. G. (1988). Problems of rainfall-runoff modeling in arid and semiarid regions. *Hydrolog. Sci. J.*, **33**, 379–400.

Reid, I. and Frostick, L. E. (1987). Flow dynamics and suspended sediment properties in arid zone flash floods. *Hydrol. Proc.* **1**, 239–53.

Schick, A. P. (1970). Desert floods. In *Results of Research on Representative and Experimental Basins*. Wallingford: International Association of Hydrological Sciences, 479–93.

Sharma, K. D. (1992). Runoff and sediment transport in an arid zone drainage basin. Ph.D. Thesis. Indian Institute of Technology, Bombay.

Sharma, K. D. and Murthy, J. S. R. (1995). Hydrologic routing of flow in arid ephemeral channels. *J. Hydr. Eng. ASCE*, **121**, 466–71.

Sharma, K. D. and Murthy, J. S. R. (1996). Ephemeral flow modeling in arid regions. *J. Arid Environ.*, **33**, 161–78.

Sharma, K. D. Vangani, N. S., and Choudhary, J. S. (1984). Sediment transport characteristics of the desert streams in India. *J. Hydrol.*, **67**, 272–81.

Sharma, K. D., Dhir, R. P., and Murthy, J. S. R. (1993). Modeling soil erosion in arid zone drainage basins. In *Sediment Problems: Strategies for Monitoring, Prediction and Control*. Wallingford: International Association of Hydrological Sciences, 269–76.

Sharma, K. D., Murthy, J. S. R., and Dhir, R. P. (1994). Stream-flow routing in the Indian arid zone. *Hydrol. Proc.*, **8**, 27–43.

Singh, V. P. (1976). *Studies on Rainfall-Runoff Modelling*. Las Cruces: New Mexico Water Resources Research Institute.

Singh, V. P. and Regl, R. R. (1983). Analytical solutions of kinematic equations for erosion on a plane. I: Rainfall of infinite duration. *Adv. Water Resour.*, **6**, 2–10.

Vangani, N. S. and Kalla, A. K. (1985). Manning's coefficient of roughness for rivers of western Rajasthan. *Ann. Arid Zone*, **24**, 258–62.

Walters, M. O. (1989). A unique flood event in an arid zone. *Hydrological Processes*, **3**, 15–24.

Wu, I-P. (1972). Recession flow in surface irrigation. *J. Irrig. Drain. Div. ASCE*, **98**, 77–90.

Yakowitz, S. (1973). A stochastic model for daily river flows in arid regions. *Water Resour. Res.*, **9**, 1271–85.

Yalin, Y. S. (1963). An expression for bed load transportation. *J. Hydr. Div. ASCE*, **89**, 221–50.

7 The Modular Modelling System (MMS): a toolbox for water- and environmental-resources management

G. H. Leavesley, S. L. Markstrom, R. J. Viger, and L. E. Hay

7.1 INTRODUCTION

Increasing demands for limited fresh-water supplies, and increasing complexity of environmental resource-management issues, present resource managers with the difficult task of achieving an equitable balance of resource allocation among a diverse group of users. Achieving such a balance is most difficult in arid and semi-arid regions. Hydrological and ecosystem models are often the tools being employed to address these resource-allocation issues.

The inter-disciplinary nature of water- and environmental-resource problems requires the use of modelling approaches that can incorporate knowledge from a broad range of scientific disciplines. Selection and application of appropriate models and tools is a function of a number of evaluation criteria, including problem objectives, data constraints, and spatial and temporal scales of application. The US Geological Survey (USGS) Modular Modelling System (MMS) (Leavesley et al., 1996b) is an integrated system of computer software that provides a research and operational framework to support the development and integration of a wide variety of hydrologic and ecosystem models, and their application to water- and environmental-resources management.

MMS supports the integration of models and tools at a variety of levels of modular design. These include individual process models, tightly coupled models, loosely coupled models, and fully integrated decision-support systems. A geographic information system (GIS) interface, the GIS Weasel, has been integrated with MMS to enable spatial delineation and characterization of basin and ecosystem features, and to provide objective parameter-estimation methods for selected models using available digital data coverages. Optimization and sensitivity-analysis tools are provided to analyze model parameters and evaluate the extent to which uncertainty in model parameters affects uncertainty in simulation results. A variety of visualization and statistical tools are also provided.

Forecasts of future climatic conditions are a key component in the application of MMS models to resource-management decisions. Forecast methods applied in MMS include a modified version of the United States National Weather Service's Ensemble Streamflow Prediction Program (ESP), and statistical and dynamical downscaling from atmospheric models.

MMS is being used to develop and apply water- and environmental-resouce management models and decision-support systems in several arid and semi-arid regions of the world. MMS provides a framework in which to collaboratively address the many complex issues associated with the design, development, and application of hydrological and environmental models in these regions. The open-source software design of the MMS facilitates the direct and indirect sharing of resources, expertise, knowledge, and costs among projects in these different regions. This paper presents an overview of the concepts and components of MMS and of applications in selected arid and semi-arid regions.

7.2 LEVELS OF MODULAR DESIGN

7.2.1 Process modules and models

MMS has a master library that contains compatible modules for simulating a variety of water, energy, and biogeochemical processes. The library may contain several modules for a given process, each representing an alternative conceptualization to simulating that process. The different conceptualizations are functions of a variety of constraints that include the types of data available and the spatial and temporal scales of application. A model for a specified application is created by coupling appropriate modules from the library. If existing modules cannot provide appropriate process algorithms, new modules can be developed and incorporated into the library.

Hydrological Modeling in Arid and Semi-Arid Areas, ed. Howard Wheater, Soroosh Sorooshian, and K. D. Sharma. Published by Cambridge University Press.
© Cambridge University Press 2008.

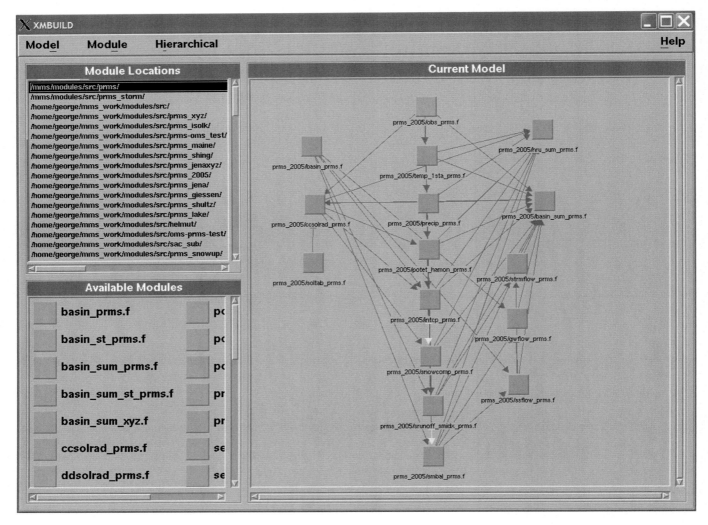

Figure 7.1 (Plate 9) The model builder interface Xmbuild

Model building in the MMS is accomplished using an interactive model-builder interface termed Xmbuild (Fig. 7.1, Plate 9). Xmbuild enables the user to select and link modules to create a model. Selecting a directory in the Module Locations window will display all the modules available in the Available Modules window. Selecting a module in this window will place it in the Current Model window. As additional modules are added to the Current Model window they are linked by using the output from one module as the input to other process modules.

Xmbuild enables users to view inputs and outputs for each module and to search the module library for all modules that provide the necessary inputs for each module. Using this search and select procedure, a user-defined model can be constructed. Module inputs and outputs include a units attribute that can be checked to ensure module compatibility. Plans include the development of an expert system to assist users in module selection based on

future research to identify the most appropriate modules for various problem objectives, data constraints, and spatial and temporal scales of application.

7.2.2 Loosely coupled models

The module-linking concept for model building applies to loosely coupled models as well. In loosely coupled models, information flow is in only one direction; output from one model is used as input to another model (Fig. 7.2). An example of a series of loosely coupled models might begin with a watershed model that simulates hillslope runoff volume and timing for input to a channel-hydraulics model. Output from the channel-hydraulics model can then be input to a fish model. The link between models is accomplished using a common database and a software component termed a "data management interface" (DMI). A DMI reformats

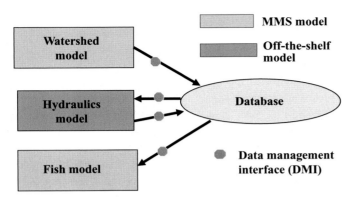

Figure 7.2 Example of loosely coupled models using a common database and DMIs

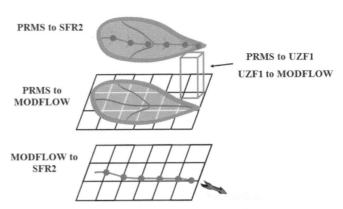

Figure 7.3 (Plate 10) Fully coupled surface-water/groundwater model

model output and writes it to the database, and reads data from the database and reformats it for input to a given model. Each DMI is unique for the database being used and for each model being applied. Writing a DMI is currently the responsibility of the user, but a library of DMIs for selected databases and models is being developed. Numerous combinations of models are possible using the loosely coupled approach. Models can be those created from the module library as well as off-the-shelf models that are not a component of MMS. The only requirement for the use of non-MMS models is that they can be executed in a batch mode.

7.2.3 Fully coupled models

The concept of linking modules to create a model can also be applied to the linking of models to create a larger integrated model. A fully coupled model refers to the coupling of individual models where there is a two-way flow of information between the models. These typically are developed to provide feedback among related processes in the linked models. Feedback is normally provided through the use of iterative computational procedures where groups of modules may be required to run multiple times to reach convergence for selected feedback processes.

A fully coupled surface-water/groundwater model is currently being developed in MMS (Fig. 7.3, Plate 10). The watershed model PRMS (Leavesley *et al.*, 1983; Leavesley and Stannard, 1995), groundwater model MODFLOW (Harbaugh, 2005), Streamflow-Routing Package SFR2 of MODFLOW (Niswonger and Prudic, 2005), and Unsaturated Flow Package UZF1 of MODFLOW (Niswonger *et al.*, 2006) are being linked to simulate the interactions among these individual models. PRMS provides the spatial and temporal recharge for MODFLOW grid cells. This recharge is passed to UZF1 for routing through the unsaturated zone to MOD-FLOW. If the water table reaches the soil surface, the UZF1 model becomes inactive in the associated grid cells, groundwater exfiltration occurs from these cells, and saturated areas are developed

in PRMS that will generate direct runoff from rain or snowmelt that occurs on these areas. Runoff from PRMS is delivered to the channel reaches in SFR2. The water-surface elevation (head) in a SFR2 reach and the head in the MODFLOW grid cell adjacent to the channel reach are then used to solve for the rate and direction of water flow between the stream and the aquifer. If the head in the aquifer is below the bottom of the channel reach, the unsaturated zone model is used to route flow from the channel to MODFLOW.

7.2.4 Decision-support systems (DSSs)

Decision-support systems are the top level of complexity for model coupling and integration. Various combinations of models from all levels of modular design can be integrated with resource-management and decision-support models to create a resource-management DSS. For example a resource-management DSS might include: (1) watershed models for simulating reservoir inflows and streamflow from unregulated basins, (2) one-dimensional and two-dimensional hydraulic models for application to selected river reaches where channel-flow characteristics may affect channel morphology or biological habitats, (3) sediment-transport and chemical-transport models to address a variety of water quality issues at the basin or reach scale, (4) agricultural models to address land-management and irrigation practices and the fate and transport of nutrients and pesticides, (5) biological and ecosystem models that address critical habitat issues, and (6) reservoir-management models to control the volume, timing, and distribution of water within a basin.

The ability to couple and integrate models for DSS development and application are provided in MMS by the object user interface (OUI) tool set. A variety of analytical, statistical, and graphical support tools are also provided to aid in the decision-support process. The capabilities of OUI and the other support tools are described in the next section.

7.3 ANALYSIS AND SUPPORT TOOLS

7.3.1 The GIS Weasel

The GIS Weasel is a geographic information system (GIS) interface for applying tools to delineate, characterize, and parameterize topographical, hydrological, and biological basin features for use in a variety of lumped- and distributed-modelling approaches. It is composed of ArcInfo (ESRI, 1992) GIS software, C language programs, and shell scripts. Distributed basin features are typically described using a concept of "model response units" (MRUs). MRUs are areas delineated within a watershed, or area of interest, which reflect a model's treatment of spatially distributed attributes, such as elevation, slope, aspect, soils, and vegetation. MRUs can be characterized using these attributes. The GIS Weasel also delineates a drainage network and computes the connectivity of MRUs with this drainage network. The location of data-collection sites can also be overlaid with the MRU map to define associations between MRUs and the data sites.

Parameter-estimation methods are implemented using ARC macro language (AML) functions. Keeping with the modular concept, a library of parameter-estimation methods is maintained in a similar fashion to the library of process modules. For a given model, a recipe file of AML functions can be created and executed to estimate a selected set of spatial parameters. This recipe file can also be modified to change the parameter-estimation method associated with a selected parameter, thus enabling the evaluation of alternative parameter-estimation methods.

Currently, methods to estimate selected spatially distributed model parameters have been developed for PRMS and TOP-MODEL (Beven *et al.*, 1995). Digital databases used for parameter estimation in the USA include: (1) USGS digital elevation models, (2) State Soils Geographic (STATSGO) 1 km gridded soils data (US Department of Agriculture, 1994); and (3) Forest Service 1 km gridded vegetation type and density data (US Department of Agriculture, 1992). Spatially distributed parameters estimated using these databases include elevation, slope, aspect, topographic index, soil type, available water-holding capacity of the soil, vegetation type, vegetation cover density, solar radiation transmission coefficient, interception-storage capacity, stream topology, and stream-reach slope and length.

7.3.2 The Object User Interface (OUI)

The Object User Interface is a Java-based, multi-purpose MMS component developed jointly by the Friedrich–Schiller University, Jena, Germany, and the USGS. OUI is a map-based interface for acquiring, organizing, browsing, and analyzing spatial and temporal data, and for executing models and analysis tools. OUI is the key component of the MMS for developing loosely coupled models and DSSs.

The functional components of OUI are a hierarchical data tree and a map window for display of, and interaction with, one or more data-tree themes (Fig. 7.4, Plate 11). The data tree provides users with access to a variety of data layers that typically include basin boundaries, model response units, stream reaches, meteorological and streamflow gauge sites, and other map-based features of interest for model application and analysis. These spatial data layers are stored in an ESRI shape file format. The display and data tree provide action buttons to initiate model applications, evaluate model results using a variety of statistical and graphical tools, analyze associated spatial and temporal data, and generate a three-dimensional animation of simulated model states.

The contents of the data tree and pull-down menus are specified using the eXtensible markup language (XML). OUI is easily applied by creating or modifying a control file called tree.xml. This file contains a variety of information, including the locations and names of all data files, format of all data, database connection parameters, locations and names of all models, and the locations and names of all associated DMIs and model management interfaces (MMIs). An MMI is a set of Java code that provides the ability to pre- and post-process data and to execute models for a user-defined set of simulations and analyses. It is, in effect, a script that creates and executes a sequence of models and analytical tools based on an established set of interface rules.

7.3.3 Optimization and sensitivity analysis tools

Optimization and sensitivity analysis tools are provided to analyze model parameters and evaluate the extent to which uncertainty in model parameters affects uncertainty in simulation results. Two optimization procedures are available to fit user-selected parameters. One is the Rosenbrock technique (Rosenbrock, 1960). The second is a hyper-tunnel method (Restrepo and Bras, 1982). The Shuffle Complex Evolution Optimization algorithm (Duan *et al.*, 1993) and the Multi-Objective COMplex Evolution algorithm (Yapo *et al.*, 1998), which is capable of solving multi-objective optimization problems, are currently being incorporated into the MMS tool set.

Sensitivity-analysis components allow the user to determine the extent to which uncertainty in the parameters results in uncertainty in the predicted runoff. Two methods of sensitivity analysis are currently available. One is the method developed for use with the original PRMS. The output of this method includes measures of the relative sensitivity, error propagation, joint and individual standard errors, and correlation among user-selected model parameters. The second method evaluates the sensitivity of any pair of parameters and develops the objective-function surface for a selected range of these two parameters. Other sensitivity-analysis methods to address the questions of parameter and prediction uncertainty are being evaluated for incorporation in MMS.

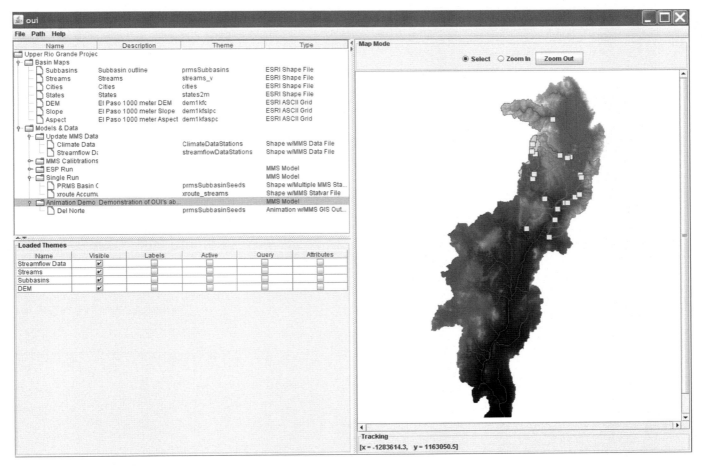

Figure 7.4 (Plate 11) The Object User Interface (OUI)

Currently the Generalized Likelihood Uncertainty Estimation (GLUE) procedure (Beven and Binley, 1992) is being added to the MMS tool set.

7.4 FORECASTING

7.4.1 Ensemble Streamflow Prediction

Forecasting capabilities are provided by the Ensemble Stream-flow Prediction (ESP) procedure (Day, 1985). ESP uses historic or synthesized meteorological data as an analog for the future. These time series are used as model input to simulate future streamflow. The initial hydrologic conditions of a watershed, for the start of a forecast period, are assumed to be those simulated by the model for that point in time. Typically, multiple hydrographs are simulated from this point in time forward, one for each year of available historic data. For each simulated hydrograph, the model is re-initialized using the watershed conditions at the starting point of the forecast period. The forecast period can vary from a few days

to an entire water year. A frequency analysis is then performed on the peaks and/or volumes of the simulated hydrograph traces, to evaluate their probabilities of exceedance. The user can view all the forecast traces and then select all or a subset of the traces for use in an associated water management model (Fig. 7.5, Plate 12). In the example shown the user selected the traces that represent the 10, 50, and 90 % probabilities of exceedance.

The ESP procedure uses historical meteorological data to represent future meteorological data. Alternative assumptions about future meteorological conditions can be made with the use of synthesized meteorological data. A few options are also available in applying the frequency analysis. One assumes that all years in the historic database have an equally likely probability of occurrence. This gives equal weight to all years. Years associated with El Niño, La Niña, ENSO neutral, Pacific Decadal Oscillation (PDO) less than −0.5, PDO greater than 0.5, and PDO neutral have also been identified in the ESP procedure. The years in these groups can be extracted separately for analysis. Alternative schemes for weighting user-defined periods, based on user assumptions or a priori information, are also being investigated.

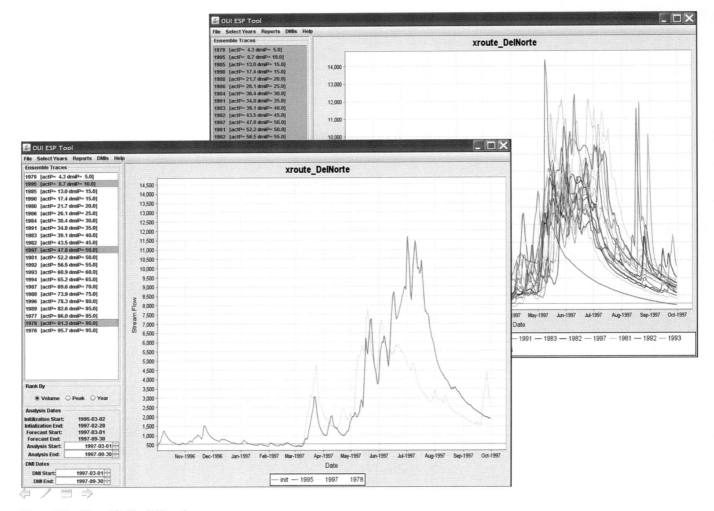

Figure 7.5 (Plate 12) The ESP tool

7.4.2 Downscaling atmospheric model output

Procedures to dynamically and statistically downscale atmospheric model output for use as input to watershed models have been developed and coupled with the MMS (Wilby *et al.*, 1999; Hay *et al.*, 2000, 2002). The dynamical downscaling procedure applies a bias correction directly to the atmospheric model grid-scale temperature and precipitation outputs. The statistical method uses a regression-based statistical downscaling model to simulate point values of daily precipitation and temperature from atmospheric-model output of grid-scale synoptic measures. The point estimates of climate variables are then spatially distributed across a basin using lapse rates and topographic information.

7.5 APPLICATIONS

The component tools in the MMS can be used individually or in various combinations to address a wide range of water- and environmental-resource management needs. A selected set of example applications in arid and semi-arid regions is presented to provide an overview of the types of resource-management issues being addressed.

7.5.1 Watershed and River System Management Program (WARSMP)

MMS is being used as the hydrological modelling and forecast component of the WARSMP (Leavesley *et al.*, 1996a). WARSMP is a cooperative effort between the USGS and the Bureau of Reclamation (BOR) to develop an operational, database-centered, DSS for application to complex, water and environmental resource-management issues (Fig. 7.6). The MMS has been coupled with the BOR RiverWare software (Fulp *et al.*, 1995) using a shared relational database. RiverWare is an object-oriented reservoir and river-system modelling framework developed to provide tools for evaluating and applying optimal water-allocation and management strategies.

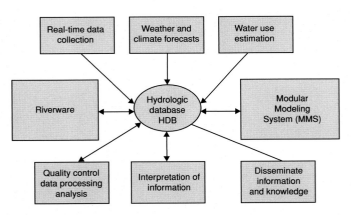

Figure 7.6 Database-centered decision-support system

MMS and RiverWare are linked using a common hydrologic database (HDB). The database used varies among the basins, with Oracle and HEC-DSS being currently supported. Real-time hydrologic and meteorological data are provided to HDB every six hours. HDB is also the database of record and so all model results and forecasts are stored there as well.

A typical exchange between RiverWare and MMS would be for RiverWare to request all inflow predictions for reservoirs in a basin. A watershed model in MMS would simulate these forecast streamflows and write them to HDB. RiverWare would read these results and evaluate alternative reservoir-management strategies. The associated reservoir releases for each strategy would be written to HDB. These strategies may have implications for environmental or flood issues. MMS could be called again to run one or more hydraulic and/or ecosystem models to evaluate the effects of the reservoir releases on selected river reaches. These results would then be written to HDB for use by RiverWare in selecting a specific management alternative.

The modelling capabilities of MMS and RiverWare include simulating watershed runoff, reservoir inflows, and the impacts of resource-management decisions on municipal, agricultural, and industrial water users, environmental concerns, power generation, and recreational interests. The WARSMP DSS is currently operational in the Gunnison River Basin, Colorado; Yakima River Basin, Washington; Rio Grande Basin in Colorado and New Mexico, and Truckee River Basin in California and Nevada.

The Upper Gunnison River Basin provides a typical example of the modelling issues that need to be considered in applying MMS. The basin is about $10\,000\,\text{km}^2$ in size. The GIS Weasel was used to partition it into subunits or MRUs, using characteristics such as slope, aspect, elevation, vegetation type, soil type, and precipitation distribution (Fig. 7.7, Plate 13). Each unit is assumed to be homogeneous with respect to its hydrologic response and to the characteristics listed. Each unit is termed a hydrologic response unit (HRU), which is just a specific type of MRU. More than 1000 HRUs were defined. The HRUs were then grouped into 15

major subbasins and 15 streamflow forecast nodes were defined. However, only 5 of the 15 subbasins had observed streamflow data. The subbasin configuration, HRU delineations, and other data layers were specified to the OUI through the tree.xml file for use in organizing the display and analysis of data and model results.

The distributed-parameter watershed model PRMS was selected for application. A set of spatial parameters was estimated for each subbasin and HRU using the GIS Weasel. An objective parameter-estimation and calibration procedure has been developed and is being tested in the MMS for the PRMS applications. In this procedure, no changes are made to the spatial parameters estimated by the GIS Weasel. Calibration is focused on the water-balance parameters affecting potential evapotranspiration (ET) and precipitation distribution, and on the subsurface and groundwater parameters affecting hydrograph shape and timing. Parameters calibrated on the five gauged basins were then transferred to the ten ungauged basins.

For streamflow simulation purposes, an MMI was created in OUI to execute the PRMS separately on each of the 15 subbasins and then to execute a channel-routing model that routes the subbasin outflows to produce a simulated streamflow hydrograph at the 15 river forecast nodes. In PRMS, a water balance and an energy balance are computed daily for each HRU. The sum of the responses of all HRUs, weighted on a unit-area basis, produces the daily subbasin streamflow. The user can view the routed streamflow at any node by activating the routing-node data layer in the OUI data tree and then clicking on the desired node in the OUI map display.

A second MMI was created to implement the ESP procedure in OUI. Here, forecast hydrographs are simulated for each subbasin, one hydrograph for each of the 24 years in the historic data record. A routed hydrograph through the entire basin is then generated for each of the 24 forecast periods. The suite of 24 hydrographs at any forecast node can then be viewed using the ESP Tool, which is a Java-based GUI in which all or a subset of the forecast hydrograph traces can be viewed (Fig. 7.5, Plate 12). For each node, a frequency analysis is computed on the suite of traces and the probability of exceedance for each trace is provided as well. The ESP Tool MMI contains the procedure to write operator-selected hydrographs to the central database HDB for use by RiverWare. The river-basin manager typically selects the hydrographs with a 10, 50, and 90 % probability of exceedance for analysis in RiverWare. The ESP procedures are typically run a few times a week to aid resource managers in making reservoir and river-system management decisions.

A major goal of the WARSMP DSS program is to effectively integrate scientific understanding into the water- and environmental-resource decision-making process. The database-centered DSS approach that combines MMS and RiverWare provides a flexible framework that can be used by scientists, resource

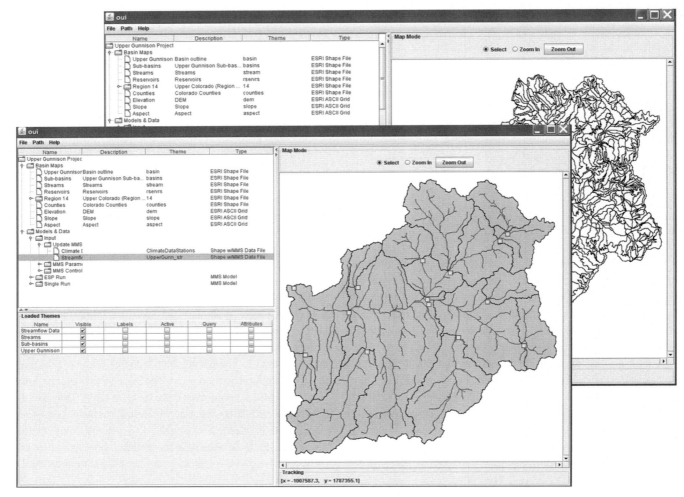

Figure 7.7 (Plate 13) Gunnison River basin delineation into HRUs, subbasins, and forecast nodes

managers, and stakeholders in this process. This approach is generic and applicable to a wide variety of basins and resource-management issues in arid and semi-arid regions.

7.5.2 River Drâa, Morocco

GLOWA is a program initiated by the German Federal Ministry of Education and Research (http://www.glowa.org/eng/home/home_objectives.htm) to focus on problems of water availability. One aim is the development of simulation tools and instruments that will aid in the development of strategies for sustainable water management at a regional level (river basins of approx. $100\,000\,km^2$), while considering global environmental changes and socioeconomic conditions. IMPETUS is program within GLOWA to investigate all aspects of the hydrological cycle in two river catchments in north-west and west Africa: the wadi Drâa in the south-east of Morocco and the River Ouémé in Benin.

MMS is being used in the IMPETUS study of the wadi Drâa. The Drâa basin forms a large transboundary catchment ($115\,000\,km^2$) stretching across Algeria, Mauretania, and Western Sahara, but the Drâa river itself is restricted to the upper parts of the catchment ($34\,609\,km^2$) located in Morocco and no longer reaches the Atlantic Ocean (Fig. 7.8).

The River Drâa drains from the High Atlas Mountains to Lac Iriki and feeds a large dam for irrigation purposes. In order to address a number of imminent problems limiting the availability and allocation of water along the wadi Drâa, 12 measurement sites were installed along a gradient of elevation and aridity. Monitoring of the thickness and the extent of snow cover in the High Atlas Mountains is essential to enable the competing water users (power generation, irrigation, domestic consumption) to have adequate supplies. In addition to seeking a better understanding and prediction of the geospheric, atmospheric, and biospheric components of the hydrological cycle, the IMPETUS

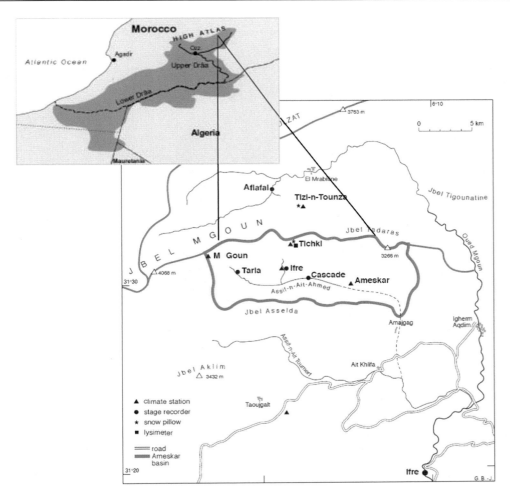

Figure 7.8 Location of the Draa River basin, Morocco

activities center, around the questions of the influence, risks, and resulting conflicts of human activities in the context of the specific social and economical structures encountered in the area.

The issues and needs for water-management DSS tools are described by de Jong *et al.* (2004, 2005). The most severe problems affecting the Drâa in terms of the natural physiography are water shortage and drought over the last five years. According to the world water-poverty index, Morocco has an index of below 45 and the Drâa belongs to the world's ten driest catchments (Ravenga *et al.*, 1998). Population, settlements, infrastructure and agriculture are concentrated around the rivers and oases. The catchment is also subject to highly variable rain- and snowfall regimes and extremely sporadic and variable discharge. The downstream reservoir Mansour Eddahbi situated near Ouarzazate, the regional capital, is strongly dependant on water input from the mountain catchments. Over the past few years, there have been large fluctuations in water input into the reservoir and its minimum capacity

is often no longer reached. The Mansour Eddahbi dam has also been subject to substantial infill by sediments and consequently rapid capacity loss. If this trend continues, the reservoir will have lost half of its capacity by the year 2030 and will require intensive study of sediment delivery to predict its entire lifetime and economical function.

MMS is currently being used to develop a DSS for the whole Drâa catchment for operational discharge forecasting. A number of physical and hydrological characteristics have been identified as having major effects on the ability to make such forecasts. These include the spatial and temporal distribution of precipitation and the determination of its form as rain or snow. Evapotranspiration and snow sublimation also play a significant role. Surface and subsurface runoff and springs are controlled by complex geomorphology and geology including limestone and basalt. Wadi river beds are highly porous and discharge is sporadic and highly variable depending on precipitation inputs and river-bed characteristics.

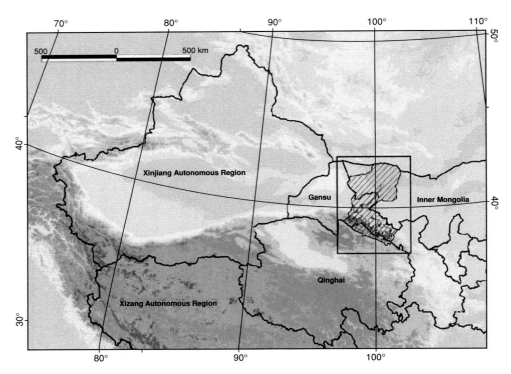

Figure 7.9   (Plate 14) Location of the Heihe River Basin, China

The initial modelling efforts have been focused on the implementation of the model PRMS to selected headwater catchments. HRU delineation and parameterization using the GIS Weasel have been based mainly on geomorphological and hydrogeological characteristics. Analysis of model results are being used to identify additional process models required to meet the water- and environmental-resource management needs on the basin. These modelling needs include the processes of snow ablation, erosion and sedimentation, channel transmission loss, surface-water/groundwater interactions, and oasis formation and use.

7.5.3   Heihe River, China

A collaborative research program has been established between the Cold and Arid Regions Environmental and Engineering Research Institute, Chinese Academy of Sciences (CAREERI/CAS) in Lahzhou, China and the USGS, for the development of integrated models and a modelling environment for inland river basins. A case study for this effort is being conducted in the Heihe River basin, Northwest China (Fig. 7.9, Plate 14). The Heihe is the second largest inland river watershed in China with an area of 140 000 km^2. The basin runs from the Qilian mountains in the south to the Alxa high plain in the north. There is a diverse set of hydrological and geological conditions in the basin that include alpine glacial, snow, grassland, forest, plain, oasis,

and desert regions. River flows originate as rainfall and snowmelt in the Qilian mountains and gradually disappear as the river flows north due to irrigation use, evapotranspiration, and channel transmission losses to groundwater.

Research is being conducted on water-resource allocation and utilization, and the integrated management of water resources, ecosystems, and economic systems in the Heihe River Basin. These research efforts include the investigation of: (1) sustainable utilization of water resources in inland river basins in arid regions under a changing environment, (2) heat and water transfer in the soil–vegetation–atmosphere system and land-surface process studies in inland river basins, (3) studies on water cycle and water-resource capacity in northwest China, and (4) the development of a basin-scale climate model for the region.

Initial collaborative research and development using MMS will focus on surface-water/groundwater issues and the development of a DSS for the Heihe Basin. The interaction of surface and groundwater in the middle and lower reaches of the Heihe River Basin is very complex, and is an obstacle to good water management. An integrated model that can quantitatively describe the spatial and temporal fluxes of surface and groundwater will be developed. Models to be evaluated include SWAT (Arnold *et al.*, 1998) and MODFLOW. The coupled model will be further linked with a water-management model. These integrated models will be part of the overall spatially explicit DSS for the Heihe River

Basin. Additional models and tools will be added to the DSS as they become identified by research results and operational needs.

7.6 SUMMARY

MMS provides a toolbox approach to model and system development. It supports multi-disciplinary model integration and the development of decision-support systems. Open-source software design allows many to share resources, expertise, knowledge, and costs. Individual modules, models, and tools can be developed by those with the relevant expertise and added to the common toolbox for use by others. Continued advances in physical and biological sciences, GIS technology, computer technology, and data resources will expand the need for a dynamic set of tools to incorporate these advances in a wide range of inter-disciplinary research and operational applications. MMS is being developed as a flexible framework in which to integrate these activities and to provide improved knowledge of hydrological and environmental processes to advance the art and science of water- and environmental-resource modelling.

MMS, the GIS Weasel, selected models and tools, and additional information on MMS can be downloaded at:

http://www.brr.cr.usgs.gov/mms/
http://www.brr.cr.usgs.gov/weasel/
http://www.brr.cr.usgs.gov/warsmp/

7.7 REFERENCES

Arnold, J. G., Srinivasin, R., Muttiah, R. S., and Williams, J. R. (1998). Large area hydrologic modeling and assessment: Part I. Model development. *JAWRA*, **34(1)**, 73–89.

Beven, K. J. and Binley, A. (1992). The future of distributed models: model calibration and uncertainty prediction. *Hydrolog. Proc.*, **6**, 279–98.

Beven, K. J., Lamb, R., Quinn, P. F., Romanowicz, R., and Freer, J. (1995). TOPMODEL. In *Computer Models of Watershed Hydrology*, ed. V. P. Singh Highlands Ranch, CO: Water Resources Publications, 627–68.

Day, G. N. (1985). Extended streamflow forecasting using NWSRFS. *J. Water Resour. Plan. Manag., ASCE*, **111**, 157–70.

de Jong, C., Machauer, R., Reichert, B., *et al.* (2004). An integrated geomorphological and hydrogeological MMS modelling framework for a semi-arid mountain basin in the High Atlas, southern Morocco. In *Complexity and Integrated Resources Management, Transactions of the 2nd Biennial Meeting of the International Environmental Modelling and Software Society*, ed. C. Pahl-Wostl, S. Schmidt, A. E. Rizzoli, and A. J. Jakeman. Manno, Switzerland: iEMSs, ISBN 88-900787-1-5, Vol. 2 736–41.

de Jong, C., Machauer, R., Leavesely, G., *et al.* (2005). Integrated hydrological modelling concepts for a peripheral mountainous semi-arid basin in southern Morocco. *EU Proceedings Geomatics for Land and Water management: Achievements and Challenges in the Euromed Context*, ed. R. Escadafal and M. L. Paracchini, Italy: Joint Research Centre, 219–27.

Duan, Q., Gupta, V. K., Sorooshian, S. (1993). A shuffled complex evolution approach for effective and efficient global optimization. *J. Optimiz. Theory Appl.*, **76(3)**, 501–21.

ESRI (Environmental Systems Research Institute) (1992). *ARC/INFO 61 User's Guide*. Redlands, CA: ESRI.

Fulp, T. J., Vickers, W. B., Williams, B., King, D. L. (1995). Decision support for water resources management in the Colorado River regions. In *Workshop on Computer Applications in Water Management*, ed. L. Ahuja, J. Leppert, K. Rojas, and E. Seely. Information Series No. 79, Fort Collins, CO: Colorado Water Resources Research Institute, 24–27.

Harbaugh, A. W. (2005). *MODFLOW-2005, The US Geological Survey Modular Groundwater Model – The Groundwater Flow Process*. US Geological Survey Techniques and Methods 6-A16.

Hay, L. E., Clark, M. P., Wilby, W. J., *et al.* (2002). Use of regional climate model output for hydrologic simulations, *J. Hydrometeor.*, **3**, 571–90.

Hay, L. E., Wilby, R. L., and Leavesley, G. H. (2000). A comparison of delta change and downscaled GCM scenarios for three mountainous basins in the United States. *J. Am. Water Resour.*, **36(2)**, 387–97.

Leavesley, G. H. and Stannard, L. G. (1995). The precipitation-runoff modelling system – PRMS. In *Computer Models of Watershed Hydrology*, ed. V. P. Singh. Highlands Ranch, CO: Water Resources Publications, 281–310.

Leavesley, G. H., Lichty, R. W., Troutman, B. M., and Saindon, L. G. (1983). *Precipitation-Runoff Modelling System – User's Manual*. US Geological Survey Water Resources Investigation Report 83–4238.

Leavesley, G. H., Markstrom, S. L., Brewer, M. S., and Viger, R. J. (1996a). The modular modelling system (MMS) – the physical process modeling component of a database-centered decision support system for water and power management. *Water Air Soil Poll.* **90**, 303–11.

Leavesley, G. H., Restrepo, P. J., Markstrom, S. L., Dixon, M., and Stannard, L. G. (1996b). The modular modeling system – MMS: user's manual. *US Geological Survey Open File Report 96–151*.

Niswonger, R. G. and Prudic, D. E. (2005). *Documentation of the Streamflow-Routing (SFR2) Package to Include Unsaturated Flow Beneath Streams – A Modification to SFR1*. US Geological Survey Techniques and Methods, 6-A13.

Niswonger, R. G., Prudic, D. E., and Regan, R. S. (2006). *Documentation of the Unsaturated Flow (UZF1) Package for Modeling Unsaturated Flow Between the Land Surface and the Water Table with MODFLOW-2005*. US Geological Survey Techniques and Methods 6-A19.

Ravenga, C. S. M., Abramovitz, J., and Hammond, A. (1998). *Watersheds of the World*. Washington DC: Water Resources Institute and Worldwatch Institute.

Restrepo, P. J. and Bras, R. L. (1982). *Automatic Parameter Estimation of a Large Conceptual Rainfall-Runoff Model: A Maximum-Likelihood Approach*. Ralph M. Parsons Laboratory Report No. 267. Cambridge, MA: Massachusetts Institute of Technology, Department of Civil Engineering.

Rosenbrock, H. H. (1960). An automatic method of finding the greatest or least value of a function. *Comp. J.*, **3**, 175–84.

US Department of Agriculture (1992). *Forest Land Distribution Data for the United States Forest Service*, http://www.epa.gov/docs/grd/forest_inventory.

US Department of Agriculture (1994). *State Soil Geographic (STATSGO) Database – Data Use Information*. Natural Resources Conservation Service, Miscellaneous Publication No. 1492.

Wilby, R. L., Hay, L. E., and Leavesley, G. H. (1999). A comparison of downscaled and raw GCM output: implications for climate change scenarios in the San Juan River Basin, Colorado. *J. Hydrol.*, **225**, 67–91.

Yapo, P. O., Gupta, H. V., and Sorooshian, S. (1998). Multi-objective global optimization for hydrologic models. *J. Hydrol.*, **204**, 83–97.

8 Calibration, uncertainty, and regional analysis of conceptual rainfall-runoff models

H. S. Wheater, N. McIntyre, and T. Wagener

8.1 INTRODUCTION

The majority of continuous-time rainfall-runoff models can be classified as conceptual, as discussed in Chapter 1 (see also Wheater *et al.*, 1993; Wheater, 2002). This type of model represents the hydrological processes that are seemingly important in the system using a simplified, conceptual representation (Fig. 8.1). These models have three notable characteristics: (a) their model structure is specified a priori, (b) the hydrological properties of the catchments are represented as parameters, which are generally assumed to be constant during each model application, (c) (at least some of) the model parameters have no direct, physical meaning and are not directly measurable. Therefore model parameters are usually estimated via calibration, using the fit of the model output time series to observed data to provide a measure of goodness of fit.

In this chapter we introduce the issues associated with calibration, and recent developments that allow the associated uncertainty to be specified. Finally the application of models to ungauged catchments (regional analysis) is discussed. For a more extensive treatment of these subjects, the reader is referred to *Rainfall-Runoff Modelling in Gauged and Ungauged Catchments* by Wagener *et al.* (2004) and also the AGU Monograph on *Calibration of Watershed Models* (Duan *et al.*, 2003).

8.2 CALIBRATION ISSUES

Two distinct approaches to calibration can be taken:

(1) A manual approach requires the user to adjust parameters interactively in successive model runs. This is a time-consuming and labour-intensive process, dependent on the insight of the modeller (although codified guidance is available for some models (Boyle *et al.*, 2000)). This dependence on the user can be seen as a strength, since it builds on accumulated experience and only intelligent steps through the parameter space will be made. However, it is also a weakness, since the process is subjective, and the parameters derived may be prone to bias. There is also no clear point at which the calibration process can be said to be complete.

In evaluating the quality of model fit, human judgement is used. The eye is a powerful integrator of different signals, and can, for example, detect that simulation performance over different parts of the hydrograph (e.g., peaks, recessions, low flows) may provide clues to guide the search for optimal values of specific parameters within the model. Alternatively the fit can be measured using one or more objective functions. These commonly aggregate the time-series residuals over the whole calibration period, for example the Nash–Sutcliffe efficiency (NSE) (Nash and Sutcliffe, 1970), which is a dimensionless form of the sum of squared errors.

(2) An automatic approach uses a computer algorithm to search the parameter space, performing multiple trials of the model. This requires that performance is specified by an objective function, and if the model has p parameters, the problem can be posed as the maximization or minimization of a $(p + 1)$-dimensional response surface. Classical optimization methods can be applied, for example gradient-based methods such as the well-known Marquard algorithm and Rosenbrock's rotating-coordinate search method (Rosenbrock, 1960), or alternatively, rule-based search methods, such as the Simplex algorithm of Nelder and Mead (1965). Current computing power has allowed more powerful variants to be developed. For example, the shuffled complex evolution method, developed at the University of Arizona (SCE-UA) (Duan *et al.*, 1992), combines the Simplex algorithm with a genetic algorithm approach. Multiple Simplexes are propagated through the parameter space, and periodically shuffled to exchange information. The automatic approach has the advantages that

Hydrological Modeling in Arid and Semi-Arid Areas, ed. Howard Wheater, Soroosh Sorooshian, and K. D. Sharma. Published by Cambridge University Press.
© Cambridge University Press 2008.

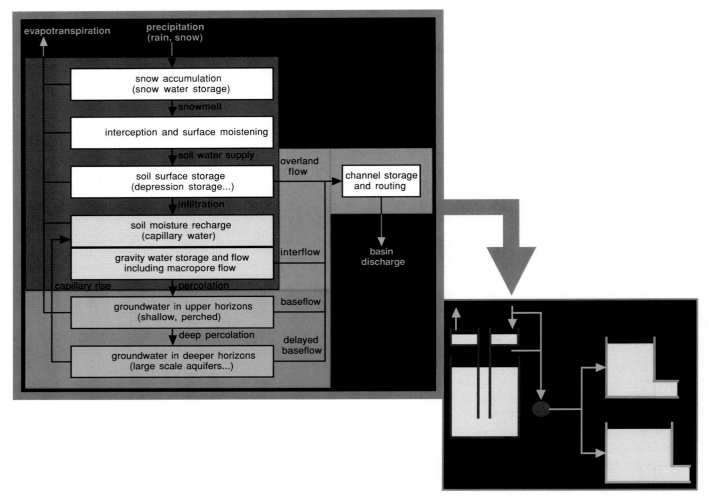

Figure 8.1 (Plate 15) Conceptual model

the computer does the hard work of exploring the parameter space, rather than the user, and that the procedure is objective; the calibration process would, through thorough searching of the parameter space, ideally be able to define a single well-identified optimum parameter set.

The development of automatic methods has allowed detailed investigation of the issues underlying the search for a global optimum set of parameter values in a way that was not possible using a manual approach. Problems in searching the parameter space have long been well-known (e.g., Ibbitt, 1970; Johnston and Pilgrim, 1976). These include:

• multiple local optima on the objective function surface;
• interdependence of parameters that gives difficulties due to the production of valleys (or ridges) in the objective function;
• insensitive directions in the parameter space, e.g., if a parameter is redundant due to a threshold value;

• search hampered by boundaries in parameter values (although restriction of the parameter space can be helpful, interactions with the boundary are problematic);
• saddle points, where first derivatives vanish but minima (or maxima) are not reached;
• different scales of parameters, which create difficulties in defining appropriate step lengths in each parameter direction.

One important result is non-uniqueness of the identified parameter sets: many combinations of parameter values provide equally good fits to the data (indeed many different model structures may also give similar objective function values (Wheater, 2002)). Beven (1993) defined this problem as "equifinality," arising from over-parameterization of models, data limitations, and structural faults in the model.

This ambiguity has serious implications for model applications. If parameters cannot be uniquely identified, then they cannot be deterministically linked to catchment characteristics. For

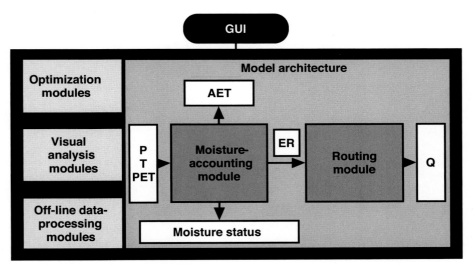

Figure 8.2 The Rainfall-Runoff Modelling Toolbox (RRMT)

example, this in principle precludes the use of models for the prediction of catchment change (where a change in catchment attributes must be represented by an appropriate change in parameter values), or for application to modelling the hydrology of ungauged catchments.

8.3 DEVELOPMENTS IN CALIBRATION METHODS

There have been three reactions to this problem of ambiguity (Wagener *et al.*, 2003a). These are discussed below, and illustrated using various methods of model performance analysis.

8.3.1 The development and analysis of parsimonious model structures

Firstly, there has been a move to simpler models, with fewer parameters, in an attempt to reduce parameter interactions and dependencies and more closely match the complexity of the model to the information content of the data. This reflects the philosophy of metric models, and hence this class of model was termed hybrid metric-conceptual (HMC) by Wheater *et al.* (1993).

An early example is the probability-distributed model (PDM) of Moore and Clarke (1981). IHACRES (see Chapter 4 by Croke and Jakeman) is another. In the PDM model, the spatial variation of soil-moisture storage capacity is represented by a prescribed probability distribution. The Pareto distribution is used here. It has the distribution function:

$$F(c) = 1 - \left(1 - \frac{c}{c \max}\right)^{b}, \qquad (8.1)$$

where c/c max is the relative soil moisture storage and the parameter b controls the degree of spatial variability. When the soil storage at any point on its distribution is filled, effective rainfall is generated. The actual evaporation is assumed to be proportional to the wetness of the catchment. This model element has two model parameters, the maximum storage capacity, c max, and the degree of spatial variability of the soil-moisture capacity in the catchment, b. The routing component of the model can take various forms. One consists of two linear reservoirs in parallel, one representing a relatively quick catchment response and the other a slower response. All the effective rainfall is split between these two reservoirs, defined by parameter %(q) (proportion of the total effective rainfall going to the quick response reservoir). The two components of outflow are aggregated into total streamflow. This model element has three model parameters, i.e., two residence times of the each reservoirs, rt(q) and rt(s), and %(q), thus giving five parameters for the model as a whole.

Models of this type can typically be considered to consist of a non-linear loss function, which accounts for rainfall losses, and a routing component, comprising a set of linear or non-linear stores in parallel. The Imperial College Rainfall-Runoff Modelling Toolbox (RRMT) (Wagener and Lees, 2001a; Wagener *et al.*, 2002), written in the MATLAB environment, supports the implementation of this type of model. It allows a wide range of alternative moisture-accounting modules and routing modules to be selected from a library and easily combined to form a chosen model structure (Fig. 8.2).

Current computing power allows such models to be run many thousands of times on an ordinary PC, and hence simple, but powerful methods to analysis model structure and performance can be easily implemented. Figure 8.3 shows typical "dotty plots"

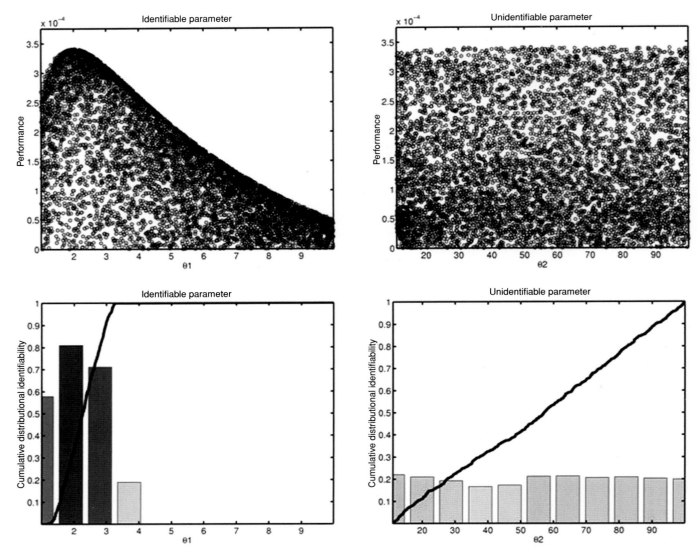

Figure 8.3 Example of a well and a poorly identified parameter. The top row shows scatter plots of parameter versus measure of performance. It has to be considered that these projections into a single parameter dimension can, however, hide some of the structure of the response surface (Beven, 1989). The bottom row shows the cumulative distribution of the best performing 10 % of parameter sets and the corresponding gradients within each segment of the parameter range

from a Monte Carlo analysis, where each dot represents a simulation. Values of the individual model parameters are drawn from pre-specified distributions by random sampling. The upper plots show the value of the objective function (which is being maximized) on the vertical axis, plotted against the associated value of a given parameter on the horizontal axis. The left-hand image shows a clearly-defined optimum parameter value; the parameter is clearly "identifiable." In contrast, the right-hand image shows that any value for the parameter can (in conjunction with different combinations of the other parameters) yield comparable results. The parameter is non-identifiable using this objective function, at least in a univariate sense.

Dynamic identifiability analysis (DYNIA) (Wagener *et al.*, 2003b) extends the analysis of parameter identifiability to investigate time dependence. A measure of parameter identifiability is required, and the lower plots in Fig. 8.3 illustrate one approach. The cumulative distribution of the best performing 10 % of parameter sets is plotted, and the corresponding gradients within each segment of the parameter range are represented by a histogram. Gray-scale coding indicates the value of the gradient – darker coloring indicates a steeper gradient, and thus greater identifiability. We can now take a moving window through the output time series (i.e., streamflow in this case) and evaluate the identifiability as a function of time. An example is shown in Fig. 8.4 for the

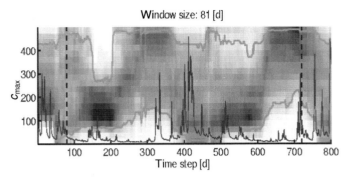

Figure 8.4 DYNamic Identifiability Analysis (DYNIA)

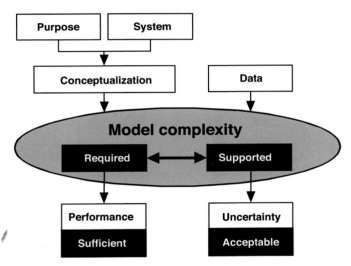

Figure 8.5 A framework for modelling

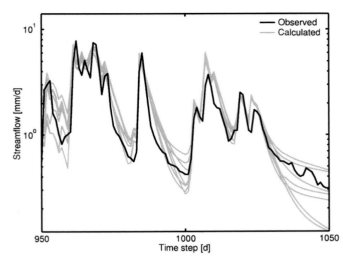

Figure 8.6 Simulated hydrographs with the same objective function value

8.3.2 The use of multiple performance criteria

Secondly, there has been a move to extract more information from the available data. Until recently, automatic optimization methods have used a single objective function (in contrast to the subjective manual assessment of goodness of fit, which intuitively inspects many different facets of model performance). The ambiguity that can arise from this is illustrated in Fig. 8.6, where a set of simulations with the same NSE value are seen to have markedly different low-flow properties. Research has shown that the information retrieved from a single objective function is sufficient to identify between three and five parameters (Beven, 1989; Jakeman and Hornberger, 1993), but it was realized that more information can be obtained from the available streamflow time series by considering different aspects of the flow time series (Wheater *et al.*, 1986; Gupta *et al.*, 1998, Boyle *et al.*, 2000).

Using multiple objective functions can provide information that improves the identifiability of certain parameters (Wheater *et al.*, 1986). It can also illustrate tensions in the model. Figure 8.7 illustrates results from a set of Monte Carlo simulations, as discussed above. Two objective functions are calculated, one (NSE* = 1−NSE) based on the sum of squared errors (and hence most strongly influenced by high-flow performance), the other (FSB) based on the errors in low-flow performance. The aim in this case is to minimize the objective functions. In the left-hand diagram it can be seen that the best low-flow performance does not coincide with the best high-flow performance, i.e., different parameter sets are required to optimize for low flows in comparison with high flows. A trade-off is needed – the user must decide where the modelling priorities lie, and choose an appropriate parameter set. In the right-hand diagram, the model structure is capable of maximizing both high- and low-flow performances with minimal trade-off required.

parameter c max (the maximum soil moisture storage capacity in the PDM model), evaluated using the NSE criterion. It can be seen that there are periods in the data series where the identifiability of the parameter increases (shown by the dark shading). It can also be seen that there are tensions in the model. There are periods where the identified parameter value is in the region of 100; there are other periods, notably where there is a significant hydrograph response following a dry period, where a higher value (around 500) is preferred.

The analysis of model structure and parameter identifiability opens the way to a new framework for modelling, in which the attributes of the model are tuned to the modelling task (Fig. 8.5). If an application is, for example, to select a model for regionalization, then it may be considered important to maximize parameter identifiability, and this may require simplification of model structure, possibly with some loss of simulation performance.

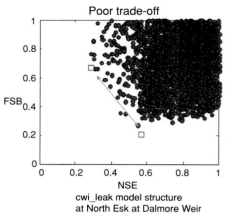 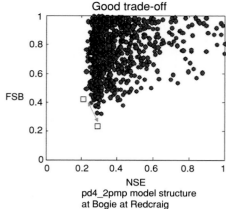

Figure 8.7 Multiple objective functions

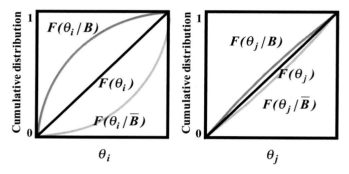

Figure 8.8 Cumulative distribution of initial $(F(\theta)_i)$, "behavioral" $(F(\theta_i|B))$ and "non-behavioural" $(F(\theta_i|\bar{B}))$ populations for a sensitive parameter θ_i, and a (conditionally) insensitive parameter θ_j

8.3.3 Abandoning the concept of a unique best-fit model

The logical conclusion from the widespread observation that there is equifinality in model parameters and in model structures is to abandon the idea that a uniquely identifiable model exists. Rather, there is a population of models (i.e., structures and parameter sets) that can be defined according to their consistency with the available data. Spear and Hornberger (1980) developed the concept of regionalized sensitivity analysis (RSA). They classified their model realizations as "behavioral" or "non-behavioral," depending on consistency with available data, and used this classification to explore parameter sensitivity. Parameters were sensitive if there was a significant difference between the set of behavioral and non-behavioral parameters. This is illustrated in Fig. 8.8. The cumulative distribution of the values of parameter θ_i for the model realisations considered "behavioural" is denoted $(F(\theta_i|B))$ and the distribution for the "non-behavioral" populations is denoted $(F(\theta_i|\bar{B}))$. In the figure, parameter θ_i is sensitive (the populations are clearly different) and parameter θ_j is insensitive.

This approach was extended by Freer *et al.* (1996); instead of two classes, the model realizations are split into ten groups of equal number, ranked according to their objective-function performance, and the cumulative distributions can be plotted to indicate parameter sensitivity.

This analysis can be extended to estimate the uncertainties that arise in model structure, parameter values, and data. A popular approach is the generalized likelihood uncertainty estimation (GLUE) procedure (Beven and Binley, 1992; Freer *et al.*, 1996). Recalling the basic assumption that for a given data set a unique best-fit model (in terms of structure and parameter values) does not exist, the likelihood that a particular model and parameter set represents the data can be estimated, using some appropriate performance criterion.

As above, for a given model structure, parameter sets are generated (based on random sampling of the feasible parameter space using a known or assumed prior distribution, e.g., a uniform distribution) and the model is run using a Monte Carlo procedure. The simulations are classified as behavioral or non-behavioral, and the latter are rejected. The objective function values of the behavioral set of simulations can be used as "likelihood" measures, which are scaled and used to weight the predictions associated with individual behavioral parameter sets. The modelling uncertainty is then propagated into the simulation results as confidence limits of any required percentile (Fig. 8.9).

There are clearly limitations in this approach. The decision on whether a simulation is behavioral or not requires judgement to define a threshold value of objective function. And the various sources of uncertainty (input- and output-data error, model-structural error, parameter-estimation error and random error) are lumped into the single entity of parameter error. Nevertheless, the ability to specify the uncertainty associated with a simulation is a fundamentally important step. Given the availability of such tools,

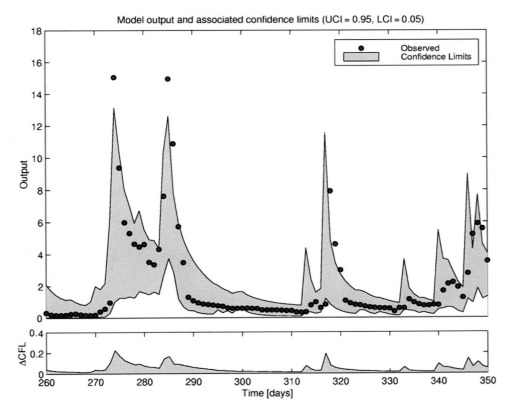

Figure 8.9 Simulation with 95 % confidence limits

it will soon no longer be adequate to present a simulation output as a single best estimate, with no attempt to specify the associated confidence intervals.

8.4 TOOLBOXES

It will be seen that important steps have been taken in our understanding of hydrological models, and that the ability to use simple Monte Carlo simulation methods has provided a powerful set of tools to support hydrological modelling through analysis of model structures, parameter identifiability and uncertainty. These techniques are now readily accessible through recently developed toolboxes.

The Imperial College Rainfall-Runoff Modelling Toolbox (RRMT) (Wagener and Lees, 2001a) has already been introduced. The toolkit contains a large selection of parsimonious conceptual rainfall-runoff models, and also optimization tools such as SCE-UA. A second Imperial College toolbox, also available in the MAT-LAB environment, is the Monte Carlo Analysis Toolbox (MCAT) (Wagener and Lees 2001b), Fig. 8.10. This supports the tools for model sensitivity and uncertainty analysis discussed above, based largely on generalized sensitivity analysis and GLUE. Recently a

capability for dynamic identifiability analysis (DYNIA) has been added to the MCAT.

8.5 REGIONALIZATION

One of the major challenges to hydrologists is the prediction of the hydrology of ungauged catchments. However, using parsimonious models and the tools described above, important progress is being made to regionalize continuous simulation rainfall-runoff models.

The procedure of regionalization can be summarized as shown in Fig. 8.11 (Wagener *et al.*, 2004). Data from gauged catchments (I) are used to calibrate a "local model," and hence derive a set of model parameters for each catchment (θ). These parameters are used, with a set of catchment characteristics (Φ), to generate a "regional model," most commonly by regression analysis. The regional model can then be simply applied to ungauged catchments – if the catchment characteristics are known, the regional model generates the parameters to run the local model structure to represent the ungauged catchment. Many issues arise, for example what is the most appropriate local model given the range of catchment types, sizes, climate characteristics; what is the most appropriate procedure for development of the regional model.

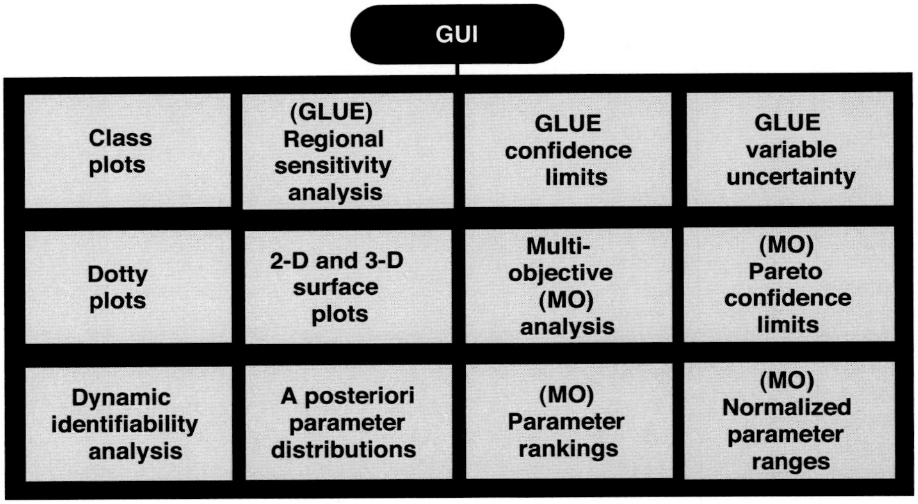

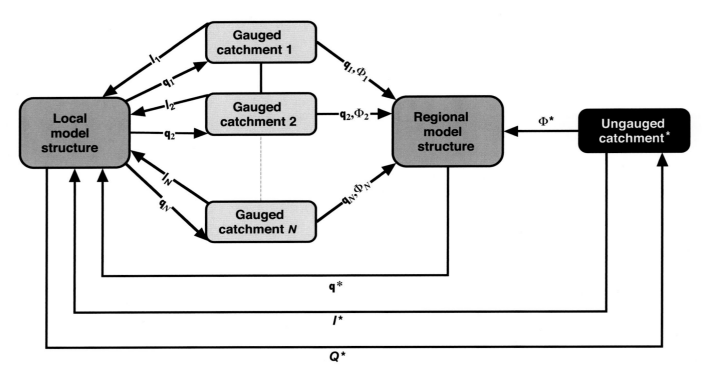

Figure 8.10 The Monte Carlo Analysis Toolbox (MCAT)

Figure 8.11 Model regionalization procedure

These are discussed in more detail by Wagener *at al.* (2004), but also provide considerable scope for new research.

A summary case study is presented below (after Wheater *et al.*, 2002), to illustrate the basic procedure, using daily modelling of 23 small to medium sized catchments from the Thames basin, UK. Mean daily discharges for the ten year period from January 1, 1990 to December 31, 1999 were used, together with mean daily precipitation (mm/d) for 36 raingauges across the study area, and weekly potential evaporation estimates. Physical catchment descriptors were available from the *Flood Estimation Handbook, FEH* (Institute of Hydrology, 1999) (see Table 8.1).

Several model structures were explored from the RRMT toolbox. Here we report results from the Catchment Wetness Index (CWI) loss function used in the IHACRES model (Jakeman *et al.*,

Table 8.1 *Description of catchment characteristics*[a]

Catchment characteristic	Unit	Description
AREA	km^2	Catchment drainage area
LDP	km	Longest drainage path
BFIHOST	–	Baseflow index derived using the HOST classification
SPRHOST	%	Standard percentage runoff derived using the HOST classification
FARL	–	Index of flood attenuation due to reservoirs and lakes
PROPWET	–	Index of proportion of time that soils are wet
DPLBAR	km	Index describing catchment size and drainage path configuration
DPSBAR	m/km	Index of catchment steepness
ASPBAR	–	Index representing the dominant aspect of catchment slopes
ASPVAR	–	Index describing the invariability in aspect of catchment slopes
RMED-1D	mm	Median annual maximum 1-day rainfall
RMED-2D	mm	Median annual maximum 2-day rainfall
RMED-1H	mm	Median annual maximum 1-hour rainfall
SAAR	mm	1961–90 standard-period average annual rainfall
SAAR$_{4170}$	mm	1941–70 standard-period average annual rainfall
URBEXT$_{1990}$	–	FEH index of fractional urban extent for 1990
URBCONC	–	Index of concentration of urban and suburban land cover
URBLOC	–	Index of location of urban and suburban land cover

[a] After Institute of Hydrology (1999)

1990), combined with a parallel linear two-store model (2PAR), to represent the quick- and slow-flow components respectively.

8.5.1 Local calibration

Table 8.2 lists the calibrated parameter sets (derived using Monte Carlo sampling using the root-mean-squared error (RMSE) criterion) and the RMSE and NSE criteria for all catchments (two catchments have been excluded due to shorter data length and calibration problems).

The resulting model fits are all considered good, with NSE values ranging from 0.73 to 0.92 (average 0.82). The reduction in performance for the top five catchments can be clearly seen and is likely to be due to the use of daily data that is relatively coarse to represent the rapid flow responses on these impermeable catchments.

8.5.2 Regional analysis

Relationships between model parameters and individual FEH catchment characteristics were investigated using simple correlation analysis. Two additional characteristics, SMDBAR (mean soil-moisture deficit) and ALTBAR (mean altitude) were only available for 12 of the 21 catchments, and were thus excluded from Table 8.1 and the full regionalization analysis. The parameter tau, the time constant of catchment wetness decline, is related to SMDBAR (correlation coefficient 0.60). The refp parameter is poorly correlated with all the catchment characteristics across most response periods. Parameter mf, which determines the difference in evapotranspiration (ET) between summer and winter, is, not surprisingly, strongly influenced by SMDBAR (correlation coefficient 0.98, see Fig. 8.12). Parameter mf also shows a relationship with BFIHOST, which may be indicative of the influence of soil type and geology on water losses from the catchment.

The residence time constant of the quick-flow reservoir, rt(q), is consistently well identified for all objective functions by BFIHOST, which reflects the underlying geology of the catchment; as BFIHOST increases, so does the residence time. rt(q) was also found to be, to some extent, related to the urban measures of concentration (URBCONC) and location (URBLOC). This may be coincidental. rt(q) is also consistently related to the catchment slope measure (DPSBAR) and the median annual maximum rainfall for a duration of one hour (RMED-1H).

The rt(s) parameter, the time constant for the slow-flow reservoir, is again strongly related to BFIHOST, here with an inverse relationship. rt(s) is also related to PROPWET, a measure of the soil moisture status over time, to the average aspect of the catchment and also to the average altitude, ALTBAR.

Finally, relationships between the percentage quickflow of total flow (%(q)) and catchment characteristics were investigated. As expected, BFIHOST shows the strongest relationship with this parameter (Fig. 8.13), which also shows some level of correlation with SMDBAR and ASPBAR.

The most significant relationships between the model parameters and catchment descriptors, indicated by the Pearson correlation coefficient in brackets, are shown in Table 8.3.

8.5.3 Multiple-regression regional analysis

A multiple-regression analysis was carried out using simple routines in MATLAB. The procedure allows the simple exclusion of outlier values, e.g., the mf values found for the three chalk catchments (C15, C16, C17). A test catchment was selected, namely

Table 8.2 *Parameter values and model fits for the CWI_2PAR model*

	Catchment number	tau	refp	mf	rt(q)	rt(s)	%(q)	RMSE	NSE
CLAY	C02	38.06	1.33	0.44	1.84	218.23	0.78	1.175	0.730
	C03	2.82	4.48	0.93	2.78	497.51	0.77	0.796	0.740
	C01	0.55	9.71	0.48	1.64	251.43	0.71	1.131	0.740
	C09	6.26	2.77	1.23	3.01	191.97	0.81	0.373	0.790
	C23	6.98	2.66	1.35	7.42	53.61	0.89	0.657	0.730
MIXED 1	C22	29.04	1.44	1.02	7.10	56.02	0.62	0.290	0.870
	C04	15.38	3.41	0.62	3.90	241.36	0.56	0.244	0.800
	C08	26.81	1.79	1.65	3.27	106.56	0.15	0.070	0.890
	C14	30.83	1.10	1.52	21.51	135.08	0.35	0.108	0.920
MIXED 2	C07	26.95	1.76	1.14	7.98	91.16	0.25	0.191	0.880
	C05	35.21	1.47	0.88	2.56	173.76	0.40	0.239	0.760
	C21	12.76	2.90	0.76	11.72	59.10	0.36	0.183	0.900
	C12	20.66	1.81	1.38	92.42	108.46	0.23	0.108	0.910
	C13	25.92	2.27	0.88	52.18	216.52	0.81	0.169	0.890
MIXED 3	C18	12.32	1.97	1.25	24.89	35.98	0.31	0.247	0.900
	C06	29.28	1.22	1.19	2.12	106.64	0.31	0.149	0.810
	C20	36.00	1.90	0.66	13.46	22.30	0.24	0.412	0.770
	C19	18.11	2.89	0.66	32.69	68.38	0.61	0.273	0.870
CHALK	C15	4.01	1.43	17.27	51.31	73.29	0.95	0.218	0.750
	C17	21.39	1.41	17.59	18.62	69.40	0.37	0.241	0.820
	C16	2.68	1.52	15.08	38.31	53.60	0.07	0.265	0.780

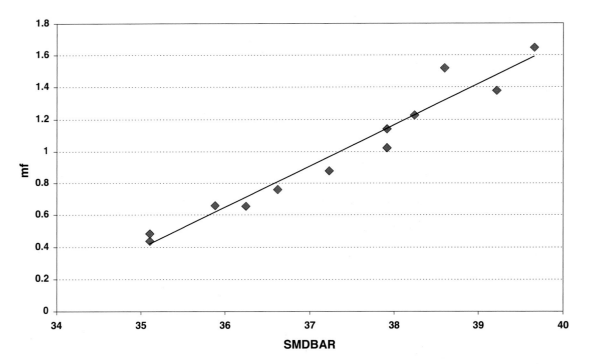

Figure 8.12 Overall RMSE best-fit mf versus SMDBAR

Table 8.3 *The most significant relationships between CWI_2PAR parameters and catchment descriptors (note – urban measures are excluded)*

Model parameter	Most significant catchment descriptors (correlation coefficient)			
tau	DPSBAR (−0.27)	BFIHOST (0.19)	RMED-1H (0.13)	ASPVAR (−0.12)
refp	BFIHOST (−0.53)	PROPWET (0.42)	FARL (−0.42)	ASPVAR (0.32)
mf	BFIHOST (0.53)	AREA (−0.32)	FARL (0.54)	PROPWET (−0.05)
rt(q)	BFIHOST (0.53)	DPSBAR (0.47)	PROPWET (−0.38)	FARL (0.27)
rt(s)	BFIHOST (−0.62)	PROPWET (−0.38)	DPSBAR (−0.38)	ASPBAR (0.37)
%(q)	BFIHOST (−0.54)	ASPBAR (0.32)	RMED-1H (−0.23)	
volc	BFIHOST (0.58)	DPSBAR (0.54)	ASPBAR (−0.44)	SAAR (0.25)

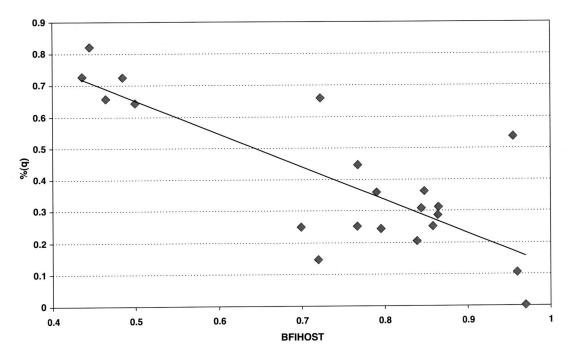

Figure 8.13 RMSE best-fit %(q) versus BFIHOST for all catchments

the River Wey at Farnham (C06), and removed from the multiple regression data. The resulting relationships are as follows:

$$\text{tau} = -0.465\,(\text{DPSBAR}) - 37.6\,(\text{ASPVAR})$$
$$- 4.26\,(\text{RMED} - 1\text{H}) + 99.6 \qquad (R^2 = 0.51)$$

$$\text{refp} = -1.20\,(\text{BFIHOST}) - 12.8\,(\text{PROPWET}) + 0.264\,(\text{ASPVAR})$$
$$- 0.269\,(\text{FARL}) + 7.05 \qquad (R^2 = 0.34)$$

$$\text{mf} = -13.3\,(\text{PROPWET}) - 0.320\,(\text{BFIHOST}) + 0.00013\,(\text{AREA})$$
$$+ 4.95\,(\text{FARL}) + 0.864 \qquad (R^2 = 0.65)$$

$$\text{rt(q)} = 29.1\,(\text{BFIHOST}) - 0.151\,(\text{DPSBAR})$$
$$- 14.1\,(\text{PROPWET}) - 3.82 \qquad (R^2 = 0.74)$$

$$\text{rt(s)} = -315\,(\text{BFIHOST}) + 1017\,(\text{PROPWET}) + 0.0149\,(\text{ASPBAR})$$
$$- 0.0619\,(\text{DPSBAR}) + 25.7 \qquad (R^2 = 0.69)$$

$$\%(q) = -1.06\,(\text{BFIHOST}) + 0.00019\,(\text{ASPBAR})$$
$$- 0.0031\,(\text{RMED} - 1\text{H}) + 1.33 \qquad (R^2 = 0.73)$$

$$\text{volc} = -0.00698\,(\text{BFIHOST}) - 0.00004\,(\text{DPSBAR})$$
$$- 0.000004\,(\text{ASPBAR}) + 0.01065 \qquad (R^2 = 0.754)$$

where volc is a factor introduced to ensure that the total volume of modelled effective rainfall equals the total volume of observed streamflow. This parameter is calculated during the calibration stage, but has to be estimated during prediction and therefore regionalization.

Table 8.4 gives a comparison of the calibrated and regionally estimated parameter values for the Farnham catchment (C06); Table 8.5 and Fig. 8.14 present the associated performance.

Table 8.4 *Calibrated and regionally estimated parameter values for the Farnham catchment (C06)*

Parameter	Calibrated	Estimated
tau	29.3	25.2
refp	1.22	1.30
mf	1.19	0.853
rt(q)	2.12	8.01
rt(s)	107	107
%(q)	0.313	0.396
volc	0.0017	0.0021

Table 8.5 *Model performance measures for the calibrated and estimated model parameters for the Farnham catchment*

Performance Measure	Calibrated	Estimated
RMSE	0.149	0.189
NSE	0.81	0.69

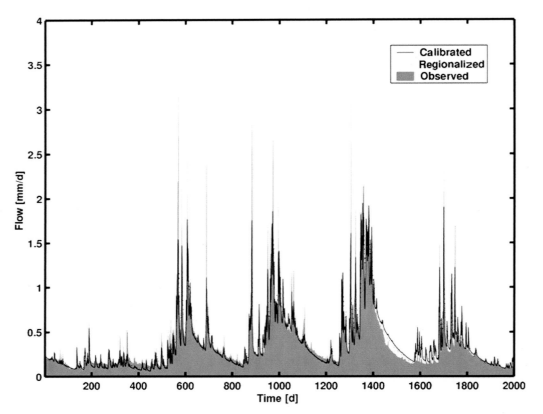

Figure 8.14 Regionalised model fits for the Farnham catchment (C06)

8.5.4   Conclusions from regionalization

This pilot investigation was of limited scope, but nevertheless provides some insight into the hydrological controls on the catchments investigated, as well as confirming the potential of the methodology for more extensive regional analysis.

The regionally estimated parameters, found using the multiple-regression method, showed encouraging results with respect to both the proximity of the estimated to the calibrated parameter values, and also in the good model fits to observed streamflow data. Only limited validation was possible given the number of catchments used. The derived regional relationships show a good level of physical justification with regard to the catchment descriptors used and the BFIHOST variable was found to be significant in many of the parameter-estimation equations, especially for the routing module parameters. This is indicative of the dominant effect of the soils and geology of the catchment, reflected in the BFIHOST value, on the catchment response characteristics.

8.6 CONCLUSIONS

This chapter has reviewed developments in our understanding of the performance of conceptual rainfall-runoff models. Since they were first developed, some 40 years ago, important progress has been made, and current computing power has enabled the development and application of a powerful range of stochastic, Monte Carlo based tools for the analysis of model performance. The limitations of these models have mainly been due to the problems of parameter inter-dependence – models have been too complex to be identified uniquely from the available data.

Various issues have been discussed. It has been shown that automatic methods of model calibration have finally caught up with the experienced user by incorporating multiple objective functions in the evaluation of model performance. For a rainfall-runoff model, different aspects of the streamflow hydrograph can be considered. For other applications, such as sediments or water quality, different output signals can be evaluated. This generally illustrates tensions in our models. The parameters which are optimal for high-flow simulation will not be the same as those which are optimal for low flows, and in a water-quality model, the hydrological parameters which are optimal for flow may not be same as those that give optimal simulations of chemistry. This provides important information to the user, who can decide what are the appropriate criteria for a particular application.

A major step forward has been the recognition that alternative models and parameters may be equally likely interpretations of the available data. This has led to the development of methods to quantify model uncertainty, which means that the modeller can provide an informed estimate of the uncertainty associated with a model simulation. This provides a radically different way of communicating information to the users of model results, and enables much more informed decisions to be made, for example about the risk associated with a given management strategy.

Finally, the ultimate challenge to the hydrologist is to predict the response of an ungauged catchment. We have seen that parsimonious models have the capability to deliver convincing results, and such regional methods can be expected to rapidly find their place in hydrological analyses (Wheater, 2002).

The modelling methods and tools discussed are available in the MATLAB-based RRMT and MCAT Toolboxes. These can be downloaded free for research users from http://ewre.cv.imperial.ac.uk. Please note that MATLAB software is required for their implementation.

8.7 REFERENCES

Beven, K. J. (1989). Changing ideas in hydrology: the case of physically based models. *J. Hydrol.*, **105**, 157–72.

Beven, K. J. (1993). Prophecy, reality and uncertainty in distributed hydrological modelling. *Adv. In Water Resourc.*, **16**, 41–51.

Beven, K. J. and Binley, A. M. (1992). The future of distributed models: model calibration and predictive uncertainty. *Hydrol. Proc.*, **6**, 279–98.

Boyle, D. P., Gupta, H. V., and Sorooshian, S. (2000). Towards improved calibration of hydrologic models: combining the strengths of manual and automatic methods. *Water Resour. Res.* **36**, 3663–74.

Duan, Q., Gupta, V. K. and Sorooshian. S. (1992). Effective and efficient global optimisation for conceptual rainfall-runoff models. *Water Resour. Res.*, **28**, 1015–31.

Duan. Q., Gupta, H. V., Sorooshian, S., Rousseau, A. N., and Turcotte, R. (eds.) (2006). *Calibration of Watershed Models*. American Geophysical Union Monograph Series, *Water Science and Application 6*. Washington, DC: AGU.

Freer, J., Beven, K., and Abroise, B. (1996). Bayesian uncertainty in runoff prediction and the value of data: an application of the GLUE approach. *Water Resour. Res.*, **32**, 2163–73.

Gupta, H. V., Sorooshian, S., and Yapo, P. O. (1998). Towards improved calibration of hydrological models: multiple and non-commensurable measures of information. *Water Resour. Res.*, **34(4)**, 751–63.

Ibbitt, R. (1970). Systematic parameter fitting for conceptual models of catchment hydrology. Unpublished Ph.D. thesis, University of London.

Institute of Hydrology (1999). *Flood Estimation Handbook*. Wallingford: Institute of Hydrology.

Jakeman, A. J. and Hornberger, G. M. (1993). How much complexity is warranted in a rainfall-runoff model? *Water Resour. Res.*, **29**, 2637–49.

Jakeman, A. J., Littlewood, I. G., and Whitehead, P. G. (1990). Computation of the instantaneous unit hydrograph and identifiable component flows with application to two small upland catchments. *J. Hydrol.*, **117**, 275–300.

Johnston, P. R. and Pilgrim, D. H. (1976). Parameter optimization for watershed models. *Water Resour. Res.*, **12(3)**, 477–86.

Moore, R. J. and Clarke, R. T. (1981). A distribution function approach to rainfall-runoff modelling, *Water Resour. Res.*, **17(5)**, 1376–82.

Nash, J. E. and Sutcliffe, J. V. (1970). River flow forecasting through conceptual models. 1: A discussion of principles. *J. Hydrol.*, **10**, 282–90.

Nelder, J. A. and Mead, R. (1965). A simplex method for function minimisation. *Comp. J.*, **7**, 308–13.

Rosenbrock, H. H. (1960). An automatic method for finding the greatest or least value of a function. *Comp. J.*, **3**, 175–84.

Spear, R. C. and Hornberger, G. M. (1980). Eutrophication in Peel inlet. II: Identification of critical uncertainties via generalised sensitivity analysis. *Water Resour. Res.*, **14**, 43–9.

Wagener, T. and Lees, M. J. (2001a). *Rainfall-Runoff Modelling Toolbox user manual*. London: Department of Civil and Environmental Engineering, Imperial College.

Wagener, T. and Lees, M. J. (2001b). *Monte-Carlo Analysis Toolbox User manual*. London: Department of Civil and Environmental Engineering, Imperial College.

Wagener, T., Lees, M. J., and Wheater, H. S. (2002). A toolkit for the development and application of parsimonious hydrological models. In *Mathematical models of large watershed hydrology*. ed. V. P. Singh and D. Frevert. Vol. 1. Highland Ranch, CO: Water Resources Publishers, 87–136.

Wagener, T., Wheater, H. S., and Gupta, H. V. (2003a). Identification and evaluation of watershed models. In *Calibration of Watershed Models. Water Science and Application*, Vol. 6, ed. Duan Qingyun, H. V. Gupta, S. Sorooshian, A. N. Rousseau, and R. Turcotte American Geophysical Union, 29–47.

Wagener, T., McIntyre, N., Lees, M. J., Wheater, H. S., and Gupta, H. V. Towards reduced uncertainty in conceptual rainfall-runoff modeling: dynamic identifiability analysis. *Hydrological Processes*, **17(2)**, 455–76.

Wagener, T., Wheater, H. S., and Gupta, H. (2004). *Rainfall-Runoff Modelling in Gauged and Ungauged Catchments*. London: Imperial College Press.

Wheater, H. S. (2002). Progress in and prospects for fluvial flood modelling. *Phil. Trans. R. Soc. Lond. A*, **360**, 1409–31.

Wheater, H. S., Bishop, K. H., and Beck, M. B. (1986). The identification of conceptual hydrological models for surface water acidification, *J. Hydrol. Proc.*, **1**, 89–109.

Wheater, H. S., Jakeman, A. J., and Beven, K. J. (1993). Progress and directions in rainfall-runoff modelling. In *Modelling Change in Environmental Systems*, ed. A. J. Jakeman, M. B. Beck, and M. J. McAleer. Chichester: Wiley, 101–32.

Wheater, H. S., Boxall, S., and Wagener, S. (2002). Regionalisation of rainfall-runoff models: an application to the Thames Basin. *Proc. British Hydrological Society 8th National Hydrology Symposium, Birmingham* London: British Hydrological Society 199–206.

9 Real-time flow forecasting

P. C. Young

9.1 INTRODUCTION

The primary objective of this chapter is to describe recent research on the design of real-time, adaptive forecasting procedures for the prediction of flow (discharge) or river level (stage) in river systems and to illustrate this with a case study based on data from a semi-arid river catchment in Western Australia. In particular, the aim is to produce an on-line, real-time approach to flow forecasting that is inherently stochastic and so able to predict not only the likely level of future flow, but also the uncertainty associated with this prediction. In this manner, the probability of a flood occurring in the near future can be quantified and this additional information can then be used as a basis for decision-making, operational management, and risk assessment in relation to the flooding of flood-prone locations.

The methodology described in subsequent sections of the chapter can be applied to the forecasting of either river flow or level. For simplicity, however, "flow" will be used here as a generic term to mean either of these two measures. Also, "flow" will be taken to mean the total flow in the river, not just the "run-off" component of total flow. In this context, the approach to forecasting described here is model-based, i.e., depending on the nature of the catchment and the forecasting objectives, the flow forecasts are based on an appropriate combination of stochastic, dynamic models for the relationships between: (i) rainfall and flow, and (ii) flow at various locations along the river, i.e. "flow routing" models. Both types of model are estimated statistically on the basis of the available rainfall-flow data using the inductive data-based mechanistic (DBM) approach to stochastic modelling (see for example, Young and Lees, 1993; Young and Beven, 1994; Young, 1998b and the references therein), where both the model structure and the associated parameters are inferred from the data with the minimum of prior assumptions.

It is important to stress that, despite this reliance on statistical inference, the DBM representation of hydrological time series is not an exercise in "black-box" modelling. Indeed, unlike many statistically based models (or alternative neural network and neuro-fuzzy black-box models), the DBM model is only considered acceptable if it has an internal structure that can be interpreted satisfactorily in a physically meaningful manner. On the other hand, the DBM models often exploit "systems" elements, such as parametrically efficient (parsimonious) transfer-function relationships that are then decomposed in a physically interpretable fashion in order to satisfy the requirements of DBM modelling. As such, DBM models are normally in an ideal form for conversion to a stochastic state-space form and can be embedded easily within a state-estimation algorithm, such as the Kalman Filter (Kalman, 1960), that then provides the main engine for state updating, data assimilation, and real-time forecasting.

This statistical approach to model-based forecasting has the virtue of being inherently stochastic and, because it is formulated in Bayesian, recursive estimation terms (see for example, Bryson and Ho, 1969; Young, 1984), it provides an ideal basis for real-time implementation and the introduction of adaptive procedures. Such adaption is motivated by a view that the rainfall-flow and flow-routing processes are inherently "non-stationary," i.e., no completely fixed model with constant parameters will be able to characterize the catchment behavior for all times into the future. As a result, it is argued that the forecasting system should be based on models that are able to adjust to any, normally small, changes in the catchment behavior not predicted accurately enough by the initially estimated model.

The chapter has another, underlying objective that is of deeper philosophical and methodological significance and is, in part, a response to the recent increased interest in the so-called "top-down" (or "holistic") approach to modelling hydrological systems (e.g., Sivapalan and Young, 2005 and Jothityangkoon *et al.*, 2001, which follow from the earlier contributions of Klemes, 1983 and Young, 2003; see also the special issue of *Hydrological Processes* devoted to this topic: **17(11)**, 2003). Interest in top-down modelling has been revived largely because the alternative "bottom-up" or "reductionist" philosophy that dominated much research

Hydrological Modeling in Arid and Semi-Arid Areas, ed. Howard Wheater, Soroosh Sorooshian, and K. D. Sharma. Published by Cambridge University Press.

during the last century has failed to solve the many problems of modelling natural environmental systems, including hydrological processes. Top-down modelling in hydrology has its parallels in the environmental (e.g., Young, 1978, 1983; Beck, 1983) and ecosystems (e.g., Silvert, 1993 and the references therein) literature of the 1970s and early 1980s.

As far as the author is aware, these latter contributions were the first to emphasize the inherent dangers of "deterministic reductionism," i.e., the widely held view that a physically based simulation model can be constructed on the basis of purely deterministic equations that reflect the modeller's perception of the physical system and that this model can be validated satisfactorily against the available data. They also presented initial thoughts on a more objective, statistical approach to modelling stochastic systems that tries to avoid such dangers as much as possible. This, in turn, led eventually to the DBM approach which is much in sympathy with the tenets of deterministc top-down modelling (e.g., Jothityangkoon et al., 2001) but is, of course, rather different in its methodological basis. The papers also adumbrated very similar anti-reductionist arguments that have appeared recently in the hydrological literature and express some of these same views within a hydrological context (Jakeman and Hornberger, 1993; Beven, 2000). Quite similar anti-reductionist views are also appearing in other areas of science: for instance, in a recent lecture (Lawton, 2001), a chief executive of the UK Natural Environment Research Council (NERC) recounted the virtues of the top-down approach to modelling ecological systems (although, for some reason, he did not appear to accept that such reasoning could also be applied to other natural systems, such as the physical environment!).

The top-down approach to modelling can be justified by the results of dominant mode analysis (DMA) (Young et al., 1996; Young, 1999a). This shows that complex systems, including large, deterministic simulation models, can be emulated, often to a remarkable degree, by much simpler models that reflect the "dominant modal dynamics" of the complex system or model, as shown in the practical examples described in these references. Given that the DBM model is based on response data from the hydrological system, it is not surprising that it reflects the dominant modal behavior of the system. Indeed, it is interesting to note that DMA exploits the same identification and estimation algorithms used in DBM modelling, as described in subsequent sections of this chapter.

Bearing the above comments in mind, a wider aim of the present chapter is to promulgate the philosophy of "inductive," dominant mode, DBM modelling as an alternative to the "hypothetico-deductive" (and often reductionist) approach that has dominated much hydrological modelling research over the last century (see Young, 2002). Other recent publications that have concentrated more centrally on this topic within this wider modelling context and can be considered as adjuncts to the present chapter are Young

(1998a,b; 1999a), Young and Pedregal (1998,1999a), Young and Parkinson (2002), Young et al. (1996), Shackley et al. (1998), Parkinson and Young, (1998), and Young et al. 2004.

Finally, it is important to stress that the present chapter concentrates on the problem of modelling rainfall-flow processes for the purpose of real-time flow forecasting. The aim is *not* to produce a detailed simulation model that is intended for "what-if" studies and planning purposes as, for example, the models considered in Beven and Binley (1992), Romanowicz et al. (1994) and Beven et al. (2000). The DBM model is, however, a special form of the hybrid metric-conceptual (HMC) model (see the next section and Wheater et al., 1993) and, if required, it can be synthesized in a form which allows for these alternative applications (e.g., Young, 2003). It is then closely related to other HMC models, such as the Bedford–Ouse (Young, 1974, Whitehead and Young, 1975) and IHACRES (Jakeman et al., 1990) models. Indeed, because of its more objective, statistical approach to model structure identification, DBM modelling can help to justify and improve the more conceptual elements in such models.

9.2 THE CATEGORIZATION OF RAINFALL-FLOW MODELS

Before considering DBM modelling in detail, it is useful to set the scene by considering the various approaches that have been used to characterize the non-linear dynamic relationship between rainfall and river flow. This is one of the most interesting modelling problems in hydrology. It has received considerable attention over the past 30 years, with mathematical and computer-based models ranging from simple black-box representations to complex, physically based catchment models. It would be impossible to review this enormous literature here. Fortunately, however, there are many books available that deal in whole, or in part, with this challenging area of science and engineering. Useful texts of this type are Anderson and Burt (1985), Shaw (1994), Singh (1995), and Beven (2001). The latter book, in particular, provides a clearly written review of the whole topic that not only deals critically with many recent developments but also provides an excellent introduction to the subject at the start of the twenty-first century.

Wheater et al. (1993) have categorized rainfall-flow models into the following four, broad types:

- *Metric models*, which are based primarily on observational data and seek to characterize the flow response largely on the basis of these data, using some form of statistical estimation or optimization (e.g., Wood and O'Connell, 1985; Young and Wallis, 1985; Young, 1986). These include purely black-box, time-series models, such as discrete and continuous-time transfer functions, neural-network and neuro-fuzzy representations (e.g., Tokar and Johnson, 1999; Jang, *et al.* 1997). Often, such models derive from, or can be related to, the earlier unit hydrograph theory, but this is not always recognized overtly.

- *Conceptual models*, which vary considerably in complexity, but are normally based on the representation of internal storages, as in the original Stanford Watershed Model of the 1960s (Crawford and Linsley, 1966). However, assumptions about catchment-scale response are not often included explicitly, notable exceptions being TOPMODEL (Beven and Kirkby, 1979) and the ARNO model (Todini, 1996). The essential feature of all these models, however, is that the model structure is specified a priori, based on the hydrologist/modeller's perception of the relative importance of the component processes at work in the catchment; and then an attempt is made to optimize the model parameters in some manner by calibration against the available rainfall and flow data.

- *Physics-based models*, in which the component processes within the models are represented in a more classical, mathematical-physics form, based on continuum mechanics solved in an approximate manner via finite difference or finite element spatio-temporal discretization methods. A well known example is the *Système Hydrologique Européen* (SHE) model (e.g., Abbot *et al.*, 1986). The main problems with such models, which they share to some degree with the larger conceptual models, are two-fold: first, the inability to measure soil physical properties at the scale of the discretization unit, particularly in relation to subsurface processes; and second, their complexity and consequent high dimensional parametrization. This latter problem makes objective optimization and calibration virtually impossible, since the model is normally so over-parameterized that the parameter values cannot be uniquely identified and estimated against the available data (see below).

- *Hybrid metric-conceptual* (HMC) *models*, in which (normally quite simple) conceptual models are identified and estimated against the available data to test hypotheses about the structure of catchment-scale hydrological storages and processes. In a very real sense, these models are an attempt to combine the ability of metric models to efficiently characterize the observational data in statistical terms (the "principle of parsimony," Box and Jenkins, 1970, or the 'Occam's Razor' of antiquity), with the advantages of conceptual models that have a prescribed physical interpretation within the current scientific paradigm.

The models in the two middle categories, above, are often characterized by a large number of unknown parameters that need to be estimated ("optimized" or "calibrated") in some manner against the observational rainfall-flow time series. Because the number of parameters is normally very large in relation to the information content of the data, however, such models are often over-parameterized and not normally identifiable, in the sense that it is impossible to estimate their parameters uniquely without imposing prior restrictions on a large subset of the parameter values prior to estimation (see for example, Young *et al.*, 1996). The author and his co-workers have addressed these problems of over-parameterization and poor identifiability associated with large environmental models many times over the past quarter century (see previous references). And recently, Beven and his co-workers (e.g., Franks *et al.*, 1997) have revisited this idea within the hydrological context, using the term "equifinality" rather than "non-identifiability" to describe the consequences of such over-parametrization: namely the existence of many different parametrizations and model structures that are all able to explain the observed data equally well, so that no unique representation of the data can be obtained within the prescribed model set.

There appear to be two main reasons for these identifiability problems. First, any limitations of the observational data can be important, since the available time series may not be sufficiently informative to allow for the estimation of a uniquely identifiable model form. In particular, the inputs to a system may not be "sufficiently exciting" (see for example, Young, 1984), in the sense that they do not perturb the system sufficiently to allow for unambiguous estimation of all the model parameters within an otherwise identifiable model structure. Secondly, even if the input does sufficiently excite the system, there are usually only a limited number of dynamic modes – the *dominant modes* of the system – that are excited to any significant extent; and the observed output of the system is dominated by their cumulative effect.

The importance of this dominant mode concept in model identification and estimation is illustrated by Appendix 1 of Young (2001b), which shows how the response of a 26th-order hydrological simulation model can be duplicated with exceptional accuracy (0.001 % error by variance) by a much simpler 7th-order dominant mode model. This is typical of most high-order linear systems and appears to carry over to non-linear systems. For example, Young *et al.* (1996), Young (1998b), and Young and Parkinson (2002) have used similar analysis to show how the responses of high-order, non-linear global carbon cycle simulation models are accurately reproduced by differential equation models of much reduced order. This is also reflected in other recent work on the simplification of global climate models (Hasselmann *et al.*, 1997; Hasselmann, 1998).

As a result of dominant modal behavior, the identifiable order is normally quite low for hydrological systems, and many previous rainfall-runoff modelling studies (e.g., Kirkby, 1976; Hornberger *et al.*, 1985; Jakeman and Hornberger, 1993; Young, 1993; 1998b; Young and Beven, 1994; Young *et al.*, 1997a,b; Ye *et al.*, 1998) suggest that a typical set of rainfall-runoff observations contain only sufficient information to estimate up to a maximum of six parameters within simple, non-linear dynamic models of dynamic order three or less. In the hydrological examples discussed later, for instance, there is clear evidence in the data of only two significant dominant modes between the effective rainfall input and the flow response (as described by a second-order transfer function model with only four or five parameters): a "quick" mode with a residence time (time constant) of a few hours; and a "slow" mode, with a residence time of many hours.

By their very nature, both the metric and HMC approaches avoid many of these "large model" problems. As a result, they provide a potentially attractive vehicle for real-time flood forecasting: they can be justified well in statistical terms and they are inherently simple in both structure and application. Such simplicity means that the forecasting system can be more easily optimized on a regular basis in order to ensure near-optimal performance. And, as we see later, it facilitates the incorporation of advanced features such as on-line state and parameter adaption. Of the two approaches, however, the attractiveness and practical utility of the basic metric model as a vehicle for flood forecasting is marred by its lack of any clearly defined internal physical interpretation.

For instance, neural-network (e.g., Tokar and Johnson, 1999) and neuro-fuzzy models have attracted a great deal of attention in recent years, but they are the epitome of black-box modelling, revealing very little of their internal structure that has any physical meaning (see the discussion in Young (2001c) on the paper by Hu *et al.* (2001), where a neuro-fuzzy model with 102 parameters can be replaced by a non-linear DBM model with only 15 parameters if the internal structure of the model is identified and taken into consideration). Neural models can also be misleading since their efficacy tends to be judged on the basis of their "one-step-ahead" prediction performance, which often obscures underlying limitations in the full "simulation" behavior of the model. Indeed, in most cases, neural models are only predictive devices and cannot be used in a simulation mode. For these and other reasons, many hydrologists tend to mistrust such black-box models as a basis for something as important as flood forecasting. Moreover, their lack of any obvious internal physical meaning means that metric models are difficult to interrogate and diagnose when errors are encountered. HMC models, on the other hand, do not suffer from these problems and, indeed, are often simpler in dynamic terms than the metric model.

Within the category of HMC models, two main approaches to modelling can be discerned; approaches which, not surprisingly, can be related to the more general deductive and inductive approaches to scientific inference that have been identified by philosophers of science from Francis Bacon (1620), to Karl Popper (1959) and Thomas Kuhn (1962). In the first "hypothetico-deductive" approach, the a priori conceptual model structure is

effectively a theory of hydrological behavior based on the perception of the hydrologist/modeller and is strongly conditioned by assumptions that derive from current hydrological paradigms (e.g., the Bedford–Ouse and IHACRES models mentioned previously). The alternative DBM approach is basically "inductive," in the sense that it tries to avoid theoretical preconceptions as much as possible in the initial stages of the analysis. In particular, the model structure is not pre-specified by the modeller but, wherever possible, it is inferred directly from the observational data in relation to a more general class of models (normally ordinary differential equations or their discrete-time equivalents). Only then is the model interpreted in a physically meaningful manner, most often (but not always) within the context of the current hydrological paradigm, e.g., the models of rainfall-flow data in Young (1993, 1998b), Young and Beven (1994), and Young *et al.* (1997a).

9.3 DATA-BASED MECHANISTIC (DBM) MODELLING

Previous publications (Beck and Young, 1975; Whitehead and Young, 1975; Young, 1978, 1983, 1992, 1993, 1998a,b; Young and Minchin, 1991; Young and Lees, 1993; Young and Beven, 1994, Young *et al.*, 1996; Young and Pedregal, 1998, 1999a; Young and Parkinson, 2002) map the evolution of the DBM philosophy and its methodological underpinning in considerable detail, and so it will suffice here to merely outline the main aspects of the approach.

The main stages in DBM model building are as follows:

(1) The important first step is to define the objectives of the modelling exercise and to consider the type of model that is most appropriate to meeting these objectives. For instance a complex spatio-temporal catchment model to be used for the purpose of "what-if" studies within a planning context is not necessarily a suitable vehicle for real-time flow forecasting, and vice versa. As a result, the prior assumptions about the form and structure of this model are kept at a minimum in order to avoid the prejudicial imposition of untested perceptions about the nature and complexity of the model needed to meet the defined objectives.

(2) The next step in DBM modelling depends upon the circumstances of the study. Prior to the acquisition of data (or in situations of insufficient data), it is advisable and often essential to develop a simulation model that represents the system in a physically meaningful manner and satisfies the defined objectives. It is also necessary to evaluate this simulation model in two main ways. First, investigate the sensitivity of the model to uncertainty, using uncertainty and sensitivity analysis (e.g., Monte Carlo Simulation (MCS) analysis, as in

Parkinson and Young, 1998). Second, using DMA (see earlier, Section 9.1), develop a reduced-order, dominant-mode model that captures the most important aspects of the model's dynamic behavior.

(3) Subsequent to the acquisition of sufficient data, an appropriate model structure is identified and its parameters estimated by a relatively objective process of statistical inference applied directly to the time-series data and based on a given general class of linear and non-linear stochastic models, such as differential equations or their discrete-time equivalents. This stage normally exploits recursive estimation methods (e.g., Young, 1984) and can involve both non-parametric and parameteric estimation, as discussed in later sections of this chapter.

(4) Regardless of whether the model is identified and estimated in linear or non-linear form, it is only accepted as a credible representation of the system if, in addition to explaining the data well, it also provides a description that has direct relevance to the physical reality of the system under study. This is a most important aspect of DBM modelling and differentiates it from more classical statistical modelling methodology and "neural" methods of black-box modelling. If necessary, the DBM model obtained at this stage should be reconciled with the dominant mode version of the simulation model considered in (2) above.

(5) Finally, the estimated model is tested in various ways to ensure that it is "conditionally valid" in the sense discussed in the next section. This involves standard statistical diagnostic tests for stochastic, dynamic models, possibly including analysis which ensures that the non-linear effects have been modelled adequately (e.g., Billings and Voon, 1986), as well as exercises in predictive validation and stochastic sensitivity analysis.

One aspect of the above DBM approach which differentiates it from alternative deterministic, top-down approaches is its inherently stochastic nature and exploitation of powerful statistical methods of model identification and parameter estimation. This means that the uncertainty in the estimated model is always quantified, so that this information can then be utilized in various ways. For instance, it allows for the application of Monte Carlo-based uncertainty and sensitivity analysis, as well as the use of the model in statistical forecasting and data assimilation algorithms, such as the Kalman Filter. The uncertainty analysis is particularly useful because it is able to evaluate how the covariance properties of the parameter estimates affect the probability distributions of physically meaningful, derived parameters, such as residence times and partition percentages in parallel hydrological pathways (see for example, Young, 1992, 1999a and the example below).

9.4 STATISTICAL IDENTIFICATION, ESTIMATION, AND VALIDATION

The statistical approach to modelling assumes that the model is stochastic: in other words, no matter how good the model and how low the noise on the observational data happens to be, a certain level of uncertainty will remain after modelling has been completed. Consequently, full stochastic modelling requires that this uncertainty, which is associated with both the model parameters and the stochastic inputs, should be quantified in some manner as an inherent part of the modelling analysis. This statistical approach involves three main stages: *identification* of an appropriate, identifiable model structure; *estimation* (optimization, calibration) of the parameters that characterize this structure, using some form of estimation or optimization; and *conditional predictive validation* of the model on data sets different to those used in the model identification and estimation. In this section, we consider these three stages in order to set the scene for the later analysis. This discussion is intentionally brief, however, since the topic is so large that a comprehensive review is not possible in the present context. It should be noted that, at Lancaster, the statistical tools for the identification, estimation, and validation of DBM models are all available in the CAPTAIN Toolbox, a novel collection of computational algorithms designed for use in the MATLAB/Simulink software environment (see http://www.es.lancs.ac.uk/cres/captain/).

9.4.1 Structure and order identification

Identification, in a statistical context, normally means the data-based inference of the most appropriate model order, as defined in dynamic system terms, although the model structure itself can be the subject of the analysis if this is also considered to be ill-defined. Most importantly, the nature of linearity and non-linearity in the model is not assumed a priori (unless there are good reasons for such assumptions based on previous data-based modelling studies), but is identified from the data using non-parametric and parametric statistical-estimation methods.

Within the hydrological "top-down" context, linear system-identification analysis is related directly to problems such as the definition of how many storage zones (conceptual "storages" or "buckets") are required to characterize the data at the scale of interest; and how these submodels are inter-connected (i.e., in series, parallel, or feedback arrangements). It must be stressed, however, that such problems arise mainly from the specification of the dynamic model order (i.e., the order of the differential equations that are used to describe the major rainfall-flow dynamics; or equivalently, here, the number of storage zones). So a parsimonious model, in this important dynamic sense, is one that has a lowest dynamic order that is consistent with the information content in the data and whose parameters are statistically significant.

Of course, the DBM model may well have other parameters that are not associated primarily with the dynamic order of the model and are not so important in identifiability terms: for instance, coefficients that parameterize any non-linearity in the system (see below). Here again, however, the presence of such parameters in the model should be justified by whether or not they are statistically significant. The statistical significance of parameter estimates can be evaluated by conventional statistical tests or, in these days of the fast digital computer, by MCS and sensitivity analysis (see for example, Chapter 6 in Saltelli *et al.*, 2000, Chapter 7 in Beven, 2001, and Ratto *et al.*, 2007).

There are a variety of statistical methods for identifying model order. Fitting criteria, such as the coefficient of determination R_T^2, based on the simulated model errors (see later, Section 9.7), can be very misleading if used on their own, since over-parameterized models can "over-fit" the data. In general, therefore, it is necessary to exploit some specific order-identification statistics, such as the correlation-based statistics popularized by Box and Jenkins (1970), the well-known Akaike Information Criterion (AIC) (Akaike, 1974), and the YIC criterion proposed by the present author (Young, 1989). In all cases, the objective is to avoid over-parametrization by identifying a model structure and order that explains the data well within a minimal parametrization, i.e., "parsimonious models" (Box and Jenkins, 1970). The time-series methods used for model order identification in the present report are outlined in Young and Lees (1993), Young and Beven (1994), Young *et al.* (1996), and Young and Parkinson (2002).

In DBM modelling, identification also includes the initial data-based estimation of the location and nature of any significant non-linear phenomena that need to be included in the model. Fortunately, the recursive estimation procedures that figure so strongly in DBM modelling allow for the estimation of any significant parameter variation in the model and this can provide information on the presence of non-stationarity and non-linearity in the system dynamics. Here, the model parameters are estimated by the application of an approach to time variable parameter (TVP) or state-dependent parameter (SDP) estimation based on recursive fixed interval smoothing (FIS), e.g., Bryson and Ho (1969), Young (1984, 1999b, 2000). In the SDP case, the temporal variation in the parameters is further related to the variation of other measured variables or "states" that characterize the system, thus identifying the presence of non-linear behavior. In effect, the FIS algorithm provides a method of non-parametric SDP estimation, with the estimates defining a graph of the SDP against the associated state variable, which then defines the nature of the non-linearity (see Young, 1993; Young and Beven, 1994; Young, 1998a, 2000, 2001a, Young *et al.*, 2001). This approach to SDP estimation is illustrated in the later practical example.

9.4.2 Estimation (optimization or calibration)

Once the model structure and order have been identified, the parameters that characterize this structure need to be estimated in some manner. There are many automatic methods of estimation or optimization available in this age of the digital computer, from the simplest, deterministic procedures, usually based on the minimization of least-squares cost functions; to more complex numerical optimization methods based on statistical concepts, such as maximum likelihood (ML). In general, the latter are more restricted, because of their underlying statistical assumptions, but they provide a more thoughtful and reliable approach to statistical inference; an approach which, when used correctly, includes the associated statistical diagnostic tests that are considered so important in statistical inference.

In DBM modelling, if the model is identified as predominantly linear or piece-wise linear, then the constant parameters that characterize the identified model structure are estimated using advanced methods of statistical estimation for linear transfer function (TF) models. As shown in Appendix 1, TF models are simply representations of linear differential equations or their discrete-time (sampled-data) equivalents. Here, the preferred methods are the refined instrumental variable (RIV/SRIV) algorithms available in the CAPTAIN Toolbox, mentioned previously in this section. These provide a robust approach to model identification and estimation that has been well tested in practical application to hydrological systems over many years. Although based on ML estimation concepts, these algorithms are more robust to assumptions about the nature of the noise and uncertainty affecting the system. Full details of the methods are provided in Young and Jakeman (1979, 1980), Jakeman and Young (1979), Young (1984, 1985). They are also outlined in Young and Beven (1994), Young *et al.* (1996), and Young and Parkinson (2002).

If non-linear phenomena have been detected and identified by SDP estimation at the previous structure identification stage of the analysis, then the non-parametric, state-dependent relationships are normally parameterized in a finite form and the resulting non-linear model is estimated using some form of numerical optimization, such as non-linear least squares or ML based on prediction-error decomposition (Schweppe, 1965; Harvey, 1989). In the present flow-forecasting context, this approach to non-linear estimation is required only to define the nature of the effective rainfall non-linearity, which appears at the input to the model, as described in subsequent sections of this chapter.

9.4.3 Conditional predictive validation

Validation is a complex process and even its definition is controversial. Some academics (e.g., Konikow and Brederhoeft, 1992, within a groundwater context; Oreskes *et al.* 1994, in relation to the whole of the earth sciences) question even the possibility of validating models. To some degree, however, these latter arguments are rather philosophical and linked, in part, to questions of semantics: what is the "truth?" What is meant by terms such as validation, verification, and confirmation? etc. Nevertheless, one specific, quantitative aspect of validation is widely accepted, namely predictive validation, in which the predictive potential of the model is evaluated on data other than that used in the identification and estimation stages of the analysis.

Predictive validation implies that, on the basis of the new measurements of the model inputs (i.e., rainfall) from the validation data set, the model produces estimates and predictions of the flow that are acceptable within the uncertainty bounds associated with the model. Note this stress on the question of the inherent uncertainty in the estimated model: one advantage of statistical estimation, of the kind considered in this chapter, is that the level of uncertainty associated with the model parameters and the stochastic inputs is quantified in the time-series analysis. Consequently, the modeller should not be looking for perfect predictability (which no-one expects anyway) but predictability which is consistent with the quantified uncertainty associated with the model, as obtained in the prior estimation stage of the analysis.

9.5 THE DBM CATCHMENT MODEL

Within the catchment modelling context, DBM models are of two types: the non-linear rainfall-flow model; and the flow-routing model, which may be linear or non-linear, depending on the nature of the catchment and the flow or stage gauging. The complete model used in flood forecasting and warning applications is composed of both types linked in a manner that reflects the physical nature of the catchment under study. A recent example of such a quasi-distributed model is described in Romanowicz *et al.* (2006). In this chapter, however, we concentrate almost completely on the rainfall-flow component, with only a brief reference to flow routing. It must be emphasized, however, that this is not because flow routing is unimportant in real-time flood forecasting. It is simply that the advances reported in this chapter relate almost entirely to rainfall-flow modelling.

9.5.1 The rainfall-flow component

The first step in DBM modelling is the consideration of the objectives. In this case, it will be assumed that this is limited to obtaining a model which explains the rainfall-flow data well on an hourly or daily basis at the whole catchment scale and, at the same time, is capable of reasonable mechanistic interpretation combined with an ability to perform well in a flood forecasting/warning context.

Note that this emphasis on the "catchment scale" is very important because the hydrological significance and interpretation of the rainfall-flow models developed below all relate to catchment-scale characteristics, such as storage and flow partitioning. These models do not relate directly to more detailed characteristics such as flow paths in the field, analysis of soil depths, etc. Note also the allusion to the "rainfall-flow" relationship, rather than the use of the more conventional "rainfall-runoff" terminology. This is to emphasize that, as discussed below, the models considered here predict both storm runoff and base-flow, which are interpreted as the major components of the total gauged flow.

Based on these objectives, the most obvious and physically meaningful model form in this hydrological context is a continuous-time, differential equation (or set of equations). Such a model is consistent, for example, with the normal formulation of conservation equations and many conventional hydrological models, e.g., conceptual models of serial and parallel connected non-linear "buckets" or "storages," as discussed, for instance, in the top-down modelling of Jothityangkoon *et al.* (2001)[1], or the consideration of such models in a stochastic context by Young (2003). However, when dealing with discrete-time, sampled data, it is often convenient to consider modelling in terms of the discrete-time equivalent of the differential equation, the discrete-time TF. Appendix 1 provides some background on TF models and shows the link between continuous and discrete-time TF models. In the rest of the chapter, we concentrate on modelling in discrete-time terms but the treatment is very similar for continuous-time differential equation models estimated from discrete-time data, as discussed recently in Young (2004) and Young and Garnier (2006).

The most common discrete-time TF models are based on the assumption that the input and output data are time series measured at a sampling interval of Δt time units. Then, a sampled variable, say the flow, is denoted by y_k, where the subscript k denotes the value of the variable at the kth sampling instant, i.e., after $k\Delta t$ time units. Using this notation and the discrete-time TF model form of Equation (A.11) in Appendix 1, previous DBM modelling of rainfall-flow data based on SDP estimation (see the references in the previous section) has confirmed many aspects of earlier hydrological research and identified the non-linear DBM model structure shown in Figure 9.1[2]. Here, the two components of the TF model are the linear component, which models the basic, underlying, hydrograph behavior; and the nonlinear component, which models the relationship between the measured rainfall r_k and the effective rainfall u_k, so controlling the magnitude of the hydrograph contribution through time.

The resulting DBM model has the form (see Appendix 1):

$$y_k = \frac{B(z^{-1})}{A(z^{-1})}u_{k-\delta} + \xi_k. \tag{9.1}$$

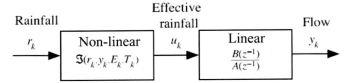

Figure 9.1 Block diagram of the generic DBM rainfall-flow model

Here, $u_{k-\delta}$ is the effective rainfall at time $(k - \delta)\Delta t$ time units previously, where δ is an "advective" time delay, i.e., the number of sampling intervals taken for the effective rainfall to have its first effect on the flow y_k. This advective time delay is normally defined as the nearest integer value of $\tau / \Delta t$, where τ is the advective time delay in time units (thus incurring a possible approximation error). The variable ξ_k represents the effects of all stochastic inputs, such as measurement inaccuracy, unmeasured stochastic disturbances, and modelling errors. In this equation, z^{-1} is the backward shift operator, i.e., $z^{-d} y_k = y_{k-d}$, while $A(z^{-1})$ and $B(z^{-1})$ are constant coefficient polynomials in z^{-1} of the following form:

$$A(z^{-1}) = 1 + a_1 z^{-1} + a_2 z^{-2} + \cdots + a_n z^{-n},$$
$$B(z^{-1}) = b_0 + b_1 z^{-1} + b_2 z^{-2} + \cdots + b_m z^{-m}. \qquad (9.2)$$

The structure (order) of the TF model Equation (9.1) is defined by the triad $[n\,m\,\delta]$ and this is normally identified from the data during the identification and estimation of the model, based on historical rainfall-flow data. This order is normally low, with $n \le 3$, $m \le 3$; while the value of δ is defined by the nature of the catchment and the location of the measurement devices, so its range is more difficult to define a priori.

The general TF model form $B(z^{-1})/A(z^{-1})$ in Equation (9.1) defines the input–output relationship between u_k and y_k and its unit impulse response is a scaled version of the underlying unit hydrograph. But, as we see later, it can also be decomposed into a parallel connection of lower-order processes. This decomposition not only makes the physical interpretation of the TF more transparent, it can also improve its performance in forecasting terms when implemented within a flood-forecasting system.

The non-linear component in Figure 9.1 takes the general form:

$$u_k = \mathcal{F}(r_k, y_k, E_k, T_k), \qquad (9.3)$$

where $\mathcal{F}(r_k, y_k, E_k, T_k)$ denotes an unknown, non-linear functional relationship defining the unobserved catchment-storage (soil-moisture) state s_k or, as we shall see later, some surrogate for this state, considered as a function of potentially important variables that may affect or be related to catchment storage. In addition to the rainfall r_k, this function may involve other relevant measured variables, such as the temperature T_k (or some function of this, such as the mean monthly temperature T_m), the potential evaporation E_k and the flow y_k, all of which could help to define the changes in soil moisture and storage if they are available.

Typically, the form of the non-linearity $\mathcal{F}(.)$ is initially identified from the rainfall-flow data through SDP estimation in non-parametric (graphical or "look-up" table) form, without any prior assumptions about its non-linear nature. This is then parameterized in some simple manner: for example, in Young (1993), Young and Beven (1994), and Young and Tomlin (2000), $\mathcal{F}(.)$ is defined as a power law $c y_k^\gamma$ in the flow y_k, with the power law exponent γ estimated from the data. In this case, the complete rainfall-flow model takes the very simple form:

$$y_k = \frac{B(z^{-1})}{A(z^{-1})} u_{k-\delta} + \xi_k, \quad u_k = c.f(y_k).r_k, \qquad (9.4)$$

where $f(y_k) = y_k^\gamma$. The fact that the non-linearity here $\mathcal{F}(.)$ is a function of the flow y_k seems very strange at first sight but the reason for this is discussed later in Section 9.6. The attraction of this SDP estimation approach is that $\mathcal{F}(.)$ is inferred from the rainfall-flow data and not assumed a priori, as in conceptual HMCs such as the Bedford–Ouse and IHACRES models, so leaving less room for unjustified over-confidence in the hypothetical definition of the non-linear model form.

The DBM model, even in the simple form of Equations (9.1) to (9.4), appears to have wide application potential. In addition to rivers in Australia (e.g., Young et al., 1997a,b) and the USA (Young, 2001b), it has been combined with an adaptive gain updating scheme in the parameter-adaptive Dumfries flood-warning system in Scotland (Lees et al., 1994), which has been operating successfully without major modification since 1991; and it is embedded within the Kalman Filter to provide a state-adaptive forecasting system for the River Hodder in north-west England (Young and Tomlin, 2000). It is also the basis for both state- and parameter-adaptive systems in Young (2002), for the River Hodder, and Romanowicz et al. (2004) for the River Severn from its source in Wales to Buildwas in England.

9.5.2 The flow-routing component

The generic flow (channel) routing model is much simpler than the rainfall-flow model since it is now widely accepted that linear TF models are often (but not always) adequate for the representation of flow dynamics in river systems. The discrete-time routing model for a single stretch of river consists of a serial connection of channel storage elements, each of which has the general form:

$$y_k^i = \frac{B_i(z^{-1})}{A_i(z^{-1})} y_{k-\delta_i}^{i-1} + \xi_k^i, \qquad i = 1, 2, \ldots, \text{nr}, \qquad (9.5)$$

where nr is the number of reaches and the i superscript denotes the reach number. Once again, it is assumed that the output of the routing model is affected by noise, here denoted by ξ_k^i. Equation (9.5) can be considered as the discrete-time equivalent of continuous-time differential equation storage equations (see Appendix 1). Normally, each of these elements is only first or second order (as defined by statistical identification and estimation based on the upstream and downstream flow data). The complete quasi-distributed catchment-routing model will consist of models such as this for the main river channels and all their tributaries within the catchment, connected accordingly, and it can involve any other measured in-flows as additions, inserted at appropriate nodal locations. The model will receive inputs from the rainfall-flow models discussed above and, in examples such as the Dumfries flood-warning model, from flow gauges far upstream which provide advance warning of impending flow changes. A typical early example of such a model is that used for studies of the Bedford–Ouse river system (Whitehead *et al.*, 1976), more recent examples are the simple River Wyre model (Young, 1986), the much more spatially complex Dumfries model (Lees *et al.*, 1994) and other models discussed in Cluckie (1993).

Note that we are restricting attention here to TF-based flow routing; this is not, of course, the only form of flow routing and other approaches are often utilized, although most of these can be considered in TF terms if this is desired (e.g., the "kinematic wave" model used in the Thames catchment model, see Greenfield, 1984, Moore and Jones, 1978). As in the case of the rainfall-flow models, flow-routing TF model parameters are normally obtained by the analysis of historical flow records using similar statistical identification and estimation methods to those used in the rainfall-flow example above.

If the catchment model is formulated in terms of the measured level (stage) rather than flow, then this not only avoids the need to utilize the stage-discharge relationship, which can introduce calibration errors, but it is also a more appropriate model form for flood forecasting, where the river level is the critical variable. In this connection, recent research has suggested that non-linear SDP transfer functions may be more appropriate than linear TFs for level–level routing, since they are better able to characterize the non-linearities associated with the changing speed (celerity) as the flood-wave moves along the river and the flood starts to inundate the flood plain.

9.6 PHYSICAL INTERPRETATION OF THE DBM MODEL

As we have stressed, an important aspect of DBM modelling is that the model can be interpreted in physically meaningful terms. In this regard, let us consider the model (9.4) which, as we shall see, is the model used in the Canning River example considered later in Section 9.8. As noted previously, the relationship $u_k = c \cdot f(y_k) \cdot r_k$ appears, at first sight, rather hard to justify in physical terms. However, it should not be taken literally and interpreted as saying that the effective rainfall is physically a function of flow, which is physically impossible. Rather, the measured flow y_k is

effectively acting here as an objectively identified surrogate for the catchment storage s_k. This seems sensible from a hydrological standpoint, since flow is clearly a function of the catchment storage and its pattern of temporal change is likely to be similar. So, the non-linear function as a whole is similar in its motivation to that used in the Bedford–Ouse and IHACRES conceptual HMC models and can be justified similarly in physical terms.

The delayed effective rainfall $u_{k-\delta}$ provides the input to the linear TF model component Equation (9.1). If this TF is greater than first-order and characterized by real eigenvalues (the roots of the $A(z^{-1})$ polynomial), as it normally will be, then the TF can be decomposed into a parallel pathway form, with first-order storage equations in each pathway (see Appendix 1 and the discussion on the physical interpretation of parallel TF models in Wallis *et al.*, 1989; Jakeman *et al.*, 1990; Young, 1992, 1993; Young and Beven, 1994; and Lees, 2000a,b). From this decomposition, it is possible to compute: the residence times (time constants); the advective time delays; the percentage partition of the flow down each of the storage pathways; and even the changing volumes associated with these pathways, all with obvious physical significance. When dealing with hourly data, there are usually two such pathways with very different dynamic characteristics. In the case of daily data, there can be three pathways, one of which represents an "instantaneous" response, i.e., a flow effect that arises within the same day of the effective rainfall occurence. For example, in the case of the Canning River example, using daily data, the best identified DBM model has a [2 3 0] TF, so that the model can be decomposed to the form:

$$y_k = 0.06 u_k + \frac{0.171 z^{-1}}{1 - 0.661 z^{-1}} u_k + \frac{0.026 z^{-1}}{1 - 0.942 z^{-1}} u_k + \xi_k.$$

(9.6)

Here, the first component on the right-hand side of the equation is the instantaneous effect, which accounts for only 5.9 % of the flow. The second component is the fast-pathway effect and this accounts for most of the flow, at 49.5 %, with a residence time of 2.4 days. The final component is the slow-pathway effect, which accounts for 44.6 % of the flow and has a residence time of 16.7 days.

Given these derived model parameters, the most obvious physical interpretation of the DBM is that the effective rainfall affects the river flow via two main pathways. First, the initial rapid rise in the hydrograph derives from the quick-flow pathway, probably as the aggregate result of the many surface and near-surface processes active in the catchment. And the long, elevated tail in the recession of the hydrograph arises from the slow-flow component, most likely the result of water displacement (probably of old water) from the storage within the groundwater system. Note that the estimate of the flow contribution of this slow-flow component is also practically useful in other ways: it provides a relatively objective estimate of the total base-flow in the river and, as such, can be utilized for base-flow quantification and removal, if this is required, as suggested by Jakeman *et al.* (1990). This contrasts with the classical unit hydrograph methods, where the base flow has to be removed rather subjectively.

Finally, it must be emphasized that the estimated TF and its decomposition are stochastic objects and so the uncertainty that is inherent in their derivation needs to be taken into consideration when interpreting the model in physically meaningful terms, as illustrated in the later Canning River example.

9.7 DATA ASSIMILATION: RECURSIVE STATE AND PARAMETER ESTIMATION

Within a flood-forecasting context, a catchment model based on rainfall-flow and flow-routing TF models should not be considered as an end in itself: rather, it is a major component of a data assimilation system that collects data from remote sensors within the catchment and "blends" these data with the model in a statistical manner to produce acceptable forecasts into the future. In this situation, the DBM model identified and estimated from the data with the objective of explaining the data may well not provide the best model for forecasting. The objective of forecasting is not simply to explain the data, but rather to maximize the forecasting performance over a specified lead time into the future. If we consider the Canning River model (9.6), for instance, it has no input time delay ($\delta = 0$) so, in order to forecast even one day ahead, it is necessary to also forecast the rainfall one day ahead which, in itself, is quite a difficult task. On the other hand, if we are able to introduce a time delay of one day into the model and still explain the data well, then there is no need for a rainfall forecast and the resulting flow forecasts are likely to be superior to those using the originally optimized model.

For illustrative purposes, let us consider the model formulation in terms of a single, second order [2 2 1] rainfall-flow model of the Canning, where there is a one day delay ($\delta = 1$) rather than the zero delay in Equation (9.6). The decomposed form of this estimated model is as follows:

$$y_k = \frac{0.185z^{-1}}{1 - 0.679z^{-1}} u_{k-1} + \frac{0.024z^{-1}}{1 - 0.946z^{-1}} u_{k-1} + \xi_k. \quad (9.7)$$

As we shall see later in Section 9.8, although this is not the best identified model, it still explains 94.9 % ($R_T^2 = 0.949^3$) of the Canning flow series over a two-year period, which is almost indistinguishable from the model (9.6), where $R_T^2 = 0.950$.

The model (9.7) could be used directly as a basis for real-time data assimilation and forecasting but, as we shall see, real-time updating is made much more straightforward and flexible if it is converted into a stochastic state-space form so that it can be embedded within a Kalman Filter (KF) state-estimation algorithm.

In the simplest situation, where the ξ_k is a white noise process e_k, with variance σ^2, the most obvious stochastic state space form of Equation (9.7) is:

$$x_k = \mathbf{F}x_{k-1} + \mathbf{G}u_{k-1} + \zeta_k, \quad (9.8)$$

$$y_k = \mathbf{h}^T x_k + e_k, \quad (9.9)$$

where:

$$\mathbf{F} = \begin{bmatrix} 0.679 & 0 \\ 0 & 0.946 \end{bmatrix} \mathbf{G} = \begin{bmatrix} 0.185 \\ 0.024 \end{bmatrix} \zeta_k = \begin{bmatrix} \zeta_{1,k} \\ \zeta_{2,k} \end{bmatrix} \mathbf{h}^T = \begin{bmatrix} 1 & 1 \end{bmatrix}. \quad (9.10)$$

In this manner, the state variables are defined as the unobserved (hidden or latent) quick and slow components of the flow, as defined by the decomposed, first-order TF relationships; and the output or "observation" Equation (9.9) combines these to form the complete flow output. For simplicity, the stochastic "system inputs" $\zeta_{1,k}$ and $\zeta_{2,k}$ are assumed to be zero mean, white-noise processes. They are introduced to allow for the inevitable uncertainty in the definition of the parallel-pathway dynamics and are an important aspect of this "state-adaptive" approach to forecasting.

For flow-forecasting purposes, the state-space model (9.8)–(9.10) is used as the basis for the implementation of the following, recursive, KF state-estimation and forecasting algorithm:

A priori prediction:

$$\hat{x}_{k|k-1} = \mathbf{F}\hat{x}_{k-1} + \mathbf{G}u_{k-1}, \quad (9.11)$$

$$\mathbf{P}_{k|k-1} = \mathbf{F}\mathbf{P}_{k-1}\mathbf{F}^T + \sigma^2 \mathbf{Q}_r, \quad (9.12)$$

$$\hat{y}_{k|k-1} = \mathbf{h}^T \hat{x}_{k|k-1}. \quad (9.13)$$

A posteriori correction:

$$\hat{x}_k = \hat{x}_{\hat{x}|k-1} + \mathbf{\Pi}_k \cdot \left\{ y_k - \hat{y}_{k|k-1} \right\}, \quad (9.14)$$

$$\mathbf{\Pi}_k = \mathbf{P}_{k|k-1}\mathbf{h}[\sigma^2 + \mathbf{h}^T\mathbf{P}_{k|k-1}\mathbf{h}]^{-1}, \quad (9.15)$$

$$\mathbf{P}_k = \mathbf{P}_{k|k-1} - \mathbf{\Pi}_k\mathbf{h}^T\mathbf{P}_{k|k-1}, \quad (9.16)$$

$$\hat{y}_k = \mathbf{h}^T \hat{x}_k. \quad (9.17)$$

In these equations, $\mathbf{P}_k$ is the error covariance matrix associated with the state estimates; and $\mathbf{Q}_r$ is the 2×2 noise variance ratio (NVR) matrix defined in the next subsection.

9.7.1 State adaption

In the above KF equations, the DBM model parameters are known initially from the model identification and estimation analysis based on the estimation data set. However, by embedding these model equations within the KF algorithm, we have introduced additional, unknown parameters, normally termed "hyper-parameters" to differentiate them from the model parameters.[4] In this example, these hyper-parameters are the elements of the noise variance ratio (NVR) matrix $\mathbf{Q}_r$ and, in practical terms, it is normally sufficient to assume that this is purely diagonal in form. These two diagonal elements are defined as $NVR_i = \sigma^2_{\zeta_i}/\sigma^2$, $i = 1, 2$; they specify the nature of the stochastic inputs to the state equations and so define the level of uncertainty in the evolution of each state (the quick- and slow-flow states respectively) relative to the measurement uncertainty. The inherent state adaption of the KF arises from the presence of the NVR parameters since these allow the estimates of the state variables to be adjusted to allow for presence and effect of the unmeasured stochastic input disturbances.

Clearly, the NVR hyper-parameters have to be estimated in some manner on the basis of the data. One well-known approach is to exploit maximum likelihood (ML) estimation based on prediction error decomposition (see Schweppe, 1964; Harvey, 1989; Young, 1999b). Another, is to assume that *all* the parameters of the state-space model (9.8, 9.9) are unknown and re-estimate them by minimizing the variance of forecasting errors over the specified forecasting interval. In effect, this optimizes the memory of the recursive estimation and forecasting algorithm (see Young and Pedregal, 1999b) in relation to the rainfall-flow data. In this numerical optimization, the multi-step-ahead forecasts $\hat{y}_{k+f|k}$, where f is the forecasting horizon, are obtained by simply repeating the prediction step in the algorithm f times, without intermediate correction. The main advantage of this latter approach is, of course, that the integrated model-forecasting algorithm is optimized directly in relation to the main objective of the forecasting system design; namely the minimization of the multi-step prediction errors.

9.7.2 Parameter adaption

Although the parameters and hyperparameters of the KF-based forecasting engine can be optimized in the above manner, we cannot be sure that the system behavior may not change sufficiently over time to require their on-line, real-time adjustment. In addition, it is well known that the measurement noise e_k is "heteroscedastic," i.e., its variance can change quite radically over time, with much higher variance occurring during storm events. For these reasons, it is wise to build some form of parameter adaption into the forecasting algorithm.

9.7.2.1 GAIN ADAPTION
It is straightforward to update *all* of the parameters in the rainfall-flow model since the estimation algorithms can be implemented in a recursive form that allows for sequential updating and the estimation of time-variable parameters (Young, 1984). However, this adds complexity to the final forecasting system and previous experience suggests that a simpler solution, involving a scalar gain adaption, is often sufficient. This is the approach that has been used successfully for some years in the Dumfries flood-warning system (Lees *et al.*, 1994) and is also discussed by Cluckie (1993), who refers to it as a "scale-factor" updating method. It involves the

recursive estimation of the gain $g(k)$ in the following relationship:

$$y_k = g_k \cdot \hat{y}_k + \varepsilon_k, \tag{9.18}$$

where ε_k is a noise term representing the lack of fit and, in the case of the model (9.4):

$$\hat{y}_k = \frac{\hat{B}(z^{-1})}{\hat{A}(z^{-1})} u_{k-\delta}, \quad u_k = c y^{\hat{\gamma}} r_k, \tag{9.19}$$

where the "hats" indicate estimated values. In other words, the time-variable scalar gain parameter g_k is introduced so that the model gain can be continually adjusted to reflect any changes in the steady-state (equilibrium) response of the catchment to the effective rainfall inputs.

The associated recursive estimation algorithm for g_k takes the usual recursive least-squares (RLS) form in the case where g_k is assumed to vary stochastically as a random-walk (RW) process (e.g., Young, 1984):

$$p_{k|k-1} = p_{k-1} + q_g, \tag{9.20}$$

$$p_k = p_{k|k-1} - \frac{p^2_{k|k-1}\hat{y}^2_k}{1 + p_{k|k-1}\hat{y}^2_k}, \tag{9.21}$$

$$\hat{g}_k = \hat{g}_{k-1} + p_k\hat{y}_k\{y_k - \hat{g}_{k-1}\hat{y}_k\}, \tag{9.22}$$

where $\hat{g}_k$ is the estimate of g_k, while q_g is the NVR defining the stochastic input to the RW process, the magnitude of which needs to be specified or optimized (see later). The adapted forecast is obtained by simply multiplying the initially computed model output $\hat{y}_k$ by $\hat{g}_k$. Note that gain adaption of this kind is quite generic and can be applied to *any* model, not just those discussed here.

9.7.2.2 VARIANCE ADAPTION
To allow for the heteroscedasticity in e_k, it is necessary to recursively estimate[5] its changing variance σ^2_k. Although a logarithmic transform might suffice, a superior approach is to use the transformation $c_k = \log(\chi^2_k) + \lambda$, where the stochastic process χ^2 is defined by:

$$\chi^2_m = \left(e^2_{2m-1} + e^2_{2m}\right)/2, \quad m = 1, \ldots N/2, \tag{9.23}$$

in which $\lambda = 0.57722$ is the Euler constant. This is motivated by Davis and Jones (1968), who showed that c_k has a theoretical distribution that is almost normal. As a result, an estimate $\hat{h}_k$ of the transformed variance can be obtained from the following recursive least-squares algorithm (cf. the above RLS algorithm for $\hat{g}_k$), where this time it is c_k that is assumed to vary stochastically as a RW process:

$$p_{k|k-1} = p_{k-1} + q_h, \tag{9.24}$$

$$p_k = p_{t|t-1} - \frac{p^2_{k|k-1}}{1 + p_{k|k-1}}, \tag{9.25}$$

$$\hat{h}_k = \hat{h}_{k-1} + p_k\{c_k - \hat{h}_{k-1}\}. \tag{9.26}$$

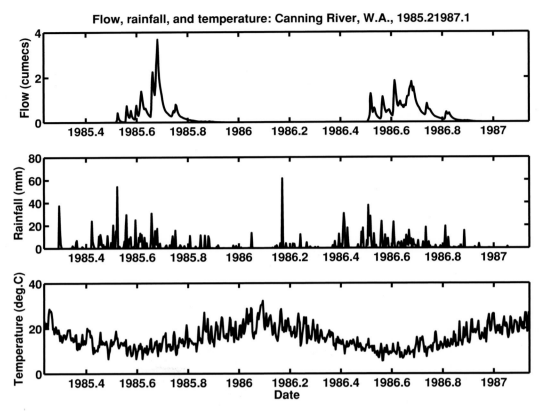

Figure 9.2 Two years of daily flow (y_k, upper panel), rainfall (r_k, middle panel) and air temperature (T_k, lower panel) for the ephemeral Canning River at Glen Eagle in south-western Australia

An estimate $\hat{\sigma}_k^2$ of σ_k^2 can then be obtained as $\hat{\sigma}_k^2 = \exp(\hat{h}_k - \lambda)$.

An alternative to algorithm (9.24)–(9.26) is the data-based identification and estimation of an SDP model for the changing variance. For instance, in Romanowicz *et al.* (2006), the variance σ_k^2 is identified as an SDP function of a simulated output, taking the following form:

$$\sigma_k^2 = \lambda_0 + \lambda_1 \hat{y}_k^2 \qquad (9.27)$$

where λ_0 and λ_1 are parameters that are optimized, together with the NVR hyper-parameter that controls the gain updating procedure discussed in Subsection 9.7.2.1. This has the advantage that the variance update is immediately effective as flow changes occur and the inherent "estimation lag" associated with the "filtering" algorithm (9.24)–(9.26) is eliminated.

9.7.2.3 HYPER-PARAMETER ESTIMATION
The above RLS estimation algorithms are, in fact, very simple examples of the KF and so it is necessary to estimate the hyper-parameters (here q_g and q_h) in some manner. Their joint estimation with the KF hyper-parameters is straightforward. However, a simpler, heuristic alternative derives from the fact that q_g and q_h control the memory of their respective RLS estimation algorithms

and the associated smoothing of the estimate (e.g., Young, 1984). Consequently, since q_g and q_h are scalar values, it is not difficult to optimize them manually to yield the best multi-step-ahead forecasts.

9.8 ILLUSTRATIVE PRACTICAL EXAMPLE

Figure 9.2 shows a portion of effective rainfall $u(t)$, flow $y(t)$, and air temperature $T(t)$ data from the Canning River at Glen Eagle in south-western Australia over the period March 23, 1985 to February 26, 1987. The Canning is a 544 km^2, semi-arid, benchmark catchment whose discharge is dominated by zero flows, with zero-flow periods occupying more than half of the recorded period. The rainfall is winter dominated, with the winter four months receiving approximately 70 % of the total annual rainfall. These data have been analyzed previously by Young *et al.* (1997a) and Ye *et al.* (1998). The latter reference uses a modified version of the IHACRES model. The present DBM modelling results are similar to those in the former reference, but the analysis has been improved in various ways, so the results are somewhat different.

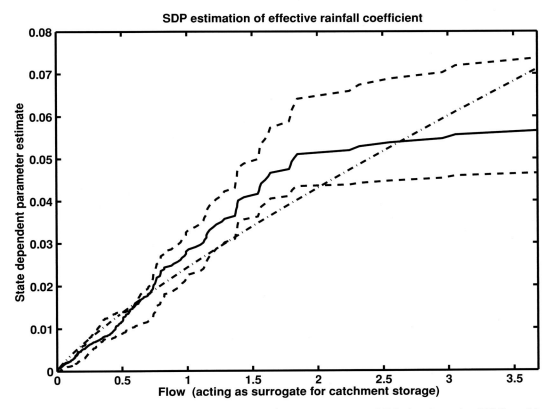

Figure 9.3 SDP non-parametric estimation of the effective rainfall nonlinearity: SDP estimate (full line) and associated 95 % confidence bounds (dashed lines). Also shown (dash-dot line) is the parametric estimate of the nonlinearity based on a power law parameterization, as obtained at the final parameter estimation phase of the analysis

9.8.1 DBM modelling

In this example, the time-series data are plentiful so the first stage in DBM modelling is model structure identification, based on the estimation data set in Fig. 9.2.[6] Initial linear identification, using the RIV estimation algorithm, suggests a [2 3 0] discrete-time TF model, but its explanation of the data is poor, implying that a non-linear model is required, as would be expected on physical grounds. TVP estimation confirms this and previous experience, showing that a time-variable parameter model is able to explain the data well. As a result, the identification proceeds further using SDP estimation, as outlined in previous sections. Figure 9.3 shows the non-parametric estimation results obtained in this manner.

The estimated shape of the effective rainfall non-linearity is similar to that obtained in most previous rainfall-flow modelling studies and the DBM model based on these non-parametric results explains 94.7 % of the flow data ($R_T^2 = 0.947$). Again, based on previous experience, a power law seems to be reasonable contender for parameterization of the SDP nonlinearity in Fig. 9.3, although it is clearly not the only possibility. Parametric estimation proceeds with this definition and yields the following parameter estimates:

$$\hat{a}_1 = -1.6031(0.012); \quad \hat{a}_2 = 0.6228(0.011); \quad \hat{\gamma} = 0.823(0.09);$$
$$\hat{b}_1 = 0.0601(0.005); \quad \hat{b}_2 = 0.100(0.010); \quad \hat{b}_3 = -0.1405(0.006);$$
$$\hat{\beta}_1 = 0.171; \quad \hat{\alpha}_1 = 0.661; \quad \hat{\beta}_2 = 0.026; \quad \hat{\alpha}_2 = 0.942;$$

where the figures in brackets are the estimated standard errors on the associated parameter estimates. The decomposed form of this model has been presented previously in Equation (9.6). Note that the derived parameters α_i, β_i, $i = 1, 2$ associated with the decomposed model have no specified standard-error bounds because they are not estimated directly. However, we consider the uncertainty in such derived parameters later. It will be noted also that, although the parametric estimate in Fig. 9.3 (dash-dot line) does not entirely reflect the shape of the non-parametric estimate (solid line), it does lie mostly within the associated 95 % confidence region. Consequently, given that the parametric estimate is also uncertain (bounds not shown to avoid confusion), it is clear that the two estimates are statistically compatible. Of course, the parametric estimate is more statistically efficient.

The simulated output of the estimated DBM model is shown as the full line in Fig. 9.4; it clearly explains the data (solid points)

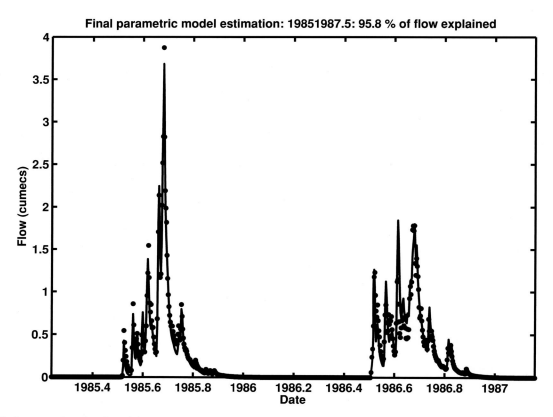

Figure 9.4 Final parametric estimation of the DBM model: comparison of the simulated DBM model output (full line) and the measured flow data (circular points).

well, consistent with the high $R_T^2 = 0.958$. The decomposition of the model has been discussed above in Section 9.6 and Fig. 9.5 shows the estimated flows through the different pathways, all plotted to the same scale. We see that the instantaneous effect (within one day) plotted in the top panel is very small, as would be expected. The quick and slow components explain much more of the flow response and it is clear that it is the latter slow pathway that leads to the extended "tail" on the associated hydrograph and can be associated with the slowly changing "base flow" in the river.

The residual noise in the estimated DBM model, ξ_k, is quite complex and heteroscedastic. On the basis of the AIC, it is identified as a 25th order autoregressive, i.e., AR(25), model. This is a rather high-order model, but it is not too unusual in rainfall-flow modelling because of the complex nature of the residuals, which tend to contain evidence of "outliers" (see below). The final residual (one-step-ahead prediction error) of the model incorporating this AR(25) noise model model, is $\sigma^2 = 0.0031$, which yields a coefficient of determination based on these residuals of $R_T^2 = 0.983$, i.e., the complete stochastic model, explains 98.3 % of the flow data. The auto and partial autocorrelation analysis of the final residuals confirms that there is no significant serial correlation left in the series and that they are sensibly white noise, as

required. However, they have significantly changing variance (heteroscedastic white noise), varying from zero when there is no flow to the highest variance when the flows are of maximum magnitude. This suggests that some small improvement in the modelling might accrue from an analysis in which this heteroscedasticity is taken into account, e.g., by modelling the heteroscedasticity and incorporating this in the estimation of the model parameters.

Other statistical diagnostic checks on the final model residuals yield reasonable results, given the nature of the data. For instance, the cross-correlation function (CCF) between the residuals and the daily temperature series T_k shows that there is no significant correlation and suggests that, in the case of the Canning River data, there is no evidence for any significant temperature-related terms in the model, either in the non-linear effective rainfall function; or as an additive term in the TF model. In other words, it would appear that using the flow as a surrogate measure for catchment storage in the effective rainfall non-linearity has removed the need to explicitly model any evapotranspirative effects based on temperature. However, the CCF between the residuals and the input rainfall r_k suggests there is some significant correlation left with the rainfall: this is quite usual with rainfall-flow models and, once more, it probably arises in this case because of the numerous

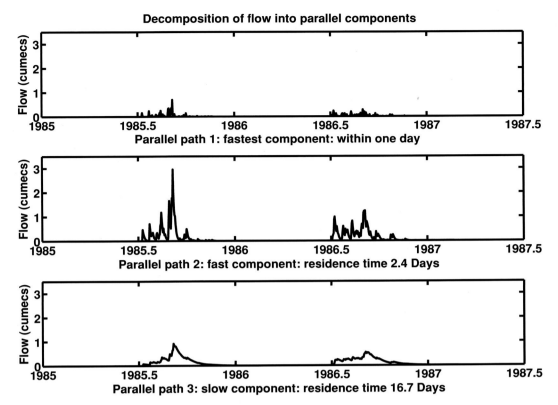

Figure 9.5 Decomposition of the Canning River flow into parallel instantaneous (top panel), quick (middle panel) and slow (bottom panel) flow components, based on the decomposition of the transfer function in the estimated DBM model

significant outliers. This problem could be obviated by simply assuming that the flow measurements at these outlier samples are missing and replacing them by their expectations. This is quite straightforward when using recursive estimation (see for example, Young, 1984), but has not been attempted in this case since it might impair the model's ability to characterize the high-flow episodes that are important in flood forecasting.

The most valuable evaluation of modelling results is predictive validation on data not used in the identification and estimation analysis. Figure 9.6 shows two typical predictive validation results obtained by using the estimated DBM model to simulate flow (from the rainfall alone) over other time periods where data are available: the left panel for the period between 1977 to 1978 and the right panel between 1979 to 1980. In the first case, the associated $R_T^2 = 0.954$, while in the latter case, this reduces to $R_T^2 = 0.928$. Clearly, these are very respectable results and they are similar to those obtained from predictive validation analysis carried out over other parts of the data set. They provide confidence in the estimated model, showing that it is robust and represents very well the short- and long-term aspects of the rainfall-flow behavior in the catchment over all of the available data.

One distinct advantage of the stochastic DBM model is the information it provides on the estimated uncertainties in both the data and the model parameters. Figures 9.7 and 9.8 illustrate how this can help in evaluating the model further. Figure 9.7 shows the estimated uncertainty in the derived residence time parameters for the two main parallel pathways, as revealed by a normalized histogram of the parameters derived from monte carlo simulation (MCS) analysis. This analysis involved 2500 stochastic realizations based on the estimated covariance matrix of the model parameters: we see that both residence times are reasonably well defined probabilistically and the distributions are roughly Gaussian in shape, with perhaps a slight skew to larger values in the case of the slow residence time in the left panel.

The same MCS analysis includes an evaluation of the predicted uncertainty associated with the model output. Here, an innovation is the inclusion of a model for the additive heteroscedastic noise (see also the next subsection). This was obtained by SDP estimation applied to the explanation of the residual variance as a function of the flow y_k and the resulting model was used to generate heteroscedastic noise with these modelled properties in the MCS realizations. We see in Fig. 9.8 that the useful effect of this, in the case of the second predictive validation data set, is to specify the standard-error bounds on the DBM model output, bounds that change in width as a function of the flow magnitude and show that the prediction of the peak flow, although in error as regards the

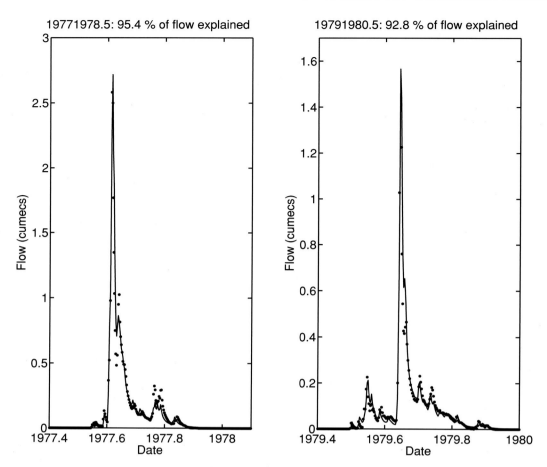

Figure 9.6 Predictive validation of the DBM model based on two sets of validation data: simulated model output (full line) compared with validation flow data (dots)

measured value (which, of course, may itself be incorrect because high-flow measurements are often subject to measurement inaccuracy) lies within the estimated uncertainty bounds.

9.8.2 Adaptive forecasting

The main objective of DBM modelling in this example is real-time forecasting. As mentioned in Section 9.7, therefore, the DBM model developed in the previous section, although best at explaining the rainfall-flow data, is not the most suitable model for this application because it requires one-day-ahead forecasting of the rainfall input. A superior alternative is to insert a "virtual" one-day delay into the DBM model, as discussed in Section 9.7. The re-estimated parameters of this DBM forecasting model, based on the estimation data set, are as follows:

$\hat{a}_1 = -1.6243(0.018);\quad \hat{a}_2 = 0.6417(0.016);\quad \hat{\gamma} = 0.777(0.096);$
$\hat{b}_1 = 0.209(0.004);\quad \hat{b}_2 = -0.191(0.004);$
$\hat{\beta}_1 = 0.185;\quad \hat{\alpha}_1 = 0.679;\quad \hat{\beta}_2 = 0.024;\quad \hat{\alpha}_2 = 0.946;$

and the decomposed form of the model has already been cited in Equation (9.7), with its most useful stochastic state space formulation in equations (9.8), (9.9), and (9.10). This model has an $R_T^2 = 0.949$, very little different to the original DBM model without the time delay. It also passes all the statistical diagnostic tests (except, as before, the CCF between the residuals and the rainfall) and, most importantly, it performs very well in real-time, one-day-ahead forecasting, as shown in Fig. 9.9. This is a plot of the one-day-ahead forecasts for the first validation data set, as obtained from a KF forecasting engine based on an adaptive version of the model.

These forecasts have a coefficient of determination, based on the one-step-ahead forecasting errors of $R_T^2 = 0.918$, which is the sort of value one would expect given the DBM modelling results. Parameter adaption is exploited, as described in Section 9.7. The optimized NVRs, based on minimizing the variance of the one-day-ahead forecasting errors associated with the adaptive parameters are $q_g = 0.0001$ and $q_h = 0.169$. The resulting adaptive gain parameter is plotted in Fig. 9.10 and we see that the

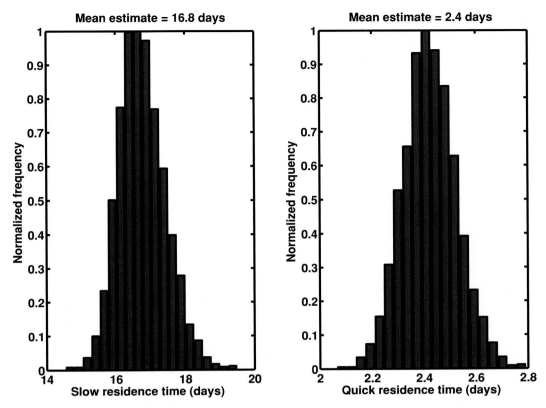

Figure 9.7 Uncertainty in the derived residence time parameters of the DBM model based on stochastic Monte Carlo analysis: slow residence time (left panel); quick residence time (right panel)

adaptive adjustment is very small (the non-adaptive value is unity) and makes very little difference in this case. However, this is fortuitous and parameter adaption should always be implemented to handle unforseen circumstances. On the other hand, the adaptive variance parameter makes considerable difference to the results, a fact that is most visible in the changing width of the 95 % confidence band in Fig. 9.9.

Finally, note that the state-space model used to obtain the above forecasting results does not contain any states associated with the identified additive noise ξ_k which, it will be recalled, is identified as an AR(25) stochastic process. This noise model could have been embedded in the state-space model and it should improve the forecasts to some extent. However, since it would also considerably enlarge the state dimension (adding 25 more stochastic state variables) and so add complexity to the forecasting system, it was not included in the present illustrative analysis.

9.9   CONCLUSIONS

This chapter describes some recent advances in stochastic modelling and forecasting that provide the basis for the implemen-

tation of real-time flow and flood-forecasting/warning systems. It argues that deterministic reductionist (or "bottom-up") simulation models are inappropriate for real-time forecasting because of the inherent uncertainty that characterizes river-catchment dynamics and the problems of model over-parametrization that are a natural consequence of the reductionist philosophy. The advantages of alternative data-based mechanistic (DBM) models, statistically identified, estimated, and validated in an inductive manner directly from rainfall-flow data, are discussed. In particular, the chapter shows how non-linear, stochastic, transfer-function models can be developed using powerful methods of recursive time-series analysis. Not only are these models able to characterize well the rainfall-flow dynamics of the catchment in a parametrically efficient manner, but, by virtue of the DBM modelling strategy, they can also be interpreted in hydrologically meaningful terms. Most importantly in the forecasting context, the models are also in an ideal form for incorporation into a data assimilation and forecasting engine based on a special, adaptive version of the Kalman Filter algorithm.

The practical example described in the chapter demonstrates how, with sufficient rainfall-flow data and no available rainfall forecasts, the approach proposed here can generate useful flow

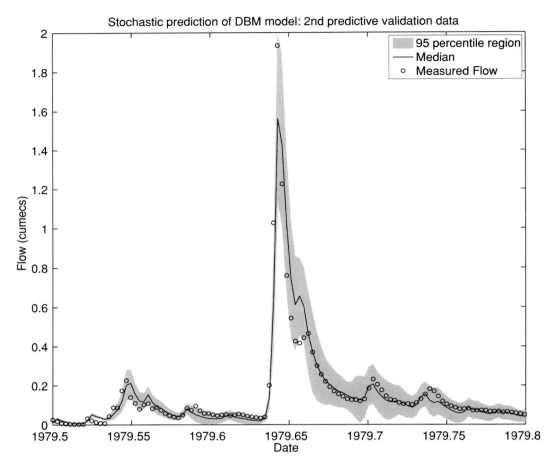

Figure 9.8 Validation of the DBM model by stochastic Monte Carlo simulation based on the 2nd validation data set: mean model output (full line), 95 % confidence bounds (gray areas) and measured flow (open circles)

forecasts for one day ahead in a semi-arid catchment; forecasts that could form the basis for flood-warning system design. Such a system is a natural development of the DBM model-based Dumfries flood-warning system (Lees *et al.*, 1994), which has been operating successfully without major modification since 1991. A more sophisticated flood forecasting system for the River Severn in the UK, based on the approach described in the present chapter, is described in Romanowicz *et al.* (2006). Both of these recursive approaches to real-time flood forecasting can be contrasted with more conventional, non-recursive, real-time forecasting procedures proposed previously. A typical example is the adaptive scheme suggested by Brath and Rosso (1993), which addresses some of the same statistical issues raised in the present chapter. However, it operates on an event basis rather than continuously; it uses repeated en-bloc optimization rather than recursive estimation; it is based on a simple conceptual model with a priori assumed structure and parameterization, and it is computationally much more demanding.

Of course, there remain a number of methodological problems still to be solved. The DBM models discussed in the chapter perform well, but they cannot be considered completely satisfactory while the model residuals retain some mildly unsatisfactory statistical characteristics. In particular, the correlation remaining between the residuals and the rainfall input shows that the model is still not fully explaining the complete rainfall-flow process (although the remaining unexplained variance represents only a small proportion of the total variance). This limitation of the current DBM models (shared, the author believes, by *all* current rainfall-flow models, whatever their type) is almost certainly due to deficiencies in the rainfall-flow data, as well as the effective rainfall non-linearity (and possibly the presence of other, smaller non-linearities in the system, as yet unquantified). There is clear need for more research on this fascinating subject and, although such research would require the analysis of a large and comprehensive rainfall-flow database covering a wide array of different catchment behavior, it would provide useful information for *all*

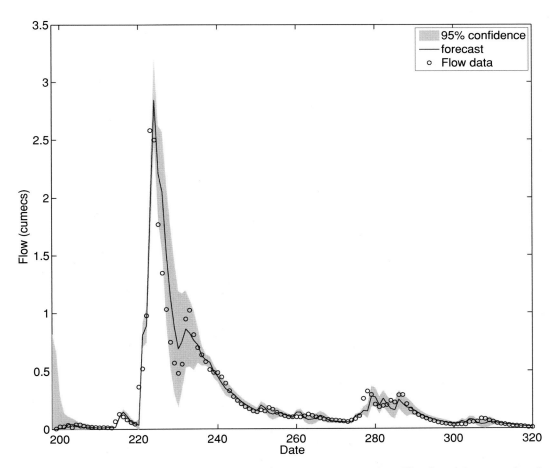

Figure 9.9 One-day-ahead forecasts for the 1st validation data set produced by the adaptive Kalman Filter forecasting system based on the DBM forecasting model: forecast (full line), 95 % confidence bounds (gray area) and flow data (open circles points)

existing rainfall-flow modelling studies, not just those discussed in this chapter.

Finally, it is interesting to compare the recursive DBM approach to hyrological modelling described in this chapter with alternative "numerical Bayesian" approaches, such as the Ensemble Kalman Filter that utilize Monte Carlo Simulation within a recursive setting, as exemplified in a number of recent papers (e.g., Moradkhani *et al.*, 2005; Thiemann *et al.*, 2001, see also the associated comment by Beven and Young, 2003). While it is clear that such methodology has the advantage of considerable generality in stochastic terms, it is computationally very intensive when compared with the DBM methodology. For instance, while the DBM calculations, as described in the present chapter, takes but a few seconds on a typical desktop computer, the Bayesian approach can occupy a much larger amount of computer time because of the need for generating many Monte Carlo realizations at each recursion. Consequently, it should only be used when a simpler alternative is not appropriate. In the case of DBM rainfall-flow and flow-

routing models, however, the identified structure is fairly simple, with the only significant non-linearity being that associated with the effective rainfall, which is situated at the input to the rainfall-flow model. For this reason, it is possible to use standard, computationally simple, recursive estimation methods, such as those described in this chapter, and a numerical Bayesian approach does not seem justified.

Also, the numerical Bayesian approach can obscure identifiability issues. For instance, it is often applied to large conceptual models that are not inherently identifiable, so that it is necessary to constrain some of the parameters so that they cannot vary too much away from the user-specified prior probability distributions (the stochastic equivalent of constraining parameters to a priori specified values). Even in the case of relatively small conceptual models, it is often necessary to constrain parameters in this manner. For instance, the Sacramento soil moisture accounting (SAC) model has 16 user-specified parameters in addition to the unit hydrograph routing (which is assumed known

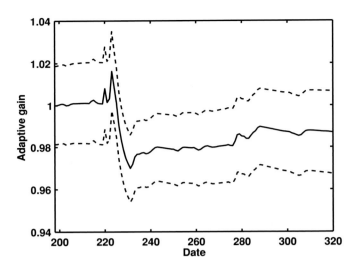

Figure 9.10 Adaptive gain estimate associated with the forecasting results in Fig. 9.9: recursive estimate of gain parameter (full line) and estimated standard error bounds (dashed lines)

beforehand). Of these 16, 13 are assumed to be known and are fixed, so that the remaining 13 can be estimated by calibration (*pers. comm*: H. Gupta (pers. comm.), see also, Thiemann *et al.*, 2001). Whilst this may be justified in a simulation modelling context, where the large model is required for "what-if" scenario studies, it is not so well justified in a forecasting context, where the parsimonious DBM model appears to be entirely adequate for forecasting and data assimilation, as shown both in the present chapter and Romanowicz *et al.*, (2006).

9.10   ACKNOWLEDGEMENTS

The author is grateful to his colleagues Professor Keith Beven, Dr. Paul McKenna and Dr. Renata Romanowicz for reading and commenting on a draft of this chapter. Naturally, the author is responsible for any errors or omissions.

APPENDIX 1

A1.1   TRANSFER FUNCTION (TF) MODELS

Most applications of TF models in hydrology tend to use discrete-time TF models because most of the literature on the identification and estimation of TF models concentrates on this form.[7] However, as we shall see, the continuous-time TF model is more transparent and immediately interpretable in a physically meaningful manner. In order to introduce TF models, therefore, let us consider first a conceptual

catchment storage equation in the form of a continuous-time, linear storage ("bucket," "tank" or "reservoir") model, see, for example, the review papers by O'Donnell, Dooge, and Young in Kraijenhoff and Moll (1986) or more comprehensive treatments, such as the recent books by Beven (2001) and Dooge and O'Kane (2003). Here, the rate of change of storage in the channel is defined in terms of water volume entering the linear storage element (e.g., river reach) in unit time, minus the volume leaving in the same time interval, i.e.:

$$\frac{dS(t)}{dt} = G Q_i(t - \tau) - Q_o(t) \tag{A.1}$$

where $Q_i(t - \tau)$ represents the input flow rate delayed by a pure time or "transport" delay of τ time units to allow for pure advection and G is a gain parameter inserted to represent gain (or loss) in the system. Making the reasonable and fairly common assumption that the outflow is proportional to the storage at any time, i.e.:

$$Q_o(t) = \rho S(t) \tag{A.2}$$

and substituting into (A.1), we obtain:

$$T \frac{dQ_o(t)}{dt} = G Q_i(t - \tau) - Q_o(t). \tag{A.3}$$

This equation is a first-order, linear differential-equation model whose response, from an initial steady-flow condition, to a unit impulsive change Q_i^{imp} of the input flow at time $t = t_0$, is given by:

$$Q_o(t) = Q_e + Q_i^{imp} e^{-(t-t_0)/T}, \tag{A.4}$$

where Q_e is the initial steady flow level. This has a typical hydrograph recession shape, with a decay time constant, T, that defines the residence time of the model. As we shall see later, combinations of two or more such first-order models, exhibit a typical unit hydrograph form (e.g., Dooge, 1959; Beven, 2001).

By introducing the derivative operator s, i.e., $s = d/dt$, and collecting like terms together, it is easy to see that Equation (A.3) can be written as:

$$(1 + Ts)Q_o(t) = G Q_i(t - \tau) \tag{A.5}$$

so that, dividing throughout by $1 + Ts$, we obtain the following continuous-time TF form of Equation (A.3):

$$Q_o(t) = \frac{G}{1 + Ts} Q_i(t - \tau), \tag{A.6}$$

where:

$$H(s) = \frac{G}{1 + Ts} \tag{A.7}$$

represents the TF in terms of the derivative operator s.[8]

A1.1.1 Physically interpretable parameters

The TF model (A.6) is characterized by three parameters: G, T, and τ. However, there are five physically interpretable model parameters associated with the model that are worth discussing. The steady-state gain (SSG), denoted by G, is obtained by setting the s operator in the TF to zero (i.e., $d/dt = 0$ in a steady state). It shows the relationship between the equilibrium output and input values when a steady input is applied. For this reason, if the input and output have similar units, G is ideal for indicating the physical losses or gains occurring in the system. In the case of a flow-routing model, for example, it indicates whether water has been added ($G > 1$) or lost ($G < 1$) between the upstream and downstream boundaries; and the percentage of water lost or gained can be defined by loss efficiency $LE = 100(1 - G)$, which will be negative if $G > 1.0$. As pointed out above, the residence time or time constant, T, is the time required for the storage element output to decay to e^{-1} or 0.3679 of its maximum value in response to an impulsive input. Finally the pure advective time delay, τ, indicates the time it takes for a flow increase upstream to be first detected downstream, and $T_k = T + \tau$ defines the travel time of the system. These five parameters typify the equilibrium and dynamic characteristics of the TF model and provide a physical interpretation of the TF model in terms of its mass transfer and dispersive characteristics.

A1.1.2 TF manipulation and block diagrams

The first order TF model (A.6) is often written in the form:

$$Q_o(t) = \frac{g_0}{s + f_1} Q_i(t - \tau), \quad \text{i.e.} \quad H(s) = \frac{g_0}{s + f_1}, \quad \text{(A.8)}$$

where:

$$g_0 = \frac{G}{T}, \qquad f_1 = \frac{1}{T}, \quad \text{(A.9)}$$

because this is the form in which the model is normally estimated (see later). Typically, a channel- or flow-routing model for a river catchment will contain a number of elemental models, such as (A.6) or (A.8), connected in a manner that relates to the structure of the catchment. For instance, a serial connection of n such elements constitutes the lag-and-route model of a single river channel (Meijer, 1941; Dooge, 1986) and, with $\tau = 0$ and all elements identical, it becomes the well-known "Nash Cascade" model (Nash, 1959). More complex river systems can be represented by a main channel of this type, with tributaries modelled in a similar manner. Also, a typical TF model between effective rainfall and flow often contains a parallel connection of two or more such storage elements (see Young, 1992, 2001a, b, 2002, 2003).

The TF formulation allows for the visual representation of a total system model in the form of a "systems block diagram." Figure A1.1 is a typical example of such a diagram that represents a catchment model consisting of an effective rainfall-flow submodel involving two different first order TFs of the form (A.6) with $\tau = 0$:

$$H_1(s) = \frac{G_1}{1 + T_1 s}, \qquad H_2(s) = \frac{G_2}{1 + T_2 s}, \quad \text{(A.10)}$$

that are connected in parallel. The flow output of this submodel forms the input to a flow-routing submodel composed of two identical, first-order TFs:

$$H_3(s) = \frac{G_3}{1 + T_3 s}, \quad \text{(A.11)}$$

again of the form (A.6) with $\tau = 0$, but this time connected in series as a Nash Cascade. The upstream input to this complete system is an effective rainfall measure $u(t)$ (see main text). The downstream output of the complete system is denoted by $y(t)$.

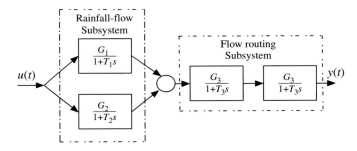

Figure A1.1 Block diagram of a hypothetical catchment model consisting of a parallel pathway, rainfall-flow subsystem in series with a two reach, flow-routing subsystem

One advantage of the TF formulation of the model shown in Fig. A1.1 is that it allows for the computation, using "block diagram algebra" of a single, multi-order TF that represents the total system. Here, TFs connected in parallel are additive; while those connected in series are multiplicative. Consequently, in this case, the two submodels can be represented by the following composite TFs:

$$H_{rf}(s) = H_1(s) + H_2(s); \qquad H_{ff}(s) = H_3(s) \cdot H_3(s) \quad \text{(A.12)}$$

So that, with the above definitions of $H_1(s)$, $H_2(s)$, and $H_3(s)$:

$$H_{rf}(s) = \frac{G_1}{1 + T_1 s} + \frac{G_2}{1 + T_2 s} = \frac{(G_1 + G_2) + (G_1 T_2 + G_2 T_1)s}{(1 + T_1 s)(1 + T_2 s)}, \quad \text{(A.13)}$$

and

$$H_{ff}(s) = \frac{(G_3)^2}{(1 + T_3 s)(1 + T_3 s)}. \quad \text{(A.14)}$$

Now, since $H_{rf}(s)$ and $H_{ff}(s)$ are connected in series, the TF of the total system $H(s)$ is obtained as the multiplication of these two composite TFs, i.e.:

$$\begin{aligned} H(s) &= H_{rf}(s) \cdot H_{ff}(s) \\ &= \frac{(G_1 + G_2) + (G_1 T_2 + G_2 T_1)s}{(1 + T_1 s)(1 + T_2 s)} \cdot \frac{(G_3)^2}{(1 + T_3 s)(1 + T_3 s)} \end{aligned} \quad \text{(A.15)}$$

Multiplying out these expressions, we see that the complete $H(s)$ is a 4th order TF that can be manipulated to the form:

$$H(s) = \frac{g_0 + g_1 s}{s^4 + f_1 s^3 + f_2 s^2 + f_3 s + f_4}, \quad \text{(A.16)}$$

where we will leave the definition of the parameters $f_i, i = 1, 2, \ldots, 4$ and $g_j, j = 0, 1$ in (A.16), as an exercise for the reader. (*Hint*: carry the above analysis with the TF representation (A.8) rather than (A.6), then it will be clear why the former representation is better for analysis, although it lacks the direct physical interpretation of the latter, which provides a better form for the block-diagram representation).

A1.1.3 The general, multi-order TF model

It is clear from the above example that, in general, serial, parallel (or even feedback[9]) connections of elemental first order TF models, such as (A.6) or (A.8), lead to a multi-order TF model that takes the general TF form:

$$x(t) = \frac{G(s)}{F(s)} u(t - \tau), \qquad (A.17)$$

where $F(s)$ and $G(s)$ are polynomials in s of the following form:

$$F(s) = s^p + f_1 s^{p-1} + f_2 s^{p-2} + \cdots + f_p, \qquad (A.18)$$

$$G(s) = g_0 s^q + g_1 s^{q-1} + g_2 s^{q-2} + \cdots + g_q, \qquad (A.19)$$

in which p and q can take on any positive integer values. Here $u(t)$ and $x(t)$ denote the deterministic input and output signals of the system at its upstream and downstream boundaries, respectively; τ is a pure advective time (transport) delay affecting the input signal $u(t)$. Finally, let us assume that the output is observed as $y(t) = x(t) + \xi(t)$, where $\xi(t)$ is a noise or stochastic disturbance signal. This noise is assumed to be independent of the input signal $u(t)$ and it represents the aggregate effect, at the downstream boundary, of all the stochastic inputs to the system, including distributed unmeasured inputs, measurement errors and modelling error. Multiplying throughout Equation (A.17) by $F(s)$ and converting the resultant equation to alternative ordinary differential equation form, we obtain:

$$\frac{d^p y(t)}{dt^p} + f_1 \frac{d^{p-1} y(t)}{dt^{p-1}} + \cdots + f_p y(t) = g_0 \frac{d^q u(t - \tau)}{dt^q}$$
$$+ \cdots + g_q u(t - \tau) + \eta(t), \quad (A.20)$$

where $\eta(t) = F(s)\xi(t)$ is a modified noise signal generated by the manipulation of the equation. Depending on the objectives of the modelling study, it may be necessary, in a complete system consisting of many subelements such as (A.6) or (A.8), to consider noise inputs *within* the system, associated with collections of subelements that have distinct physical meaning, e.g., stochastic lateral inflows. The structure of this model, in either form (A.17) or (A.20), is defined by the triad $[pq\tau]$.

A1.1.4 Discrete-time, sampled data TF models

To date, the most popular form of TF modelling has been carried out using the discrete-time (DT) equivalents of the models (A.8) and (A.17). In the case of Equation (A.8), this discrete-time TF model takes the form:

$$Q_{o,k} = \frac{b_0}{1 + a_1 z^{-1}} Q_{i,k-\delta} \qquad (A.21)$$

Here, $Q_{o,k}$ is the downstream flow measured at the kth sampling instant, that is at time $k\Delta t$, where Δt is the sampling interval in time units. $Q_{i,k-\delta}$ is the input flow at time $(k - \delta)\Delta t$ time units previously, where δ is the advective time delay, normally defined as the nearest integer value of $\tau/\Delta t$ (thus incurring a possible approximation error), and z^{-1} is the backward shift operator, i.e., $z^{-r} Q_{o,k} = Q_{o,k-r}$. Of course, this model can be written in its "difference equation" form, namely:

$$Q_{o,k} = -a_1 Q_{o,k-1} + b_0 Q_{i,k-\delta}, \qquad (A.22)$$

which is obtained by simple cross multiplication and application of the z^{-1} operator. This reveals that the flow $Q_{o,k}$ at the kth sampling instant is a proportion $-a_1$ (note that, in the present context, a_1 will be a negative number less than unity, so that this is a positive proportion) of its value $Q_{o,k-1}$ at the previous $(k - 1)$th sampling instant, plus a proportion b_0 of the delayed upstream flow input $Q_{i,k-\delta}$ measured δ sampling instants previously.

The values of the parameters a_1 and b_0 in Equations (A.21) and (A.22) can be related to the parameters of the model (A.8) in various ways depending upon how the input flow $Q_i(t)$ is assumed to change over the sampling interval between the measurement of $Q_{i,k-1}$ and $Q_{i,k}$ (since it is not measured over this interval). The simplest and most common assumption is that it remains constant over this interval (the so-called zero-order hold, ZOH, assumption), in which case the relationships are as follows:

$$a_1 = -\exp(-f_1 \Delta t) \quad b_1 = \frac{g_0}{f_1}\{1 - \exp(-f_1 \Delta t)\} \qquad (A.23)$$

Note that, because these relationships are functions of the sampling interval Δt, for every unique CT model such as (A.8), there are infinitely many DT equivalents (A.21), depending on the choice of Δt, all with different parameter values defined in (A.23). Following from the definition of this first-order DT model, the general multi-order DT equivalent of the continuous-time TF relationship in (A.17), with the added observation noise ξ_k (the discrete-time equivalent of $\xi(t)$), is defined as follows:

$$x_k = \frac{B(z^{-1})}{A(z^{-1})} u_{k-\delta}, \qquad y_k = x_k + \xi_k, \qquad (A.24)$$

where:

$$A(z^{-1}) = 1 + a_1 z^{-1} + a_2 z^{-2} + \cdots + a_n z^{-n}, \qquad (A.25)$$

$$B(z^{-1}) = b_0 + b_1 z^{-1} + b_2 z^{-2} + \cdots + b_m z^{-m}. \qquad (A.26)$$

Normally $n = p$ but m may be equal or greater than q. The structure of this DT model is defined by the triad $[nm\delta]$. Note finally that the TFs in both (A.24) and (A.17) are composed of ratios of polynomials (in z^{-1} and s respectively), so they are sometimes referred to as rational transfer functions.

A1.1.5 The unit hydrograph and finite impulse response TF models

The reader can verify that the division of the numerator polynomial $B(z^{-1})$ by the denominator polynomial $A(z^{-1})$ in the TF model (A.24) normally results in an infinite dimensional polynomial:

$$G(z^{-1}) = g_0 + g_1 z^{-1} + g_2 z^{-2} + \cdots + g_\infty z^{-\infty}, \qquad (A.27)$$

so that the equation can also be written in the alternative form:

$$y_k = \sum_{i=\delta}^{\infty} g_i u_{k-i} + \xi_k, \quad g_j, j = 0, 1, \cdots, \delta - 1 = 0 \qquad (A.28)$$

This will be recognized as the discrete-time form of the convolution integral equation associated with the solution of differential equations (here with a pure time delay of δ sampling intervals or $\delta \Delta t$ time units) and, once again, we see that the TF model is simply a discrete-time equivalent of a continuous-time differential-equation model.

Equation (A.28) has important connotations in hydrology because, if the noise $\xi_k = 0$ for all t, then its unit impulse response, i.e., the response of the equation to a unit impulse input ($u_k = 1.0$ for $t = 1, u_k = 0$ for all $t > 0$), is equivalent to the hydrological unit hydrograph. Sometimes, indeed, the Equation (A.28) is referred to as a TF model, although its infinite dimensional nature is an obvious restriction. Despite this disadvantage, some hydrologists have used the equation directly in RF modelling by considering it in a finite impulse response (FIR) form, in which the upper limit of the summation is set to some value $r < \infty$ where it is considered that the ordinates of the impulse response have become small enough to be ignored.

Unfortunately, it can be shown that the number of FIR model parameters (or "weights"), g_i, $i = 1, 2, \ldots r$, is nearly always much larger than the number of parameters in the rational TF form (A.24). As a result, the statistical estimates of these parameters will normally have unacceptably high variance and have to be constrained in some manner. For instance, Natale and Todini (1976) use quadratic programming to compute the constrained least-squares estimates of the FIR parameters in order to ensure that they are all positive and, if necessary for conservation purposes, they all sum to unity (i.e., the model has unity steady-state gain). Even with this approach, however, it is difficult to recommend the FIR model because the rational TF model (A.24), normally with far fewer parameters, is not only entirely equivalent to the infinite dimensional IR model, without any approximation, it is also easier to estimate from rainfall-flow data using the methodology discussed in this chapter.

ENDNOTES

1 Although these authors discuss modelling at annual, monthly, and daily scales, the conceptual arguments are similar.

2 This model is similar in concept to the variable-gain factor model suggested by Ahsan and O'Connor (1993), although its identification, estimation, and implementation is quite different.

3 R_T^2 is defined as $1 - [\sigma_{se}^2 / \sigma_y^2]$, where σ_{se}^2 is the variance of the simulated model errors and σ_y^2 is the variance of the output variable. Note that this not the same as the standard coefficient of determination R^2, which is based on the one-step-ahead prediction errors. R_T^2 is very similar to the "Nash–Sutcliffe efficiency" in the hydrological literature (Nash and Sutcliffe, 1970).

4 Of course this differentiation is rather arbitrary since the model is inherently stochastic and so these parameters are simply additional parameters introduced to define the stochastic inputs to the model when it is formulated in this state-space form.

5 A non-recursive ML formulation of this heteroscedasticity problem is given by Sorooshian (1985).

6 This example is utilized as a demonstration example in the CAPTAIN Toolbox for MATLAB ™ (see previous reference in Section 9.4).

7 This Appendix is taken in part from Young (2005), which also describes the historical context of TF models and their use in hydrology.

8 The same letter s (or sometimes p) is used to represent the related Laplace transform operator. Considered in these Laplace transform terms, it is possible to utilize Laplace transform methods to handle initial conditions on the variables in the model and analytically compute its response (here $Q_o(t)$) to variations in input variable (here $Q_i(t)$). However, this is not essential in the present context, although interested reader should find the Appendix is a good primer for the study of Laplace transform methods (e.g., Schwarzenbach and Gill, 1979).

9 The block diagram algebra for a feedback connection is a little more complicated but, since it is not particularly relevant in the current context, the interested reader should consult a standard text on the subject (e.g., Schwarzenbach and Gill, 1979) to find the details.

REFERENCES

Abbot, M. B., Bathurst, J. C., Cunge, J. A., O'Connell, P. E., and Rasmussen, J. (1986). An introduction to the European Hydrological System – Système Hydrologique Européen, SHE. *J. Hydrol.*, **87**, 45–59.

Ahsan, M. and O'Connor, K. M. (1993). A simple non-linear rainfall-runoff model with a variable gain factor. *J. Hydrol.*, **155**, 151–83.

Akaike, H. (1974). A new look at statistical model identification. *IEEE Trans. Auto. Control*, AC **19**, 716–23.

Anderson, M. G. and Burt, T. P. (eds.) (1985). *Hydrological Forecasting*. Chichester: Wiley.

Bacon, F. (1620). *Novum Organum*: see for example Basil Montague (ed. and trans.) (1854), *The Works*, 3 vols. Philadelphia: Parry & MacMillan, 343–71.

Beck, M. B. (1983). Uncertainty, system identification and the prediction of water quality. In *Uncertainty and Forecasting of Water Quality*, ed. M. B. Beck and G. Van Straten. Berlin: Springer-Verlag, 3–68.

Beck, M. B. and Young, P. C. (1975). A dynamic model for BOD-DO relationships in a non-tidal stream, *Water Res.*, **9**, 769–76.

Beven, K. J. (2000). Uniqueness of place and process representations in hydrological modelling, *Hydrol. Earth Sys. Sci.*, **4**, 203–13.

Beven, K. J. (2001). *Rainfall-Runoff Modelling: The Primer*. Chichester: Wiley.

Beven, K. J. and Binley, A. M. (1992). The future of distributed models: model calibration and uncertainty prediction, *Hydrol. Proc.*, **6**, 279–98.

Beven, K. J. and Kirkby, M. J. (1979). A physically based variable contributing area model of basin hydrology. *Hydrol. Sci. J.*, **24**, 43–69.

Beven, K. J., Romanowicz, R., and Hankin, B. (2000). Mapping the probability of flood inundation (even in real time). In *Flood Forecasting: What Does Current Research Offer the Practitioner?* ed. M. J. Lees and P. Walsh, 56–63. BHS Occasional paper No. 12. London: produced by the Centre for Ecology and Hydrology on behalf of the British Hydrological Society.

Beven, K. J. and Young, P. C. (2003). Comment on "Bayesian recursive parameter estimation for hydrologic models" by M. Thiemann *et al.*, *Water Resour. Res.*, **39(5)**, 1116.

Billings, S. A. and Voon, W. S. F. (1986). Correlation based model validity tests for non-linear models, *Int. J. Contr.*, **44**, 235–44.

Box, G. E. P. and Jenkins, G. M (1970). *Time-Series Analysis, Forecasting and Control*. San Francisco: Holden-Day.

Brath, A. and Rosso, R. (1993). Adaptive calibration of a conceptual model for flash flood forecasting. *Water Resour. Res.*, **29**, 2561–72.

Bryson, A. E. and Ho, Y.-C. (1969). *Applied Optimal Control*. Massachusetts: Blaisdell Publishing.

Cluckie, I. D. (1993). Real-time flood forecasting using weather radar. In *Concise Encyclopedia of Environmental Systems*, ed. P. C. Young. Oxford: Pergamon Press, 291–8.

Crawford, N. H. and Linsley, R. K. (1966). *Digital simulation in hydrology: the Stanford Watershed Model IV*, Tech. Report no. 39, Stanford University, California.

Davis, H. T. and Jones, R. H. (1968). Estimation of the innovations variance of a stationary time series, *J. Am. Stat. Assoc.*, **63**, 141–9.

Dooge, J. C. I. (1959). A general theory of the unit hydrograph. *J. Geophys. Res.*, **64**. 241–56.

Dooge, J. C. I. (1986). Theory of flood routing. In *River Flow Modelling and Forecasting*, ed. D. A. Kraijenhoff and J. R. Moll. Dordrecht: D. Reidel, 39–65.

Dooge, J. C. I. and O'Kane, J. P. (2003). *Deterministic Methods in Systems Hydrology*. Lisse: Balkema.

Franks, S., Beven, K. J., Quinn, P. F., and Wright, I. (1997). On the sensitivity of soil–vegetation–atmosphere transfer (SVAT) schemes: equifinality and the problem of robust calibration. *Agric. Forest Meteorol.*, **86**, 63–75.

Greenfield, B. J. (1984). *The Thames Water Catchment Model*, Internal Report, Technology Development Division, Thames Water, UK.

Harvey, A. C. (1989). *Forecasting, Structural Time Series Models and the Kalman Filter*. Cambridge: Cambridge University Press.

Hasselmann, K. (1998). Climate-change research after Kyoto. *Nature*, **390**, 225–6.

Hasselmann, K., Hasselmann, S., Giering, R., Ocana, V., and Storch, H. von (1997). Sensitivity study of optimal emission paths using a simplified structural integrated assessment (SIAM). *Clim. Change*, **37**, 345–86.

Hornberger, G. M., Beven, K. J., Cosby, B. J., and Sappington, D. E. (1985). Shenandoah watershed study: calibration of a topography-based variable contributing area hydrological model for a small forested catchment. *Water Resour. Res.*, **21**, 1841–50.

Hu, J., Kumamaru, K., and Hirasawa, K. (2001). A quasi-ARMAX approach to modelling nonlinear systems. *Int. J. Cont.*, **74**, 1754–66.

Jakeman, A. J. and Hornberger, G. M. (1993). How much complexity is warranted in a rainfall-runoff model? *Water Resour. Res.*, **29**, 2637–49.

Jakeman, A. J., Littlewood, I.G., and Whitehead, P. G. (1990). Computation of the instantaneous unit hydrograph and identifiable component flows with application to two small upland catchments, *J. Hydrol.*, **117**, 275–90.

Jakeman, A. J. and Young, P.C. (1979). Refined instrumental variable methods of recursive time-series analysis. Part II: multivariable systems. *Int. J. Contr.*, **29**, 621–44.

Jang, J.-S. R., Sun, C.-T., and Mizutani, E. (1997). *Neuro-Fuzzy and Soft Computing*. Upper Daddle River, NJ: Prentice Hall.

Jothityangkoon, C., Sivapalan, M., and Farmer, D. L. (2001). Process controls of water balance variability in a large semi-arid catchment: downward approach to hydrological model development. *J. Hydrol.*, **254**, 174–98.

Kalman, R. E. (1960). A new approach to linear filtering and prediction problems. *ASME Trans., J. Basic Eng.*, **82-D**, 35–45.

Kirkby, M. J. (1976). Hydrograph modelling strategies. In *Processes in Physical and Human Geography*, ed. R. J. Peel, M. Chisholm, and P. Haggett. London: Academic Press, 69–90.

Klemes, V. (1983). Conceptualisation and scale in hydrology, *J. Hydrol.*, **65**, 1–23.

Konikow, L. F. and Bredehoeft, J. D. (1992). Groundwater models cannot be validated. *Adv. Water Resour.*, **15**, 75–83.

Kraijenhoff, D. A. and Moll, J. R. (1986). *River Flow Modelling and Forecasting*. Dordrecht: D. Reidel.

Kuhn, T. (1962). *The Structure of Scientific Revolutions*. Chicago: University of Chicago.

Lawton, J. (2001). Understanding and prediction in ecology, Institute of Environmental and Natural Sciences, Lancaster University, Distinguished Scientist Lecture.

Lees, M. J. (2000a). Advances in transfer function based flood forecasting. In *Flood Forecasting: What Does Current Research Offer the Practitioner?* ed. M. J. Lees and P. Walsh. BHS Occasional paper No. 12. London: produced by the Centre for Ecology and Hydrology on behalf of the British Hydrological Society, 41–55.

Lees, M. J. (2000b). Data-based mechanistic modelling and forecasting of hydrological systems. *J. Hydroinfo.*, **2**, 15–34

Lees, M., Young, P. C., Beven, K. J., Ferguson, S., and Burns, J. (1994). An adaptive flood warning system for the River Nith at Dumfries. In *River Flood Hydraulics*, ed. W. R. White and J. Watts. Wallingford: Institute of Hydrology, 65–75.

Meijer, H. (1941). Simplified flood routing. *Civ. Eng.*, **11**, 306–7.

Moore, R. J. and Jones, D. A. (1978). An adaptive finite difference approach to real-time channel flow routing. In *Modeling, Identification and Control in Environmental Systems*, ed. G. C. Vansteenkiste. Amsterdam: North Holland, 153–70.

Moradkhani, H., Sorooshian, S., Gupta, H. V., and Houser, P. R. (2005). Dual stateparameter estimation of hydrological models using ensemble Kalman Filter. *Adv. Water Resour.*, **28**, 135–47.

Nash, J. E. (1959). Systematic determination of unit hydrograph parameters. *J. Geophys. Res.*, **64**, 111–15.

Nash, J. E. and Sutcliffe, J. V. (1970). River flow forecasting through conceptual models: discussion of principles. *J. Hydrol.*, **10**, 282–90.

Natale, L. and Todini, E. (1976). A stable estimator for linear models. 2: Real world hydrologic applications. *Water Resour. Res.*, **12**, 672–6.

Oreskes, N., Shrader-Frechette, K., and Belitz, K. (1994). Verification, validation, and confirmation of numerical models in the earth sciences. *Science*, **263**, 641–6.

Parkinson, S. D. and Young, P. C. (1998). Uncertainty and sensitivity in global carbon cycle modelling. *Clim. Res.*, **9**, 157–74.

Popper, K. (1959). *The Logic of Scientific Discovery*. London: Hutchinson.

Ratto, M., Young, P. C., Romanowicz, R., Pappenberger, F., Saltelli A., and Pagano, A. (2007). Uncertainty, sensitivity analysis and the role of data based mechanistic modelling in hydrology. *Hydrology and Earth System Sciences (HESS), European Geophysical Union*, **11**, 1249–66 (www.hydrol-earth-syst-sci.net/11/1249/2007/).

Romanowicz R. J., Beven, K. J., and Tawn, J. A. (1994). Evaluation of predictive uncertainty in non-linear hydrological models using a Bayesian Approach. In *Statistics for the Environment (2), Water Related Issues*, ed. V. Barnett and F. Turkman. Chichester: Wiley, 297–318.

Romanowicz, R., Young, P. C., and Beven, K. J. (2004). Data assimilation in the identification of flood inundation models: derivation of on-line, multi-step predictions of flows. In *Hydrology: Science & Practice for the 21st Century*, ed. B. Webb, Vol. 1. London: British Hydrological Society, 348–53.

Romanowicz, R. J., Young, P. C., and Beven, K. J. (2006). Data assimilation and adaptive forecasting of water levels in the River Severn catchment. *Water Resour. Res.*, **42** (doi10.1029/2005WR004373).

Saltelli, A., Chan, K., and Scott, E. M. (2000). *Sensitivity Analysis*. Chichester: Wiley.

Schwarzenbach, J. and Gill, K. F. (1979). *Systems Modelling and Control*. London: Edward Arnold.

Schweppe, F. (1965). Evaluation of likelihood functions for Gaussian signals. *IEEE Trans. Info. Theory*, **11**, 61–70.

Shackley, S., Young, P. C., Parkinson, S. D., and Wynne, B. (1998). Uncertainty, complexity and concepts of good science in climate change modelling: are GCMs the best tools? *Clim. Change*, **38**, 159–205.

Shaw, E. M. (1994). *Hydrology in Practice*, 3rd edn. London: Chapman & Hall.

Silvert, W. (1993). Top-down modelling in ecology. In *Concise Encyclopedia of Environmental Systems*, ed. P. C. Young. Oxford: Pergamon Press, 60.

Singh, V. P. (ed.) (1995). *Computer Models of Watershed Hydrology*. Colorado: Water Resources Publications.

Sivapalan, M. and Young, P. C. (2005). Downward approach to hydrological model development. In *Encyclopedia of Hydrological Sciences*, ed. M. G. Anderson, Vol. 3, part II. Hoboken, NJ: Wiley, 2081–98.

Sorooshian, S (1985). Development of rainfall-runoff models. *Appl. Math. Comp.*, **17**, 279–98.

Thiemann, M., Trosset, M., Gupta, H., and Sorooshian, S. (2001). Bayesian recursive parameter estimation for hydrologic models. *Water Resour. Res.*, **37**, 2521–35.

Todini, E. (1996). The ARNO rainfall-runoff model. *J. Hydrol.*, **175**, 339–82.

Tokar, A. S. and Johnson, P. A. (1999). Rainfall-runoff modeling using artificial neural networks. *ASCE J. Hydrol. Eng.*, **4**, 232–9.

Wallis, S. G., Young, P. C., and Beven, K. J. (1989). Experimental investigation of the aggregated dead zone (ADZ) model for longitudinal solute transport in stream channels, *Proc. Inst. Civ. Engrs: Part 2*, **87**, 1–22.

Wheater, H. S., Jakeman, A. J., and Beven. K. J. (1993). Progress and directions in rainfall-run-off modelling. In *Modelling Change in Environmental Systems*, ed. A. J. Jakeman, M. B. Beck, and M. J. McAleer, Chichester: Wiley, 101–32.

Whitehead, P. G. and Young, P. C. (1975). A dynamic-stochastic model for water quality in part of the Bedford–Ouse River system. In *Computer Simulation of Water Resources Systems*, ed. G. C. Vansteenkiste. Amsterdam: North Holland, 417–38.

Whitehead, P. G., Young, P. C., and Hornberger G. H. (1976). A systems model of stream flow and water quality in the Bedford–Ouse river. Part l: stream flow modeling, *Water Res.*, **13**, 1155–69.

Wood, E. F. and O'Connell, P. E. (1985). Real-time forecasting. In *Hydrological Forecasting*, ed. M. G. Anderson and T. P. Burt. Chichester: Wiley, 505–58.

Ye, W., Jakeman, A. J., and Young, P. C. (1998). Identification of improved rainfall-runoff models for an ephemeral low yielding Australian catchment, *Environ. Mod. Software*, **13**, 59–74.

Young, P. C. (1974). Recursive approaches to time-series analysis. *Bull. Inst. Maths Appl.*, **10**, 209–24.

Young, P. C. (1978). A general theory of modeling for badly defined dynamic systems. In *Modeling, Identification and Control in Environmental Systems*, ed. G. C. Vansteenkiste. Amsterdam: North Holland, 103–35.

Young, P. C. (1983). The validity and credibility of models for badly defined systems. In *Uncertainty and Forecasting of Water Quality*, ed. M. B. Beck and G. Van Straten. Berlin: Springer-Verlag, 69–100.

Young, P. C. (1984). *Recursive Estimation and Time-Series Analysis*. Berlin: Springer-Verlag.

Young, P. C. (1985). The instrumental variable method: a practical approach to identification and system parameter estimation. In *Identification and System Parameter Estimation*, ed. H. A. Barker and P. C. Young. Oxford: Pergamon Press, 1–16.

Young, P. C. (1986). Time-series methods and recursive estimation in hydrological systems analysis. In *River Flow Modelling and Forecasting*, ed. D. A. Kraijenhoff and J. R. Moll. Dordrecht: D. Reidel, 129–80.

Young, P. C. (1989). Recursive estimation, forecasting and adaptive control. In *Control and Dynamic Systems: Advances in Theory and Applications*, Vol. 30, ed. C. T. Leondes. San Diego: Academic Press, 119–66.

Young, P. C. (1992). Parallel processes in hydrology and water quality: a unified time series approach. *Jnl. Inst. Water and Env. Man.*, **6**, 598–612.

Young, P. C. (1993). Time variable and state dependent modelling of nonstationary and non-linear time series. In *Developments in Time Series Analysis*, ed. T. Subba Rao. London: Chapman & Hall, 374–413.

Young, P. C. (1998a). Data-based mechanistic modelling of engineering systems. *J. Vibr. Contr.*, **4**, 5–28.

Young, P. C. (1998b). Data-based mechanistic modelling of environmental, ecological, economic and engineering systems. *Environ. Mod. Software*, **13**, 105–22.

Young, P. C. (1999a). Data-based mechanistic modelling, generalised sensitivity and dominant mode analysis. *Comp. Phys. Comm.*, **117**, 113–29.

Young, P. C. (1999b). Non-stationary time series analysis and forecasting. *Prog. Environ. Sci.*, **1**, 3–48.

Young, P. C. (2000). Stochastic, dynamic modelling and signal processing: time variable and state dependent parameter estimation. In *Non-stationary and Non-linear Signal Processing*, ed. W. J. Fitzgerald, A. Walden, R. Smith, and P. C. Young. Cambridge: Cambridge University Press, 74–114.

Young, P. C. (2001a). The identification and estimation of non-linear stochastic systems. In *Non-linear Dynamics and Statistics*, ed. A. I. Mees. Boston: Birkhauser, 127–66.

Young, P. C. (2001b). Data-based mechanistic modelling and validation of rainfall-flow processes. In *Model Validation: Perspectives in Hydrological Science*, ed. M. G. Anderson and P. D. Bates. Chichester: Wiley, 117–61.

Young, P. C. (2001c). Comment on "A quasi-ARMAX approach to the modelling of nonlinear systems" by J. Hu *et al.*, *Int. J. Contr.*, **74**, 1767–71.

Young, P. C. (2002). Advances in real-time flood forecasting. *Phil. Trans. R. Soc. Lond.*, **360**, 1433–50.

Young, P. C. (2003). Top-down and data-based mechanistic modelling of rainfall-flow dynamics at the catchment scale, *Hydrol. Proc.*, **17**, 2195–217.

Young, P. C. (2004). Identification and estimation of continuous-time hydrological models from discrete-time data. In *Hydrology: Science & Practice for the 21st Century*, Vol. 1, ed. B. Webb. London: British Hydrological Society, 406–13.

Young, P. C. (2005). Rainfall-runoff modeling: transfer function models. In *Encyclopedia of Hydrological Sciences*, Vol. 3, Part II, ed. M. G. Anderson. Hoboken, NJ: Wiley, 1985–2000.

Young, P. C. and Beven, K. J. (1994). Data-based mechanistic modelling and the rainfall-flow nonlinearity, *Environmet.*, **5**, 335–63.

Young, P. C. and Garnier, H. (2006). Identification and estimation of continuous-time rainfall-flow models. In *Proc. 14th International Federation on Automatic Control (IFAC) Symposium on System Identification SYSID06*, Newcastle, NSW, Australia, Oxford: Pergamon, 1276–81.

Young, P. C. and Jakeman, A. J. (1979). Refined instrumental variable methods of recursive time-series analysis. Part I: single input, single output systems. *Int. J. Contr.*, **29**, 1–30.

Young, P. C. and Jakeman, A. J. (1980). Refined instrumental variable methods of recursive time-series analysis. Part III: extensions. *Int. J. Contr.*, **31**, 741–64.

Young, P. C. and Lees, M. J. (1993). The active mixing volume: a new concept in modelling environmental systems. In *Statistics for the Environment*, ed. V. Barnett and K. F. Turkman. Chichester: Wiley, 3–43.

Young, P. C. and Minchin, P. (1991). Environmetric time-series analysis: modelling natural systems from experimental time-series data. *Int. Jnl. Biol. Macromol.*, **13**, 190–201.

Young, P. C. and Parkinson, S. D. (2002). Simplicity out of complexity. In *Environmental Foresight and Models*, ed. M. B. Beck. Amsterdam: Elsevier, 251–301.

Young, P. C. and Pedregal, D. (1998). Data-based mechanistic modelling. In *System Dynamics in Economic and Financial Models*, ed. C. Heij and H. Schumacher. Chichester: Wiley, 169–213.

Young, P. C. and Pedregal, D. (1999a). Macro-economic relativity: government spending, private investment and unemployment in the USA 1948–1998. *Struct. Changes Econ. Dynam.*, **10**, 359–80.

Young, P. C. and Pedregal, D. (1999b). Recursive and en-bloc approaches to signal extraction. *J. App. Stats.*, **26**, 103–28.

Young, P. C. and Tomlin, C. M. (2000). Data-based mechanistic modelling and adaptive flow forecasting. In *Flood Forecasting: What Does Current Research Offer the Practitioner?*, ed. M. J. Lees and P. Walsh. BHS Occasional paper No. 12 London: produced by the Centre for Ecology and Hydrology on behalf of the British Hydrological Society.

Young, P. C. and Wallis S. G. (1985). Recursive estimation: a unified approach to the identification, estimation and forecasting of hydrological systems. *Appl. Math. Compu.*, **17**, 299–334.

Young, P. C., Chotai, A., and Beven, K. J. (2004). Data-based mechanistic modelling and the simplification of environmental systems. In *Environmental Modelling: Finding Simplicity in Complexity*, ed. J. Wainright and M. Mulligan. Chichester: Wiley, 371–388.

Young, P. C., Jakeman, A. J., and Post, D. A. (1997a). Recent advances in data-based modelling and analysis of hydrological systems. *Water Sci. Tech.*, **36**, 99–116.

Young, P. C., McKenna, P., and Bruun, J. (2001). Identification of non-linear stochastic systems by state dependent parameter estimation. *Int. J. Contr.*, **74**, 1837–57.

Young, P. C., Parkinson, S. D., and Lees, M. (1996). Simplicity out of complexity in environmental systems: Occam's Razor revisited. *J. Appl. Stat.*, **23**, 165–210.

Young, P. C., Schreider, S. Y., and Jakeman, A. J. (1997b). A streamflow forecasting algorithm and results for the Upper Murray Basin. In *Proc. MODSIM 97 Congress on Modelling and Simulation, Hobart, Tasmania*, ed. A. D McDonald and M. McAleer. Canberra: The Modelling and Simulation Society of Australia, Inc.

10 Real-time flood forecasting: Indian experiences

R. D. Singh

10.1 INTRODUCTION

Since time immemorial, floods have been responsible for loss of crops and valuable property, and untold human misery in the world. India has been no exception. An area of more than 40 million ha in India has been identified as flood prone. India, which is traversed by a large number of river systems, experiences seasonal floods. It has been the experience that floods occur almost every year in one part or the other of the country. The rivers of north and central India are prone to frequent floods during the south-west Monsoon season, particularly in the months of July, August, and September. In the Brahmaputra river basin, floods have often been experienced as early as in late May while in southern rivers floods continue till November. On average, floods in India have affected about 33 million persons between 1953 and 2000. There is every possibility that this figure may increase due to population growth and development in the flood plains.

Floods occur due to natural as well as man-made causes. Major causes of floods in India include intense precipitation, inadequate capacity within riverbanks to contain high flows, and silting of riverbeds. In addition, other factors are land slides leading to obstruction of flow and changes in the river course, retardation of flow due to tidal and backwater effects, poor natural drainage, cyclone and heavy rainstorms/cloud bursts, snowmelt and glacial outbursts, and dam-break flow.

For minimizing the losses due to floods, various flood control measures are adopted. Such measures – which should more correctly be termed as "flood management" – can be planned either through structural engineering measures or non-structural measures. Wise application of engineering science has afforded ways of mitigating the ravages due to floods and providing a reasonable measure of protection to life and property. Such measures comprise multi-purpose reservoirs and retarding structures which store flood waters, channel improvements which increase the flood carrying capacity of the river, embankments and levees which keep the water away from flood-prone areas, detention basins which retard and absorb some flood water, flood-ways which divert flood flows from one channel to another, and overall improvement in the drainage system. Various engineering measures could ultimately protect a large portion of such flood-prone areas. Up until 1985 about 13 million ha of land had been covered by some flood-protection measures. The large back-log of unconstructed, though economically feasible, flood-control projects will take quite some time to be cleared in view of the shortage of funds allocated and allocatable to the flood-control sector. Further, no systematic study has been taken up in India to examine the efficacy of these structures on river-flow regimes during the period of flood. This means that many areas will remain unprotected for a considerable time, thus calling for measures to ensure, meanwhile, flood-loss mitigation.

It has also been recognized that permanent protection of all flood-prone areas for all magnitude of floods by such structural means is neither possible nor feasible because of various factors such as financial constraints, cost–benefit criteria or topographic limitations of the region. Real-time flood forecasting and flood-plain zoning are some of the important non-structural measures adopted for the management of floods. The process of estimating the future stages or flows and their time sequence at selected vulnerable points along the river course during floods may be called "real-time flood forecasting." Real-time flood-forecasting systems are formulated for issuing flood warnings in order to prepare for the arrival of the flood, including ultimately evacuation plans. Experience has shown that loss of human life and property, etc. can be reduced to a considerable extent by giving reliable advance information about the coming floods. People could be moved to safer places in an organized manner as soon as the flood warnings are received. Valuable moveable property and cattle could be saved by transferring them to places of safety. The effectiveness of real-time flood-forecasting systems in reducing flood damage depends upon how accurately the estimation of future stages or discharges of the incoming flood, and their time sequence at selected points along the river, can be predicted. The rivers of alluvial

Hydrological Modeling in Arid and Semi-Arid Areas, ed. Howard Wheater, Soroosh Sorooshian, and K. D. Sharma. Published by Cambridge University Press.
© Cambridge University Press 2008.

plains exhibit meandering, shifting of the course, and unstable cross-sections due to the problem of sediment transport. These hydraulic changes in the river behavior complicate the issue of adopting suitable measures for flood management.

The magnitude and severity of the floods, caused by excessive rainfall in the river catchments, depend upon the nature and extent of rainfall and the characteristics of the specific watersheds. For example, intense, shorter-duration rainfall or cloud bursts in small, steep catchments (e.g., having a basin area < 1200 km^2 and time-to-peak < 6 hours) results in flash floods, whereas heavy rainfall of longer duration in the large catchments may generate floods which are sustained for longer periods. Suitable real-time flood-forecasting techniques (Wood and O'Connell, 1985) are required to forecast such floods. Accurate real-time flood forecasts are required to be issued well in advance in order to provide sufficient time, known as lead time, for evacuating the people from the areas likely to be affected. The lead times available for flash-flood forecasts are small, which makes the implementation of evacuation plans very difficult.

The techniques available for real-time flood forecasting may be broadly classified in four groups: (i) deterministic modelling, (ii) stochastic modelling, (iii) statistical modelling, and (iv) computational techniques like artificial neural network (ANN) and fuzzy logic. Deterministic models are based on either index-catchment models or conceptual-catchment models. Such models tend to simulate the basin response to hydrological events based on historical data and do not fully utilize information collected during an event. Further, the deterministic models were originally developed for design studies, and their formulation has not been influenced by the need to incorporate hydrological information in real time. As a consequence, these models cannot readily be updated, and may prove difficult to re-initialize following telemetry or computer breakdowns. For example, a forecast from a conceptual model is often made of contributions from a number of stores (reservoirs), the contents of which have to be specified in order to initialize the model, which is clearly not practical in a real-time context (unless the model is run continuously). Similar difficulties are also encountered in unit hydrograph models (which were primarily developed to meet design objectives) as the necessary separation of base flow and storm runoff is difficult to invoke in real time. The inherent weaknesses of deterministic hydrological models for real-time applications have led to interest in stochastic time-series models with structures more suited to real-time forecasting. The time-series models and their applications to real-time flow forecasting evolved in the 1970s and have been successfully applied for real-time forecasting. The statistical models involve the development of the relationships correlating the flood characteristics of forecasting stations and upstream gauging stations considering the various other factors influencing the floods. The ANN and fuzzy-logic based models, which have the potential for real-time flood forecasting, are capable of considering the inherent non-linear interactions in the rainfall-runoff process.

Depending on the availability of hydrological and hydro-meteorological data, basin characteristics, computational facilities available at the forecasting stations, warning time required, and the purpose of the forecast, different flood-forecasting techniques are being used in India. Some of the commonly used techniques include: (i) simple relations developed correlating the stage-discharge data, (ii) co-axial correlation diagrams developed utilizing the stage, discharge and rainfall data, etc., (iii) event-based hydrological system models for small to moderate-sized catchments, (iv) network models consisting of the subbasins and subreaches for the large-sized catchments, and (v) deterministic hydrologic models (at selected places). Stochastic models have been mostly applied by researchers and academicians for real-time flood forecasting, and their applications have been restricted to a few places. The application of more recent computing techniques such as ANN and fuzzy logic are currently in the development stage and being mostly used by academicians and researchers.

Thus, in India the statistical approach is most widely used to formulate real-time flood forecasts. Event-based network models and multi-parameter hydrological models are applied for some pilot projects during the implementation of international projects. Flash floods occur in the arid and semi-arid regions. However, as yet no effective system has been implemented for flash-flood forecasting for these regions. Some flood-warning arrangements have existed, but these were largely aimed at transmitting limited information on flood levels from upstream points to the areas lower down. Such warnings have limited utility in as much as they do not indicate the likely levels and the time of arrival of floods at the vulnerable places. Further, they do not often give adequate advance notice.

10.2 FLOOD PROBLEMS IN INDIA

The main problems in India with respect to floods are inundation, drainage congestion due to urbanization, and bank erosion. The problems depend on the river system, topography of the place, and flow phenomenon. Being a vast country, the flood problems in India may be visualized on a regional basis. However, for the sake of simplicity, India may be broadly divided into four zones of flooding, namely: (i) Brahmaputra River Basin, (ii) Ganga River Basin, (iii) North-West Rivers Basin, and (iv) Central India and Deccan Rivers Basin. Flooding in these zones is presented in the following subsections.

10.2.1 Brahmaputra River Basin

The first zone belongs to the basins of the rivers Brahmaputra and Barak with their tributaries. It covers the States of Assam,

Arunachal Pradesh, Meghalaya, Mizoram, northern parts of West Bengal, Manipur, Sikkim, Tripura, and Nagaland. The catchments of these rivers receive large amounts of rainfall. As a result of this, floods in this region are frequent and are severe by nature. The general tectonic up-wrapping of the north-east region has a significant effect on the river Brahmaputra. Almost all northern tributaries of the Brahmaputra are affected by landslides in the upper catchment. Further, the rocks in the hills where these rivers originate are friable and susceptible to erosion and thereby cause exceptionally high silt loads in the rivers. In addition, the region is subject to severe and frequent earthquakes, causing numerous landslides in the hills, which upsets the regime of the rivers. Important problems in this region are flood inundation due to spilling of banks, drainage congestion due to natural as well as man-made structures, and change of river flow. In recent years, erosion along the banks of the Brahmaputra has been enormous and has become a serious concern among water-resource engineers.

The main problems of flooding in Assam are inundation caused by spilling of the rivers Brahmaputra and Barak as well as their tributaries. In northern parts of West Bengal, the rivers Teesta, Torsa, and Jaldakha are in flood every year and inundate large areas. During flooding, these rivers carry large amounts of silt and have a tendency to change their courses. The rivers in Manipur spill over their banks frequently. The lakes in the territory are filled up during the Monsoon and spread to inundate large marginal areas. In Tripura, flood problems are spilling and erosion by rivers.

10.2.2   Ganga River Basin

The Ganga and its many tributaries (the Yamuna, the Sone, the Ghaghra, the Gandak, the Kosi, and the Mahananda) constitute the second zone. This zone covers Uttaranchal, Uttar Pradesh, Bihar, south and central parts of West Bengal, parts of Haryana, Himachal Pradesh, Rajasthan, Madhya Pradesh, and Delhi. The normal annual rainfall of this region varies from about 60 cm to 190 cm, of which more than 80 % occurs during the south-west Monsoon. The rainfall increases from west to east and from south to north.

The flood problem is mostly confined to the areas on the northern bank of the Ganga River. The damage is caused by the northern tributaries of the Ganga spilling over their banks and changing their courses. Though the Ganga carryies huge discharges (57 000 to 85 000 m³/s), the inundation and erosion problems are confined to specific places only. In general, the flood problem increases from west to east and from south to north. In the north-western parts of the region, there is the problem of drainage congestion. The drainage problem also exists in the southern parts of West Bengal. The problem becomes acute when the main river, in which the water is to be drained, already has a high water level. The flooding and erosion problem is serious in Uttar Pradesh, Bihar, and West

Bengal. In Rajasthan and Madhya Pradesh, the problem is not so serious. In Bihar, the floods are largely confined to the rivers of North Bihar and are an annual feature. Most of the rivers (e.g., the Burhi Gandak, the Bagmati, the Kamla Balan, other smaller rivers of the Adhwra Group, the Kosi in the lower reaches, and the Mahananda at the eastern end) spill over their banks, causing considerable damage to crops and dislocating traffic. High floods occur in the Ganga, occasionally causing considerable inundation of the marginal areas in Bihar.

In Uttar Pradesh, the flooding is frequent in the eastern districts, mainly due to spilling of the Rapti, the Sarada, the Ghaghra, and the Gandak. The problem of drainage congestion exists in the western and north-western areas of Uttar Pradesh, particularly in the Agra, Mathura, and Meerut districts. Erosion is experienced in some places on the left bank of Ganga, on the right bank of the Ghaghra, and on the right bank of the Gandak. In Haryana, flooding takes place in the marginal areas along the Yamuna and the problem of poor drainage exists in some of the south-western districts. In Delhi, a small area along the banks of the Yamuna is subject to flooding by river spills. In addition, local drainage congestion is experienced in some of the developing colonies during heavy rains. In the south and central parts of West Bengal, the Mahananda, the Bhagirathi, the Ajoy (GFCC, 1986), and the Damodar cause flooding due to inadequacy in river channels and the tidal effect. There is also the problem of erosion of the banks of rivers and on the left and right banks of the Ganga, upstream and downstream, respectively, of the Farakka barrage.

10.2.3   North-West River Basins

This is the third zone and comprises the basins of the north-west rivers such as the Sutlej, Ravi, Beas, Jhelum, and Ghaggar. In comparison to the two zones mentioned above, the flood problem in this zone is relatively less. The major problem is that of inadequate surface drainage, which causes inundation and water logging.

Another cause of flooding has been water logging in the irrigated areas and changes in river regimes due to increased ground-water levels. At present, the problems in Haryana and Punjab are mostly of drainage congestion and water logging. The Ghaggar River used to disappear in the sand dunes of Rajasthan after flowing through Punjab and Haryana. The Jhelum floods occur frequently in Kashmir causing a rise in the level of the Wullar Lake, thereby submerging marginal areas of the lake.

10.2.4   Central India and Deccan Rivers Basin

Important rivers in the fourth zone are the Narmada, the Tapi, the Mahanadi, the Godavari, the Krishna, and the Cauvery. These rivers have mostly well-defined stable courses. They have adequate capacity within the natural banks to carry the flood discharge except in their lower reaches and in the delta area, where the average bed slope is very flat. The lower reaches of the important rivers on the east coast have been embanked.

This region covers all the southern States namely Andhra Pradesh, Chhattisgarh, Karnataka, Tamil Nadu, Kerala, Orissa, Maharashtra, Gujarat, and parts of Madhya Pradesh. The region does not have very serious problems except for some of the rivers of Orissa (the Brahmani, the Baitarni, and the Subarnarekha). The delta areas of the Mahanadi, Godavari, and the Krishna rivers on the east coast periodically face flood and drainage problems, in the wake of cyclonic storms.

The Tapi and the Narmada are occasionally in high flood, affecting areas in the lower reaches of Gujarat. Flood problems in Andhra Pradesh are confined to spilling by the smaller rivers and the submergence of marginal areas along the Kolleru Lake. Rivers like Budameru and the Thammileru not only overflow their banks along their courses to Kolleru Lake but also cause a rise in the lake level resulting in inundation of adjoining lands.

In Orissa, damage due to floods is caused by the Mahanadi, the Brahmani, and the Baitarani, which have a common delta. Water from these rivers inter-mingles in the delta and results in a very high water level, which causes severe flooding in the region. The coastal districts are densely populated and receive heavy precipitation in the Eastern Ghat region. The silt deposited constantly by these rivers in the delta area raises the flood-water level, and the rivers often overflow their banks or break through new channels causing heavy damage. The lower reaches of the Subarnarekha are affected by floods and drainage congestion. The small rivers of Kerala when in high flood often cause considerable damage. In addition, there is also the problem of mud flow from the hills, which results in severe losses.

10.3 SPECIAL FLOOD PROBLEMS

In this section, some special flood problems in India are presented.

10.3.1 Problem of Tal areas

Natural depressions where water gets deposited during the Monsoon and retained for a longer period are known as Tal areas. Generally, they hamper normal activity, affecting the Kharif crop. The Mokama group of Tals in Bihar is known for its flood problem. Water accumulated in these areas during the Monsoon remains stagnant up to September. Similar problems are also seen in the Ghaggar detention basin in Rajasthan, and depressions available at Ottu, Bhindawas, and the Kotla lakes in Haryana. Flood problems in such areas are of a special nature and need to be treated separately.

10.3.2 River bank/bed erosion

All natural rivers have mobile beds. Therefore, depending on the flow phenomenon, there may be aggradation and/or degradation in the river banks and beds. A river erodes its banks due to various reasons causing considerable loss of land, deterioration of the river regime, and sometimes this accounts for huge losses during floods. Rivers in the Brahmaputra–Barak and Ganga Basins are prone to severe erosion. The erosion is governed by the discharge in the river, bed slope, sediment flow, and composition of bed and bank materials. Deforestation of upper catchments and hills leads to increased sediment load in rivers. Effects of seismic disturbances, and the topographic conditions of land, also contribute to the erosion problem. River erosion causes a loss of land resources. The river behavior causes new riverine landmass to be built up, but this only becomes productive after many years and cannot compensate for the land-loss due to erosion. Erosion in the Majuli island, the largest river island in the world, is the most appropriate example to state the severity of the problem. The Brahmaputra Board has estimated that in the Majuli Island, the annual loss of land due to erosion could be about $3.9 \, km^2$ and an economic loss of about Rs 31.5 million per annum (IWRS, 2001).

10.3.3 Sediment transport by rivers

One of the problems associated with the floods in India is the transport of sediments by rivers during floods. Himalayan rivers originating from Nepal bring a lot of sediment during floods to the alluvial plains in the valley. The transport of sediments (suspended and bed load) has a major role on river behavior and river morphology. Thus, the flood problem and its management measures depend a lot on sediment transport. Several observation sites for sediment transport in rivers are maintained by the Central Water Commission (CWC). With the help of the data (on average suspended-sediment load), the pattern of increase/decrease of silt load can be examined to assess the morphological changing trend of the river. As recorded by CWC, the total live and gross storage capacity created in India is about 177 and $217 \, km^3$, respectively. Based on the sedimentation data of 144 reservoirs, the weighted average annual loss in gross storage due to silting is computed to be 0.44 %. Thus, the likely annual loss in the total gross storage of $217 \, km^3$ is $0.95 \, km^3$. Similarly, the annual loss in live storage is 0.31 % based on the data of 42 reservoirs. Thus the likely annual loss in total live storage is $0.55 \, km^3$. Considering the average density of 1.137 tonnes/m^3, based on the data of 13 reservoirs, the weight of the total sediment deposits in all the reservoirs in India is 1080 million tons annually.

10.3.4 Dam-break flows

Flooding due to dam breaks can be a mega-disaster as it is often associated with huge loss of life and property. An unusual high peak in a short duration and the presence of a moving hydraulic shock/bore make it a different problem as compared to other natural floods. In India, historical occurrences (e.g., the failure of the Machhu Dam, Panshet Dam, Nanak Saagar Dam) of dam-break floods are common (Palaniappan, 1997). Sometimes, blockage of water due to deposits caused by landslides takes place. When this natural blockage fails due to the increased amount of water upstream, huge flooding can occur. The behavior is similar to that of dam-break floods.

10.3.5   Urban drainage

Flooding of cities in India is a common and annual event. Due to encroachment of the flood-plain areas, the presence of several structures, and the absence of proper regulations for maintenance, an artificial flood is created. Therefore, proper drainage networks for a city have to be developed.

10.3.6   Flash floods

Flash floods are characterized by the sudden rise and recession of a flow of a small volume and high discharge which causes damage because of its suddenness. They generally take place in hilly regions where the bed slope is very steep. Typical examples are the flash floods of Arunachal Pradesh and of Satluj in 2000. A large reservoir downstream of flood-prone areas can absorb the flood wave. Flash floods are also experienced in arid and semi-arid regions due to the intense and short duration rainfall in the small catchments of the region.

10.3.7   Floods due to snowmelt

Snowmelt is a gradual process. Usually, it does not produce floods. However, sometimes, glaciers hold large quantities of retained water. When released suddenly, this can cause severe flooding. The rivers originating from the Himalayas in the north are fed by snowmelt from glaciers. In 1929, the outburst of the Chong Khundam Glacier (Karakoram) caused a flood peak of over $22\,000\,\text{m}^3/\text{s}$ at Attock.

10.3.8   Floods in coastal areas

Floods in Indian river basins are also caused by cyclones. Coastal areas of Andhra Pradesh, Orissa, Tamilnadu, and West Bengal experience heavy floods regularly. The flood due to a super cyclone combined with heavy rainfall during October 1999 in the coastal region of Orissa is an example. During the past 110 years (1891–2000), over 1000 tropical cyclones and depressions, originating in the Bay of Bengal and Arabian Sea, moved across India. The passage of such storms over a river basin leads to severe floods.

10.4   THE NEED FOR FLOOD FORECASTING

Warning of the approaching floods provides sufficient time for the authorities:

(1) To evacuate the affected people to the safer places.
(2) To make an intense patrol of flood protection works such as embankments so as to save them from breaches, failures, etc.
(3) To regulate the floods through barrages and reservoirs, so that the safety of these structures can be taken care of during the higher return-period floods.
(4) To operate the multi-purpose reservoirs in such a way that an encroachment into the power and water-conservation storage can be made to control the incoming flood.
(5) To operate the city drains (outfalling into the river) to prevent bank flow and flooding of the areas drained by them.

10.5   DEVELOPMENT OF FLOOD FORECASTING IN INDIA

In 1969, the Government of India created a Central Flood Forecasting Directorate headed by a Superintending Engineer. In 1970, under the Member (Floods), six flood forecasting divisions were set up on inter-state river basins. These covered the flood prone basins/subbasins of the Ganga, the Brahmaputra, the Narmada, the Tapti, the Teesta, and the coastal rivers of Orissa. By the year 1977, the Central Flood Forecasting Organisation comprised one Chief Engineer's Office, three circles and 11 divisions.

In most States of India, there are arrangements for the issue of flood warnings from the upstream stations to the downstream stations. These warnings include:

(1) Whether the river is rising above a certain specified level, known as danger level or not.
(2) Whether the river is rising or falling.
(3) Whether the stage of the river is "low," "medium," or "high."

The above warnings, issued by telegrams, telephone, or wireless systems are purely qualitative in nature and they give only an indication of the nature of the flood. Such procedures are at present being followed in West Bengal, Andhra Pradesh, and Bihar states.

After the completion of certain multi-purpose projects like the Hirakud in Orissa, DVC Projects in Bihar/Bengal and Bhakra in Punjab, forecasting techniques have been evolved using the data of rainfall and stream gauges in the catchment upstream of the dam. Correlation diagrams have been prepared using historical data to predict the inflow into the reservoir. Based upon this, the reservoir operations are determined. Such flood forecasting systems have also been set up for Yamuna in Delhi, Koshi in Bihar, and Krishna and Godavari in Andhra Pradesh.

The Central Water Commission has established a network of more than 147 flood forecasting and warning sites on various inter-state rivers (CWC, 1989). The data from the river gauges and the rainfall are transmitted to the flood-forecasting centres from all the key stations by means of wireless or telegrams. Based on these data and the correlation curves already developed with previous data, the forecasts are issued daily to the concerned authorities so that they can take the appropriate measures.

10.6 DATA REQUIREMENTS

Basically gauge (i.e., water-level or stage) or discharge data and/or rainfall data are required for flood-forecasting purposes. The number of reporting stations depends upon hydrologic need and availability of observers and communications.

The number of raingauge stations in the basin should be such that:

(1) The areal rainfall in the catchment can be estimated with the desired accuracy, and
(2) The variation in the areal distribution as well as time distribution can be identified.

For network design of river gauges, the following points should be kept in mind:

(1) Wherever the forecast is being issued on the basis of gauge correlation, the base station and forecasting station must be equipped with gauges.
(2) In the case where more than one tributary joins the main stream and the forecast is based on a multiple coaxial diagram, there should be at least one gauge on each of the tributaries. The location of gauges on the tributaries should be such that the time of the travel from base station to forecasting station in respect of tributaries as well as main stream is constant.
(3) Where a routing model forms the basis of the formulation of a forecast, the reach has to be divided into various subreaches. For each subreach, in addition to the gauge reading, discharge observation should also be carried out.
(4) For incorporating the effect of an intermediate catchment area, at least one gauge has to be installed. If the channel is not well defined, it will be imperative to install an adequate number of rain gauges to account for the contribution from the intermediate catchment.

10.7 METHODOLOGY EMPLOYED FOR ISSUE OF FLOOD FORECASTS

The various operational steps involved before the issue of forecasts and warning are as follows:

(1) Observation and collection of hydrological and meteorological data.
(2) Transmission/communication of data to the forecasting centres.
(3) Analysis of data and formulation of forecasts.
(4) Dissemination of forecasts and warning to the administrative and engineering authorities of the states.

The above phases are described briefly in the following paragraphs:

10.7.1 Data collection

Observation and collection of hydrological data is done by the Hydrological Observation and Flood-Forecasting Organisation (HO & FFO), Central Water Commission. The Flood Meteorological Offices (FMO) collect and transmit the meteorological data. The former is responsible for planning of river-gauge/discharge network, collection of gauges and discharge data and communication of the data to its flood-forecasting centers, while the latter is responsible for planning of rain-gauge networks in consultation with HO & FFO and for collection and transmission of rainfall data to the flood-forecasting centers. The FMO provides information regarding the general meteorological situation, rainfall amounts in the last 24 hours and heavy rainfall warnings for the next 24 hours for different regions to the concerned flood-forecasting centers of the HO & FF Organisation.

At present, data from nearly 380 hydrological and 500 hydro-meteorological stations are collected every day and utilized by flood-forecasting centres for formulation of forecasts during the Monsoon period. Similarly, the meteorological data which include rainfall amounts, heavy rainfall warnings, the general synoptic situation, and weather forecasts are supplied by FMO to the concerned Divisions and Subdivisions/Control Rooms of HO & FFO daily by telephone, failing which, the information is collected by a special messenger of the HO & FF Organisation from the FMO office. Nowadays, the data of operational sites are mostly transmitted to the forecasting centers over wireless networks of the HO & FFO, most of which are 15 W SSB sets.

10.7.2 Data transmission

Transmission of data on a real-time basis from the hydrological and hydro-meteorological sites to the flood-forecasting control rooms is a vital factor in flood forecasting. Transmission of data should be as quick as possible to issue forecasts as much in advance as possible in order to enable organization of relief measures. Transmission of the observed data on a real-time basis is, therefore, essential for an efficient flood-forecasting system.

Land-line communication, i.e., by telephone/telegram, was the earliest and very commonly used mode for data transmission in flood-forecasting services till 1970. This system had the following major drawbacks:

(1) The telegraph offices were not always located very close to the data-observation sites and consequently a lot of time was wasted in performing the journey between the site and the telegraph office.

(2) During heavy rainfall periods, when timely requirement of the data becomes essential, the telegraph/telephone system became frequently out of order.

The communication system was improved by installing VHF/HF wireless (mostly 100 W/15 W HF sets) at the data-collection sites. The wireless stations are generally operated by wireless operators for transmission of data to the control rooms. Provision of wireless mechanics has also been made for repairs of the sets and their maintenance. Planning, operations, maintenance, and improvement of the communication network is looked after by officers and supporting staff. In some of the pilot projects, automatic data-transmission systems like telemetry systems have been used. However, the maintenance of such automatic instruments in the field is one of the major problems which requires immediate attention.

10.7.3 Data analysis and forecast formulation

After receipt of the hydrological and hydro-meteorological data at the control room, the data are compiled, scrutinized, and analyzed by engineers/hydro-meteorologists engaged in this work. A system of data processing before use in forecast formulation has been introduced to reduce chances of error. Many forecasting centers have been provided with microcomputer facilities for data processing.

The next important step is the formulation of the forecast. In fact, the analysis of data and formulation of forecast is the most important stage in the process of forecasting.

The various flood-forecasting centers use different forecasting models, based on the availability of hydrological and hydro-meteorological data, the basin characteristics, computational facilities available at forecasting centres, warning time required, and purpose of forecast. However, some of the common methods being used by various centres are given below:

(1) Simple correlation – based on stage-discharge data.
(2) Co-axial correlation – based on stage, discharge, and rainfall data, etc.
(3) Routing by Muskingum method.
(4) Successive routing through subreaches.

The forecasts obtained from the correlation diagrams or mathematical models, etc., are modified as necessary to arrive at a final forecast based on the prevailing conditions in the river. This requires intimate knowledge of the river by the forecaster. Forecasts once issued are further modified and revised forecasts are issued if necessary if additional information is received after the initial forecasts are made.

10.7.4 Dissemination

The final forecasts are communicated to the concerned administrative and engineering authorities of the state and other agencies connected with flood-control and management work by telephone or by special messenger/telegram/wireless depending upon local factors like the vulnerability of the area and the availability of communication facilities, etc.

On receipt of flood forecasts, the above agencies disseminate flood warnings to the officers concerned and people likely to be affected and take necessary measures like strengthening of the flood-protection and control works and evacuation of people to safer places before they are engulfed by floods. Generally, the state governments set up control rooms at state and district headquarters which receive forecasts and then further disseminate the flood warning to the affected areas and organize relief as well as rescue operations. Flood forecasts are also passed on to the All India Radio, Doordarshan, and the local newspapers for wider publicity in the public interest.

10.8 METHODS FOR FORMULATING REAL-TIME FLOOD FORECASTS

The range of methods available for formulating real-time flood forecasts is summarized in Section 10.1. The methods mainly used in India may be categorized under two groups: (i) statistical methods and (ii) deterministic methods.

10.8.1 Statistical methods

Methods based on the statistical approach make use of statistical techniques based on analysis of historical data to develop methods for the formulation of flood forecasts. The methods thus developed can be presented either in the form of graphical relations or mathematical equations. A large number of data, covering a wide range of conditions are analyzed to derive the relationships which *inter alia* include gauge-to-gauge relationships, with or without additional parameters, and rainfall-to-peak stage relationships. These methods are most commonly used in India. The Central Water Commission is the central authority for the issue of real-time flood forecasts in India. Thus the methods presented herein are derived from CWC (1989).

For developing the correlation between upstream and downstream level gauges/discharges the following parameters required to be considered:

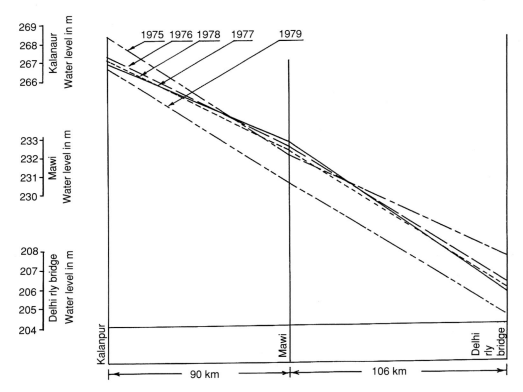

Figure 10.1 Flood profile chart of river Yamuna for Kalanaur Delhi reach

(1) Stage and discharge of the base station.

(2) Stage and discharge of the forecasting station.

(3) Change in stage and discharge of the base station.

(4) Travel time at various stages.

(5) The rainfall (amount, intensity, and duration) in the intervening catchment.

(6) Topography, nature of vegetation, type of soil, land use, density of population, depth of groundwater table, soil moisture deficiency, etc. of the intervening catchment.

(7) The atmospheric and climatic conditions.

(8) Stage and discharge of any important tributary joining the main stream between the base station and the forecasting station.

Factors (1) to (4) are basic parameters used in developing the correlation curves. Factors (5) and (6) are taken into account by introducing the rainfall and measures such as the antecedent precipitation index (API). However, factor (7) is not very important for Indian rivers as most of the floods occur during Monsoon period only.

One of the most simple and useful graphical relations is the "flood profile nomogram." This diagram indicates the peak stage at each station along a river for a storm. A number of such lines are drawn for various conditions of storm. The various lines should be drawn in different inks and the specific meteorological conditions,

such as heavy concentrated rainfall, or other conditions, such as breach of embankment, etc., should be mentioned. Such a nomogram for the Yamuna river is shown in Fig. 10.1. Although the diagram does not help in accurate forecast formulation, it serves as a useful guide in checking the formulated forecast.

The various type of graphs which are used in forecast formulation can be classified as:

(1) Direct correlation between gauge stages or discharges of the upstream (u/s) and downstream (d/s) stations.

(2) Correlation between gauge stages or discharges at u/s and d/s stations with additional parameters.

Some of the correlation diagrams which are commonly in use are discussed below.

DIRECT CORRELATION BETWEEN GAUGE STAGE AND DISCHARGE AT UPSTREAM AND DOWNSTREAM LOCATIONS

In such graphs, gauge stage, and discharge data from forecasting stations and base stations are utilized in different forms. The following types of correlation are generally used:

(1) Correlation between the Nth hour stage of base station and $(N+T)$th hour stage of forecasting stations, where T is the travel time of flood wave between the base station and

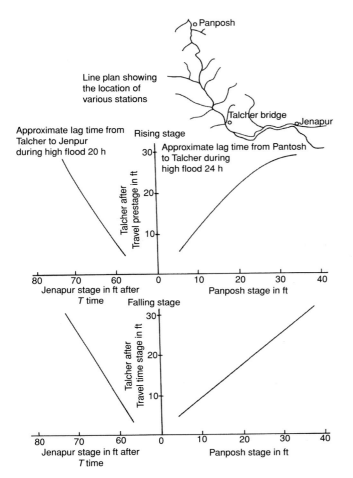

Line plan showing the location of various stations

Figure 10.2 Correlation diagram of River Brahmani between gauge at Panposh, Talcher, and Jenapur

forecasting station. Figure 10.2 shows one such graph which is used for forecasting the river stage in the River Brahmini in Orissa.

This type of graph can be developed and used for reaches of the river where: (i) there is no major tributary with considerable discharge, (ii) the catchment between the two stations is small so that the effect of rain is negligible, and (iii) the travel time from base station to the forecasting station is fairly constant for various stages. However, in most cases the travel time is not constant and varies with water level. Apart from this, such relations give considerable errors under different conditions. These relations can be considerably improved if the following aspects are taken into account:

(i) The variation in travel time.

(ii) Varying conditions during rising and falling stages of the flood (Fig. 10.2).

(iii) Antecedent conditions of the stream (this can be roughly taken into account by drawing two different sets of

curves, one for the few initial flood waves and the other for remaining flood waves).

(iv) Downstream boundary conditions (the data of the stages for high tides or larger rivers may be used to consider backwater effects).

(2) Direct correlation between the peak stage/discharge at forecasting station and base station. The gauge (or peak) at the base station and the gauge (or peak) at the forecasting station for the various intensities of flood are plotted. The travel time at various intensities of flood is also plotted corresponding to peak. Such graphs have been successfully used for the Subernarekha River in Orissa (Fig. 10.3). The warning time available is about 24–30 h.

(3) Correlation between the change in stage of the base station and the change in the stage of forecasting stations during T hours (T = time of travel of flood wave between the base station and forecasting site). Such a method obviates errors, to some extent, due to aggradation or degradation in the river section, depending upon flows. This correlation has been found more suited to large rivers with more uniform change in levels and discharges between the base stations and the forecasting stations. Such a graph developed for the river Ganga between Dighaghat and Dandhighat is shown in Fig. 10.4. Separate graphs have been developed for rising stage and falling stage.

(4) Correlation between the Nth hour and ($N+T$)th hour stages of the forecasting station with change in stage at the base station during the past T hours as variable. Different sets of graphs are drawn for rising and falling conditions of the river. Such graphs are used for forecasting river stages at a number of sites. One such correlation used for forecasting the Dalmau stage on the River Ganga, under the Lucknow Division, is shown in Fig. 10.5. If there is a large fluctuation in the stage of the base station, the parameter of average stage within the past T hours at the base station is introduced in the 1st quadrant instead of change in stage of the base station. In the 2nd quadrant the Nth hour stage of the base station is also included to account for the intensity of flood. This has been found suitable when the base station is d/s of a control structure on the river, through which the flows are released with wide fluctuations. Such a correlation is shown in Fig. 10.6.

(5) In rivers having wide fluctuation in u/s stages and, relatively, much reduced fluctuations in lower reaches due to large-scale inundation/valley storage in between the two points, a tendency effect is considered. This is done by correlating Nth and ($N+T$)th hour stages of the forecasting site over the past T hours as a variable in the 1st quadrant. Then in the 2nd quadrant, the average stage of the base station is considered as a variable. This type of graph has proved quite useful in the Bagmati and Adhwara group of rivers of Bihar in the Ganga

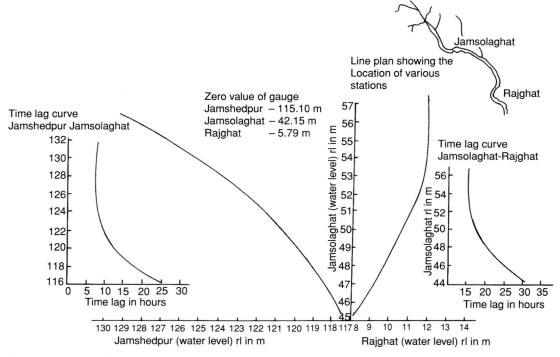

Figure 10.3 Correlation diagram of River Subernarekha between the gauges at Jamshedpur, Jamsholaghat, and Rajghat

Basin. One such graph developed for the Kamtaul site of River Adhwara is shown in Fig. 10.7.

(6) Gauge-to-gauge correlation in coastal rivers. The coastal rivers pose special problems in regard to formulation of forecasts because of the tidal effect. The simple gauge-to-gauge relation will not yield satisfactory results. Before developing gauge–gauge correlation charts, it is considered imperative to

analyze the various cases arising in coastal rivers separately and develop a different set of curves for the formulation of forecasts. The various cases encountered in the coastal rivers, are discussed below in brief:

(i) The river is in high stage and there is no tidal effect.
(ii) The river is in low stage and there is a tidal effect. As a result of this there will be backwater effects in the lower reaches of the river. The backwater effect has to be incorporated in the formulation of the forecast.
(iii) The river is in high stage and there is a tidal effect. The river will not be able to drain freely and there will be

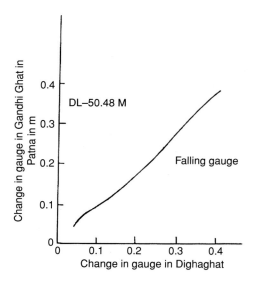

Figure 10.4 Correlation graph for Dighaghat Patna

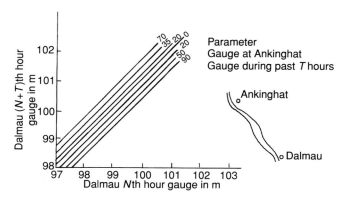

Figure 10.5 Correlation graph for Dalmau site on the River Ganga

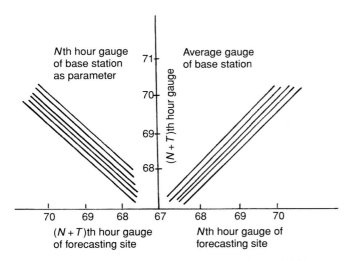

Figure 10.6 Correlation between the Nth hour and $(N+T)$th hour stages for the forecasting station with change in stages at the base station during past T hours as variable

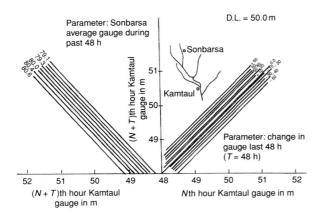

Figure 10.7 Correlation graph for the site Kamtaul River, Adhwara

a locking effect which will affect the forecast significantly. This aspect has to be considered, while developing the charts, etc., for the formulation of the forecast. Thus, it is seen that the three cases, mentioned in the preceding sections, call for the development of different sets of charts to be utilized for the formulation of forecasts.

Realising the necessity of considering the tidal effect in the formulation of forecasts for the area below Akhupada in the Baitarni river basin, a tidal gauge has been recently installed at Chandbali. The data from the tidal gauge in conjunction with the annual tidal-table (published yearly by the Survey of India) forecast will be quite useful in determining the tidal influence and backwater effects on the lower reaches of the river. This will be of particular importance when tides are enhanced by storm surges accompanying the movement of tropical cyclones on shore. The data obtained from the

tidal gauges will be useful in the development of charts for forecast formulation, considering the tidal effect.

(7) Mathematical representations. Some of the correlation diagrams developed using the u/s and d/s stages have been discussed above. Besides these the discharges at u/s and d/s stations are also used for formulation of the forecast and some mathematical equations have been developed and are in use. A few of them are discussed below.

Gauge-discharge Poanta–Tajewala model (for River Yamuna)
The travel time from Poanta to Tajewala being two hours, the Poanta gauge height in ft at t hours, $G_P(t)$ is correlated to Tajewala discharge in cusecs, $Q_T(t+2)$ for the rising limb:

$$Q_T(t+2) = 7.4045 \times 10^5 G_P(t)^{4.333}. \qquad (10.1)$$

Another relation has been developed for the falling limb when the travel time of three hours is found to be more appropriate and the relationship is:

$$Q_T(t+3) = 1.819 \times 10^{-6} G_P(t)^{5.555}. \qquad (10.2)$$

The graphical representation is shown in Fig. 10.8.

Gauge-rise models for various reaches of Yamuna
The height of the flood wave at the d/s section is related to its height at the u/s section:

$$(G_{DP} - G_{DO}) = a(G_{UP} - G_{UO}) + b, \qquad (10.3)$$

where G_{DP} and G_{UP} are the peak gauge at the d/s and u/s sections, G_{DO} and G_{UO} are the estimated stages at the time of recorded peak, had the recession prior to the start of the flood wave continued, a and b are constants, to be evaluated on the basis of past flood data. One such equation developed for the Kalanur and Delhi reach of the Yamuna is shown in the Fig. 10.9.

Discharge-rise models
The discharge rise due to a flood wave at a d/s section is related to that at an u/s section, if the effect of the intermediate catchment contribution is not significant:

$$(Q_{DP} - Q_{DO}) = m(Q_{UP} - Q_{UO}) + n, \qquad (10.4)$$

where Q_{DP} and Q_{UP} are the discharges at the d/s and u/s sections, Q_{DO} and Q_{UO} are the estimated discharges at the time of recorded peak, had the recession prior to the start of flood wave continued, and m and n are constants to be evaluated on the basis of past flood data. One such relation developed for the Kalanaur–Mawi reach of the Yamuna is shown in Fig. 10.10.

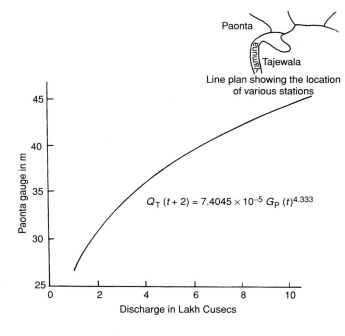

Line plan showing the location of various stations

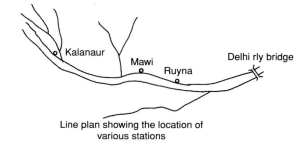

Line plan showing the location of various stations

$$Q_T\,(t+2) = 7.4045 \times 10^{-5}\,G_P\,(t)^{4.333}$$

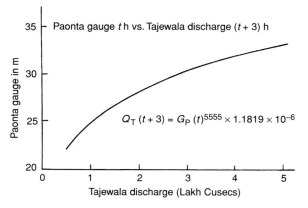

$$Q_T\,(t+3) = G_P\,(t)^{5555} \times 1.1819 \times 10^{-6}$$

Figure 10.8 Poanta gauge vs Tajewala discharge (with 2 h lag) rising limb

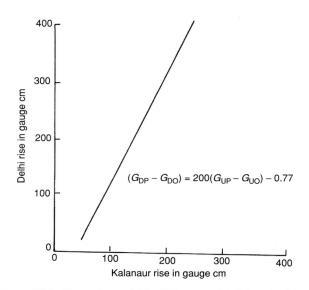

$$(G_{DP} - G_{DO}) = 200(G_{UP} - G_{UO}) - 0.77$$

Figure 10.9 Gauge-rise model for Kalanaur and Delhi reach of the River Yamuna

CORRELATION BETWEEN GAUGES AT UPSTREAM AND DOWNSTREAM WITH ADDITIONAL PARAMETERS

When the direct gauge-to-gauge correlation is not successful because of appreciable contribution due to rainfall in the reach catchment, intermediate tributaries, or the varying soil moisture condition, etc., then the introduction of additional parameters of discharge of the tributary, average rainfall over the intercepting catchment, API, etc., becomes necessary and gives better results.

With the increasing availability of data and introduction of better data transmission facilities, correlation diagrams are being developed with additional parameters. The various parameters are introduced in different quadrants.

Some such diagrams which are at present under use are discussed below in brief:

(1) Correlation between the Nth hour, and $(N+T)$th hour gauge of forecasting station with change in the level of a tributary during the past T_1 hours and change in level of the base station during the past T hours.
(2) Correlation between Nth hour and $(N+T)$th hour gauge of forecasting station with the following parameters:
    (i) Rise/fall at u/s base station.
    (ii) Rainfall observed at the u/s base station.

When a number of tributaries affect the water level at the forecasting station, then the change in the base station on the main river as well as base stations on the tributaries can be considered as additional parameters.

Multi tributary model

A discrete, linear, time-invariant model has been developed for operational flood forecasting of the river Brahmaputra at

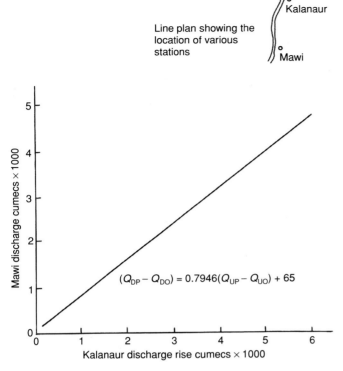

Figure 10.10 Correlation between rise in discharge at Kalanaur and Mawi

Dibrugarh. This model is based on the difference between the gauge reading at the forecasting station and the upstream base station in the tributary. The use of differences in gauge readings as input in the model takes care of the aggradation or degradation of the river bed of the tributary and the main river. The model in general is expressed as:

$$g_{(i+T),i} = A_{1i} g_{i,(i-T)} + \sum_{j=1}^{m} A_{2,j} h_{(i-T+T_j),(i+T)_j}$$

$$+ \sum_{j=1}^{m} A_{3,j} h_{(i-T_j),(i-T_j-T)}, \qquad (10.5)$$

where

m = number of tributaries (three in this case)

T = forecasting time

T_j = lag time between the forecasting stations $(T \leq T_j)$

$h_{(i-T_j+T),(i-T_j)}$ = difference in gauges at the u/s station on the tributary between the $(i - T_j + T)$th and $(i - T_j)$th instants.

$h_{(i-T_j),(i-T_j-T)}$ = difference in gauges at the u/s station on the tributary between $(i - T_j)$th and $(i - T_j - T)$th instants.

$g_{i,(i-T)}$ = difference in gauge at ith and $(i - T)$th instants of time at the forecasting station.

$g_{(i-T),i}$ = difference in gauge at $(i - T)$th and ith instants of time at the forecasting station, i.e., the forecast value.

$A_{1,j}$, $A_{2,j}$, $A_{3,j}$ are the parameters which are to be determined and can be estimated by the method of least squares. In this case the forecast of Dibrugarh is formulated with the help of observed gauge data on three major upstream tributaries, namely Dihang, Debang, and Lohit.

Rainfall-stage method

The relationship for estimating the peak discharge or the peak stage with the help of rainfall data is of great operational significance in the sense that it enables one to find the expected peak discharge or stage, which is one of the important requirements in flood warning. In its simplest form it is the relation between the average rainfall over the catchment and the peak stage. This relation may be either graphical or mathematical and can be very easily established by using the statistical technique. The results can be further improved by incorporating other parameters such as API, etc. These relations are used for many places with quite good results but the deficiency in this method is that the time of occurrence of the peak or the full shape of the hydrograph cannot be forecast.

One such relation has been developed for the Anandpur site on the Baitarni river where the peak discharge is estimated by using the relation:

$$Q_{\max} = 1.451 - 0.1678 + 0.0129x^2, \qquad (10.6)$$

where

$Q_{\max}$ = peak discharge at Anandpur in cusec

$x = x_1 + x_2$

x_1 = weighted storm rainfall over the catchment in cm

x_2 = effective antecedent rainfall in cm.

The weighted storm rainfall over the catchment is estimated by assigning certain weights to the various stations, depending upon the area and geographical condition:

$$X = 10.2A + 0.7B + C + 0.6D, \qquad (10.7)$$

where A, B, C, and D represent the rainfall at various stations in cm.

The effective antecedent rainfall is taken as a certain percentage of the antecedent rains. Table 10.1 has been assumed and used in all calculations.

The relation has been developed by analyzing data from 23 previous storms and the results are quite satisfactory.

10.8.2 Deterministic methods

One of the important areas in hydrology pertains to the study of the transformation of the time distribution of rainfall on the catchment to the time distribution of runoff. This transformation is often studied by first relating the volume of rainfall to the volume of direct

Table 10.1

Weighted antecedent rainfall in mm	Percentage of antecedent rainfall to be taken as effective
0–15	Nil
15–20	20 %
20–30	25 %
40–60	30 %
60–80	35 %
80–100	40 %
100–120	45 %
120–140	50 %
140–160	60 %
160–180	70 %
180–200	80 %
More than 200	90 %

surface runoff, thus determining the time distribution of rainfall excess (the component responsible for direct surface runoff on the catchment) and then transforming it to the time distribution of direct runoff though a discrete or continuous mathematical model. The first step decides the volume of the input to the catchment and therefore any error in its determination is directly transmitted through the second step to the time distribution of direct runoff. A number of watershed conceptual models find this component for each time-step through a number of stores representing various processes on the catchment. The parameters of these models including those in the functional relationships are determined from the historical record and their performance is tested by simulating some of the rainfall-runoff events which have not been used in the parameter estimation process. The models generally need to be run continuously so that the status of various stores is available at all times. One of the operational uses of these models is in the area of real-time flood forecasting required for real-time operation of the reservoir. In such a situation these models are run by inputting the rainfall and forecasts are issued assuming no rainfall beyond the time of forecast value of the rainfall in the future.

The infiltration part of these models and their context decide the volume of input. At the time of calculation the catchment is also performing the transformation operation to produce the direct runoff at the gauging station. Since the model is simulating the action of the catchment, it would be appropriate to make use of this information in finding out the contribution which the rainfall is going to make to the direct runoff on the catchment. However, the complexity of these models does not lend itself to this exercise during the event. The SSARR (stream-flow synthesis and reservoir regulation) model, Sacramento model and NAM-System 11 FF model are some of the watershed conceptual models used for formulating real-time flood forecasts. In India, real-time flood forecasts have been formulated in some pilot projects using these

models. However, these conceptual models are not being utilized because of inadequacy of data and problems associated with the proper calibration of the models.

Of late, methods based on the unit hydrograph approach have been formulated for real-time forecasting which overcome the difficulties associated with complex hydrological models. The unit-hydrograph method has long been recognized as a useful tool for converting excess rainfall to direct surface runoff by linear transformation. The assumptions underlying this method and their limitations with regard to areal size, linearity, and uniform spatial and temporal distribution of rainfall have been discussed in most of the text books and many research papers. In India, the applications of the unit hydrograph technique are restricted to catchments of sizes less than $5000 \, km^2$. However, for larger catchments, a network model has been developed. In this model the catchment is divided into subcatchments and the main river is divided into subreaches, considering two consecutive nodes. The nodes are the points where the tributaries of the subcatchments join the main river. The principle of the unit hydrograph is applied for converting the excess rainfall to direct surface runoff for each subcatchment considered as lateral flow to the river, and the flood-routing technique is used for routing the direct surface runoff at the upstream node through the river subreaches to the downstream node. The computations are performed for the network model structure to estimate the direct surface runoff at the outlet of the catchment. The HEC-1 model is one example that has an option of network simulation using these concepts.

The Hydrologic Engineering Centre (HEC) of the US Army Corps of Engineers has developed a computer model HEC-1F, a modification of model HEC-1, for the purpose of real-time forecasting. The HEC-1F model uses the unit-hydrograph technique with constant loss rate to forecast the runoff. Forecasting by the HEC-1F model is accomplished by re-estimating the unit-hydrograph parameters and the loss-rate parameters, as additional rainfall-runoff data are reported, and using these updated parameters the future flows are estimated for forecasting. The Snyder's synthetic unit hydrograph described by two parameters is used as the unit-hydrograph model. For the estimation of the unit hydrograph and constant loss-rate parameters, the model uses a univariate search technique of optimization. HEC(1984) provides complete details of the model. In India, the HEC-1 F model is applied for formulating the flood forecasts of a pilot basin.

10.9 FLOOD-FORECASTING STATUS IN ARID AND SEMI-ARID REGIONS OF INDIA

The arid and semi-arid regions of India receive less rainfall as compared to the other regions. The occurrence of floods in these regions is not as frequent as elsewhere. However, in the past,

REAL-TIME FLOOD FORECASTING

flash floods have been occasionally experienced in the regions due to high intensity, short duration rainfall causing loss of human lives and property. As such there is no system of forecasting flash floods because the lead period is very short. There is a need to develop a real-time rainfall-forecasting system to provide flash-flood forecasts well in advance so that emergency action can be taken to evacuate the people so that their lives may be saved and the loss of property may be minimized. In this regard, provision may be made for radars and other automatic instruments to measure and forecast the rainfall together with the advance communication and dissemination system in this region.

10.10 MODERNIZATION OF FLOOD-FORECASTING SERVICES

The flood-forecasting services provided by the Government of India through the Central Water Commission, which has over 25 years of experience in the field and commendable performance to its credit, is presently poised for a big leap forward by application of modern technology in the field of communication, and introduction of high-speed computers for forecast formulation.

In order to improve the warning time and the accuracy of the forecast, it was considered necessary to adopt the latest technology for real-time collection and transmission of hydrological and hydro-meteorological data and application of the high-speed computer, using hydrological and mathematical models.

In order to achieve these objectives, a pilot project to establish a fully automatic operational river- and flood-forecasting system in the country with the assistance of the World Meteorological Organisation (WMO)/United Nations Development Programme (UNDP) has been implemented in the Yamuna Basin up to Delhi. The experience gained from this project is being applied to modernize other forecasting centers in the country.

Another scheme, the CWC–DHI (Central Water Commission and Danish Hydraulic Institute, Denmark) collaboration project, with the Damodar River Basin as focus project, has also been implemented and computerized mathematical models developed by DHI are being used for inflow forecasting for the formulation of a flood-control scheme. A scheme on similar lines has been taken up for the Godavari Basin.

The availability of hydrological and meteorological time-series data as well as spatial data is somewhat limited. Because of this, it is difficult to calibrate and validate the multi-parameter hydrological models for formulating real-time flood forecasts in the field. Recently, a World Bank aided Hydrology Project, Phase I, has been completed for the nine states of India, namely Andhra Pradesh, Maharashtra, Gujarat, Kerala, Tamilnadu, Karnataka, Madhya Pradesh, Chattishgarh, and Orissa. Under this project, the observation networks for monitoring the hydro-meteorological variables have been strengthened for different river basins located

in these states. A hydrological information system (HIS) has been developed for the processing, storing, and retrieving of data collected from the field. State data centers are linked with the national data centers and a system has been evolved for providing the data to user agencies through the Internet. It has improved the data availability for these states. Another World Bank Assisted Hydrology Project, Phase II, is likely to start soon wherein the hydrological analysis would be carried out using the processed data available for HIS. Under this project, the development of a decision-support system (DSS) for real-time flood forecasting of the Bhakhara–Beas reservoir system has been envisaged.

10.11 RESEARCH DEVELOPMENTS

Computing techniques such as stochastic models, artificial neural networks (ANN) and fuzzy logic are being applied for real-time flood forecasting. However, such applications are mostly limited to research and development at academic and research institutions of the country. Operational uses of such techniques are yet to be established and encouraged.

10.11.1 Stochastic models for real-time flood forecasting

Several stochastic/time-series models have been proposed for modelling hydrological time series and generating synthetic stream flows. These include the Box–Jenkins class of models (Box and Jenkins, 1970, 1976; Salas *et al.*, 1980). The time-series models are considered to be most suited for real-time forecasting as on-line updating of model forecasts and parameters can be achieved using various updating algorithm. It has been observed that the dynamic stochastic time-series models are most suitable for on-line forecasting of floods (Kalman, 1960; Sage and Husa, 1969; Eykhoff, 1974; Kashyap and Rao, 1975; Kumar, 1980; O'Connell, 1980; Chander *et al.*, 1980, 1984; Chander and Prasad, 1981). These models also provide a means for the quantification of the forecast error, which may be used to calculate the risks involved in the decisions based upon these forecasts. Further, these models can be operated even with interrupted sequences of data and are easy to implement on computer and other computing devices.

10.11.2 Artificial neural-network models for real-time flood forecasting

The formulation of real-time flood forecasting using statistical and stochastic models is based upon the assumption of linearity, whereas the quantity of runoff resulting from a given rainfall event depends upon a number of factors and is dominantly non-linear. Recently, another class of black-box models in the form of artificial neural networks (ANN) has been introduced in modelling real-time problems, wherein the non-linear relationship between the rainfall and runoff process is modelled. The ANN model has wide applicability in civil engineering applications and many research papers have been published on its application. The use of ANN in real-time flood forecasting is of very recent origin and is still in the evolution stage. Recently Xiong and O'Connor (2002) studied four updating models for real-time flood forecasting, in which the authors have shown that the use of an ANN model as a forecast error update model has in fact not improved the real-time flow forecasting efficiencies over that of the standard AR model.

10.11.3 Fuzzy-logic techniques for real-time flood forecasting

The emergence of a flood and thus its forecast depend elementarily on the discharge process in the natural catchment area of the river. This process is rather complex and its mapping into a suitable process model for an automated flood forecast is accordingly difficult. Hydrologic models are useful only to the degree that they represent processes in the world. Mathematical models have been developed based either on physical considerations or on a statistical analysis to estimate floods from small- as well as large-size catchments. Existing flood-forecasting models are highly data specific and complex. Unlike mathematical models that require precise knowledge of all the contributing variables, fuzzy logic offers a more flexible, less assumption dependent and self-adaptive approach to modelling flood processes, which by their nature are inherently complex, non-linear, and dynamic. Fuzzy logic-based models can be used to model process behavior even with incomplete information. Fuzzy logic is widely regarded as a potentially effective approach for effectively handling non-linearity inherently present in the hydrological processes (Zadeh, 1965). Moreover, this technique can be used for modelling systems on a real-time basis (Lohani *et al.*, 2005). Other advantages include the potential for improved performance, faster model development and execution times and therefore reduced costs, the capability to plug fuzzy logic directly into conventional models, and the ability to provide a measure of prediction certainty. Fuzzy logic-based procedures may be used when conventional procedures are getting rather complex and expensive, or vague and imprecise information flows directly into the modelling process. With fuzzy logic it is possible to describe available knowledge directly in linguistic terms and according to rules. Quantitative and qualitative features can be combined directly in a fuzzy model. This leads to a modelling process which is often simpler, more easily manageable, and closer to the human way of thinking compared with conventional approaches.

The use of fuzzy logic in the field of hydrological forecasting is a relatively new area of research, and the potential to enhance flood forecasting by incorporating other soft computing technologies into a hybrid solution still remains to be exploited. Recently use of fuzzy set theory has been introduced to interrelate variables in hydrologic process calculations and modelling the aggregate behavior. Further, the concepts of fuzzy decision-making (Bellman and Zadeh, 1970) and fuzzy mathematical programming have great potential application in flood-management models to provide meaningful decisions in the face of conflicting objectives.

10.12 CONCLUDING REMARKS

In India, most of the techniques for formulating real-time flood forecasts are based on statistical approaches. For some pilot projects, network models and multi-parameter hydrological models are used. Conventional systems of communication are normally used for transmitting the data in real time. Automatic systems of data communication like telemetry systems are used in pilot projects on a limited scale. In arid and semi-arid areas, flash-floods are usually experienced. There is no system for formulating the flash-flood forecasts in these regions. It results in heavy loss of lives and property.

There is a need for significant improvement of the real-time flood forecasting systems in India. Efficient automatic communication systems are required to be established for transmitting the data in real time. Forecasting techniques such as deterministic models, stochastic models, ANN, and fuzzy logic techniques, etc. require study and a suitable method may be recommended for field applications based on the performance evaluation criteria and considering the data availability and purpose of the forecast. The information about a flood has to be disseminated well in advance to the people likely to be affected so that an emergency evacuation plan may be prepared and properly implemented. A hydrological information system is required to be developed for all the river basins of India. Techniques based on deterministic approaches have to be developed and applied to real-time flood forecasting. The pilot project proposed under the World Bank aided Hydrology Project, Phase II, for the development of decision-support system (DSS) for the Bhakhra–Beas reservoir system would provide useful information to decision-makers in real time for taking necessary flood management measures downstream of the reservoir. Furthermore, the technology to be acquired under the project would be very useful for developing such systems in other flood-prone river basins. Flash-flood forecasting is another important area which requires immediate attention.

REFERENCES

Bellman, R. E. and Zadeh, L. A. (1970). Decision-making in a fuzzy environment. *Manag. Sci.*, **17(4)** B141.
Box, G. E. P. and Jenkins, G. M. (1970). *Time Series Analysis: Forecasting and Control*, 1st edn. San Francisco, CA: Holden Day.

Box, G. E. P., and Jenkins, G. M. (1976). *Time Series Analysis: Forecasting and Control*, revised edition. San Francisco, CA: Holden Day.

Central Water Commission (CWC) (1989). *Manual on Flood Forecasting.* New Delhi: River Management Wing.

Chander, S. and Prasad, T. (1981). Forecasting and prediction of floods. In *Indo-US Workshop on Flood Mitigation.* Unpublished proceedings.

Chander, S., Kapoor, P. N. and Natarajan, P. (1984). *Newer Techniques in High Flow Range Forecasting.* Delhi: Civil Engineering Department, Indian Institute of Technology (Indian National Committee for the International Hydrological Programme).

Chander, S., Spolia, S. K., and Kumar, A. (1980). Flood stage forecasting in rivers. In *Monsoon Dynamics*, ed. M. J. Lighthill. Cambridge: Cambridge University Press.

Eykhoff, P. (1974). *System Identification, Parameter and State Estimation.* New York: Wiley.

GFCC (1986), *Comprehensive Plan of Flood Control for the Ajay River System.* **Vol II/8, March**.

Hydrologic Engineering Centre (HEC) (1984). *HEC1 Flood Hydrograph Package, User's Manual*, US Army Corps of Engineers.

IWRS (2001). Theme Paper on *Management of Floods and Droughts.* Roorkee, India: Indian Water Resources Society, 4–12.

Kalman, R. E. (1960). A new approach to linear filtering and prediction problem transportation. *ASME J. Basic Eng.*, **82**, 35–44.

Kashyap, R. L., and A. R. Rao (1975). *Dynamic Stochastic Models from Empirical Data.* New York: Academic Press.

Kumar, A. (1980). Prediction and real time hydrological forecasting. Ph.D. Thesis, Indian Institute of Technology, Delhi, India.

Lohani A. K., Goel, N. K., and Bahtia, K. K. S. (2005). *Real time flood forecasting using fuzzy logic.* International Conference on Hydrological Perspectives for Sustainable Development (HYPESED-2005), Roorkee, India.

O'Connell, P. E. (1980). *Real Time Hydrological Forecasting and Control.* Wallingford: Institute of Hydrology, 264.

Palaniappan, A. B. (1997). Dams and their failures. In *Lecture Notes on Dam Break Flow Analysis.* Roorkee: National Institute of Hydrology, 126–71.

Sage, A. P. and Husa, G. W. (1969). Adaptive filtering with unknown prior statistics. *Proc. Joint Automatic Control Conference.* Boulder, CO, 760–9.

Salas, J. D., Delleur, J. W., Yevjevich, V., and Lane, W. L. (1980). *Applied Modeling of Hydrologic Time Series.* Littleton, CO: Water Resources Publications.

Wood, E. F. and O'Connell, P. E. (1985). Real time forecasting in hydrological forecasting. In *Hydrological Forecasting*, ed. M. G. Anderson and T. P. Burt. Chichester: John Wiley Sons, 505–58.

Xiong, L. and O'Connor, K. M. (2002). Comparison of four updating models for real time river flow forecasting. *Hydrolog. Sci. J.*, **47**(4), 621–40.

Zadeh, L. A. (1965). Fuzzy sets, *Info. Cont.*, **8**, 338–53.

11 Groundwater flow modelling in hard-rock terrain in semi-arid areas: experience from India

S. Ahmed, J.-C. Maréchal, E. Ledoux, and G. de Marsily

11.1 INTRODUCTION

Across the world, the concern for water resources is growing as a result of population growth, climate change, and alarming signs that in some areas of the world, groundwater resources are being depleted at an unsustainable rate. This has prompted a re-examination of the world's water resources and the relationship between water and the environment. According to a United Nations survey, scarcity of fresh water is, in some areas, considered to be the world's most pressing concern (UN, 1987; El-Shibini and El-Kady, 2002). In many countries, to meet the increased demand for water, groundwater resources must be tapped. However, to ensure sustainability, much greater emphasis must be put on groundwater management than on exploration for new groundwater resources, as most productive aquifers have already been identified. Groundwater is particularly important in arid and semi-arid regions that lack perennial sources of surface water due to low rainfall and high evapotranspiration. This article focuses on groundwater management in hard-rock areas in semi-arid climates, where aquifers exist in the upper weathered-fissured section of the system; these aquifers receive little recharge, and have different and more complex characteristics than in classical sedimentary media. Specialized techniques are thus required to characterize and manage them.

Groundwater modelling has produced answers to many difficult questions that arise in the course of hydrogeological investigations. At present, it has become an indispensable tool in understanding and effectively managing aquifer systems. Some of the important tasks that a groundwater model can be used for are: interpreting the measured head and concentration data, estimating natural recharge, understanding the mechanisms of flow and transport, identifying the effect of boundary conditions, understanding the interaction between aquifers and surface water bodies, estimating the extent of contamination, and predicting the head and concentration distribution in space and time as well as fore-casting the effects of various management schemes or remedial measures.

This chapter first gives a general description of aquifer systems in hard-rock terrain in semi-arid regions of India, including their resources and exploitation. There follows an introduction to groundwater modelling principles, the various steps involved in the development of groundwater models and data requirements, and the use of geostatistical methods in modelling and uncertainty estimation. The chapter ends with an Indian case study.

11.2 AQUIFERS AND WATER RESOURCES IN (SEMI)-ARID REGIONS OF INDIA

Low, erratic rainfall, nutrient-poor soils, high temperatures, and high solar radiation characterize semi-arid regions. Water-resource problems are among the most important issues facing the people living in these regions, and as noted above, groundwater is important and often the main resource. The rapid increase in groundwater exploitation in arid and semi-arid regions has had adverse effects on both water quantity and quality. Overexploitation of resources has been recorded in recent studies – e.g., in the Ogallala aquifer in the United States, the Coastal and Souss aquifers in Morocco, and the Hermosillo aquifer in Mexico, see for example, Kromm and White (1992) Oroz (2001) Custodio (2002) Belden and Osborn (2002) Younes and Razack (2003).

Irrigation is a particular problem. In India and northern China, for example, the irrigated area grew from about 25 % of the cultivated surface in the 1960s to well over 50 % in the 1990s. The sources of irrigation water vary widely between countries, depending on the climatic and hydrogeological conditions and the development strategies of the nations. Over 50 % of India's irrigated area is supplied by groundwater, with around 80 % in semi-arid regions; this percentage is 43 % in the USA, 27 % in China, and 25 % in Pakistan (World Bank, 1999). As a result,

Hydrological Modeling in Arid and Semi-Arid Areas, ed. Howard Wheater, Soroosh Sorooshian, and K. D. Sharma. Published by Cambridge University Press. © Cambridge University Press 2008.

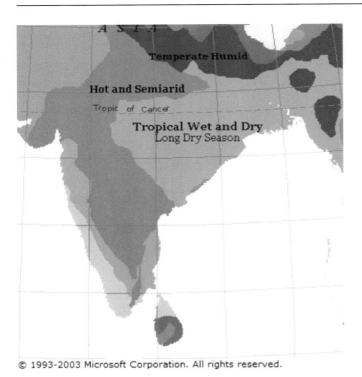

Figure 11.1 Semi-arid regions in India

the limit of groundwater development for irrigation in the world's major "breadbasket" countries may be reached in a few years. In India this will be before the end of the present decade; soon only the renewable annual recharge to the aquifers will be exploitable, together with artificial recharge, if implemented rapidly. The exploitation of groundwater resources in India, which was the source of the "green revolution," has therefore to be slowed down in the coming decades. At present, however, the groundwater aquifers are poorly managed; only the unreliable power supplies are currently constraining the growth of irrigated agriculture.

The water crisis in India is more pronounced in the western and peninsular areas where acute water shortages are predicted to occur in the coming 25 years. Climatologically, the major part of the southern Indian peninsula has semi-arid, bordering on arid, conditions with seasonal rainfall (500 mm) and high potential evapotranspiration (1800 mm). In India as a whole, 48 % of the land is semi-arid and 25 % is arid, which includes about 5 % of desert (Fig. 11.1). In semi-arid India, the natural hydrological and hydrogeological systems are extremely diverse. The resources include rainfall, surface water, and groundwater. Surface water from ephemeral streams, perennial rivers, and reservoirs provides an important source of groundwater recharge and groundwater provides the base flow of streams and rivers; therefore groundwater and surface water resources need to be treated together in planning and management.

In India, groundwater is withdrawn from traditional large-diameter dug wells (which can also store water or be used for artificial recharge) and from bore wells, whose numbers have greatly increased in the last 30 years. Comparable resources above ground include local tanks (small reservoirs), hill or check-dams, and canals to store and distribute surface water from large dams. Major irrigation in India depends on surface reservoirs and groundwater.

11.2.1 Rainfall

Natural precipitation is the most important source of water for farming in India. Rainfall depends to different degrees on the south-west Monsoon, the north-east Monsoon, shallow cyclonic depressions and disturbances, and violent local storms. India receives most of its rainfall from the south-west Monsoon originating in the Indian Ocean. About 75 % of the rainfall is received in four months i.e., June to September, but it is characterized by uneven geographical and seasonal distribution and frequent departures from the normal pattern (Fig. 11.2).

The south-west Monsoon rainfall received during the months of June–July is critical and the success of the Kharif crop (i.e., the crops that need excess water, e.g., rice) depends greatly on the distribution and amount of rain during these two months. The south-west Monsoon is responsible for 75–80 % or more of the total annual rainfall in the country. The north-east monsoon occurs during October–November when cyclonic storms form in the Bay of Bengal and, when they strike coastal Andhra Pradesh or the Coromandel coast, they bring heavy rain to these areas. About 11 % of the total rainfall in the country is received during this season.

11.2.2 Hard-rock aquifers

In arid and semi-arid regions, due to a lack of surface-water resources, shallow groundwater has historically been exploited for irrigation. This includes hard-rock aquifers, which are very often found in arid regions.

The hard rocks form a significant group and occupy about 65 % of India's total land area (Fig. 11.3). Stratigraphically they range from Archean to recent (laterites). A hydrogeological province is defined as a subsurface groundwater basin, with the same suite of rocks of the same age and geochemical characteristics. By this definition, all rocks containing aquifers with secondary porosity can be grouped into a single hydrological province, namely "hard rocks." They include the Deccan Traps, Precambrian sandstone, quartzite in the Vindhyans, and Cuddapah formations and granite, gneiss, schist, and phyllite of Archean age. The crystalline, quartzite, sandstone, and many schistose rocks formed as a result of metamorphism are also called hard rocks (Radhakrishna, 1970).

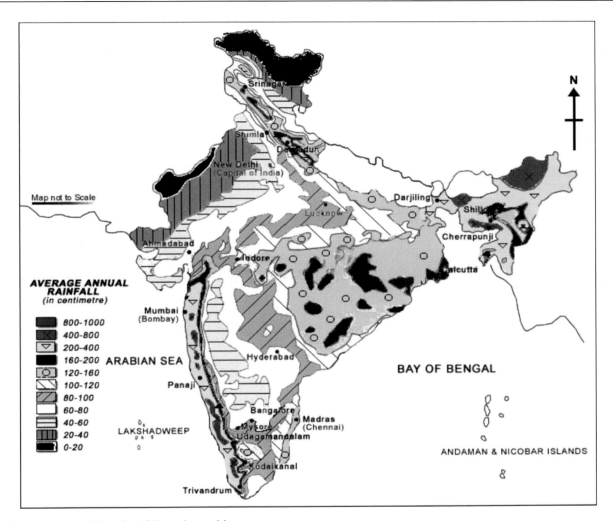

Figure 11.2 Spatial variability of rainfall: south-west Monsoon

Groundwater occurs in the upper weathered mantle of the underlying hard-rock formations and in the fractures and joint planes. The faults and shear zones also contain large supplies of groundwater.

Unweathered massive hard rocks have low permeability (Fig. 11.4), but groundwater occurs in the: (i) weathered zone (Fig. 11.5) and (ii) fractured, fissured and jointed zones (Figs. 11.4 and 11.6), which form the water-bearing strata. Minor fractures and joints are inter-connected as well as connected to the major joints that in some cases function as conduits to produce large yields in the wells.

The weathering of hard rocks in India's warm climate is mostly due to the weathering of minerals by the infiltrating rainfall. In a crystalline rock, the minerals such as feldspar and mica are mostly chloritized or transformed into clay. The quartz is less weathered, and remains as sand grains. The weathered zone is therefore made up of a mixture of sand and clay, with some remains of other minerals from the original rock. But the weathering process is very slow: to transform 10 m of granite into a weathered zone, hundreds of thousands or millions of years are required. But the Indian subcontinent has been stable and the outcropping hard rocks subjected to weathering for many hundreds of millions of years. Therefore, the thickness of the weathered zone can reach more than 30 m in some areas. Its permeability is, in general, rather low, much lower than in alluvial deposits, for example.

The fact that a fracture network exists only in the upper part of the subsurface profile and that its characteristics are relatively homogeneous indicate that a superficial and laterally continuous process is at the origin of the fractures. In the absence of glaciers and conditions favoring high erosion as is the case in India, another process than decompression must be invoked to explain the genesis of the fractures. Wyns *et al.* (2004) have shown that

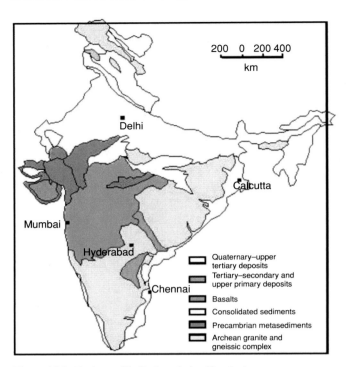

Figure 11.3 Geology of India (broad classification)

Figure 11.4 Unweathered and massive granites. (Subvertical and horizontal jointing causing weathering)

fracturing in hard-rock areas depends not only on the decompression processes, but also on mineral weathering. The weathering of the micaceous minerals, particularly biotites, into clays causes an increase in volume, which induces cracks in the rock that initiate fracturing. Where the rock texture is isotropic (e.g., in granite), the fractures are mostly subparallel to the contemporaneous weathering surface (Maréchal *et al.*, 2003b). In highly foliated rocks (metamorphic rocks), the orientation of fractures can also be controlled by the rock structure (Pye, 1986). Resulting non-horizontal fractures are less affected by the closing effect of lithostatic constraints. This could explain the less rapid vertical decrease in the permeability of these rocks compared with granitic rocks as reported by Havlík and Krásný (1998) and Krásný (1999). At a regional (and borehole) scale, the weathering process is much more homogeneous than tectonic phenomena (local deep-fault zones).

Below the weathered zone lies the weathered-fractured zone. In fact, the transition from the sandy weathered zone to the fissured zone is progressive, and the thickness of the weathered-fractured zone is on the order of 50–100 m, in general. There usually is a good hydraulic connection between the weathered zone and the weathered-fractured zone. Only in very few cases is a low-permeability layer inter-bedded between the two, see for example, Pistre (1993). Two conductive fracture sets – a horizontal set and a subvertical one – can be observed on outcrops (Fig. 11.6). The latter connects the horizontal network, ensuring a good con-

nectivity in the aquifer. The weathered-fractured zone of the hard-rock aquifer has, in general, the highest permeability, higher than that of the weathered zone. Nevertheless, the subvertical set of fractures is less permeable than the horizontal one, introducing a horizontal-to-vertical anisotropy ratio for the permeability due to the supremacy of horizontal fractures (Maréchal *et al.*, 2003a,b; Maréchal *et al.*, 2004).

The vertical fractures are the result of tectonic stresses and deformation that have affected the rock since its formation. They may extend much deeper than 100 m, but in general, below a certain depth, these fractures are closed due to the lithostatic constraint of the overlying layer and to clogging by clay or other minerals. Studies of litho-logs and depth-yield characteristics of the wells in hard-rock systems show that the water-bearing fracture systems tend, in general, to become less abundant with depth. The permeability decreases with depth (Maréchal *et al.*, 2004). Thus, the hard rock is very rarely permeable below 100 m, except in some particular settings, e.g., the Ploemeur site in Brittany, France, where high-yield wells are found deeper than 120 m (Le Borgne *et al.*, 2006).

The ability of hard-rock aquifers to store water is due to the porosity of the rocks, which mainly consists of the pores between the grains in the weathered zone and the openings of the fractures in the weathered-fractured zone. The unweathered blocks of rock between the fractures in this zone have, in general, a very low porosity and cannot store much water. The major part of the

Figure 11.5 The weathered zone in an abandoned dug well

Figure 11.6 Fractures, fissures, and joints in granite as potential aquifers

water storage is in the weathered zone, the volume of the fracture openings is, in general, small.

The rock type, degree, and genesis of secondary openings, and capacity to hold and transmit water under normal conditions are some of the characteristics that need to be evaluated with a fair degree of accuracy in hard-rock aquifers. In most areas, immediately after the first few showers of the Monsoon, there is a considerable rise in the groundwater levels, indicating that infiltrated water has percolated through the weathered rock and filled the void space (fractures or pores). The infiltration process is not instantaneous, and the upper part of the weathered zone can hold water that is slowly migrating downward. During subsequent rains, runoff may be significant if there is no gap between the rainfall events, because the void space of the upper layer is filled with water. This runoff may be stored in local reservoirs (tanks) and can infiltrate into the aquifer over long periods of time, if the tank bottom is not silted. Traditionally, local communities would clean each tank by hand every year, to increase the groundwater recharge, but this traditional practice has largely been lost and most tanks are no longer a source of recharge for the aquifers. This could be improved by mechanically removing the silt, or by artificially injecting the water stored in the tanks into the aquifer, e.g., by using abandoned dug wells. The weathered-zone thickness of the hard-rock aquifers is a limiting factor for the recharge. In addition, the heterogeneity of the hard-rock aquifers may also restrict the recharge because of impervious detached blocks of unweathered rocks within the weathered zone.

In very arid regions, the functioning of the recharge is similar, but instead of occurring every year, it may only occur once every 10 to 30 years. In the Sahara desert, for instance, it does not rain for many years until a heavy storm brings very large amounts of rain (e.g., 300 mm in one event), part of which may infiltrate. See for example, Besbes (2003).

11.2.3 Groundwater exploitation

The data show that in India, prior to 1975, groundwater was withdrawn mostly from shallow, open dug wells. They were more than 10 m in diameter and reached depths of 12 to 15 m, tapping mostly the weathered zone, which was fully saturated with water; fluctuations in water levels typically occurred about 3 m below ground level. This scenario has changed totally because of well drilling from 1975 to 1985. Bore wells 15 cm in diameter started tapping the fractured aquifer. The depths of the bores initially ranged from 25 to 30 m in 1970–80. These wells tapped both the weathered and weathered-fractured zones and the dug wells were also used as reservoirs to store water. Due to increasing demands commensurate with a large increase in the rural population, bore wells were drilled to depths of more than 60 to 70 m in the period 1980–90. The average yield of wells in these rocks ranges from 10 to 100 m^3/d. The net result of the drilling has been that a large part of the weathered zone has dried up. The dug wells have become defunct due to water-level decline, and, presently, most of them are abandoned. Historical data of Indian water levels for the past two decades indicate that they have typically declined by 6 to 8 m in the discharge zones, and by 12 to 15 m on average and up to 25 m in the withdrawal areas (Subrahmanyam *et al.*, 2000; Radhakrishna, 2004).

11.3 THEORETICAL BACKGROUND OF NUMERICAL AQUIFER MODELLING

The formulation of the general groundwater-flow equation in porous media is based on mass conservation and Darcy's law, and can be represented as follows in three dimensions (see for example, Bear, 1979; Mercer and Faust, 1981; Marsily, 1986):

$$\frac{\partial}{\partial x}\left(K_x\frac{\partial h}{\partial x}\right) + \frac{\partial}{\partial y}\left(K_y\frac{\partial h}{\partial y}\right) + \frac{\partial}{\partial z}\left(K_z\frac{\partial h}{\partial z}\right) = \pm R' + S_s\frac{\partial h}{\partial t}$$

(11.1)

where K_x, K_y, and $K_z(LT^{-1})$ are the hydraulic conductivities (or permeability) along the x, y and z coordinates, respectively, assumed to be the principal directions of anisotropy, h (L) is the hydraulic head given by $h = p/\rho g + z$ (p is the pressure, ρ the density of water, g the gravity constant, and z the elevation); h is also the water level measured in piezometers; R' (T^{-1}) is an external volumetric flux per unit volume entering or exiting the system and S_s (L^{-1}) is the specific storage coefficient.

Equation (11.1) can be integrated over the vertical (two-dimensional flow) and simplified when the flow takes place in confined conditions in the aquifer, and the head is assumed constant over the vertical:

$$\frac{\partial h}{\partial x}\left(T_x\frac{\partial h}{\partial x}\right) + \frac{\partial h}{\partial y}\left(T_y\frac{\partial h}{\partial y}\right) = \pm R + S\frac{\partial h}{\partial t},$$

(11.2)

where T_x and T_y $(L^2 T^{-1})$ are the transmissivities (product of the hydraulic conductivity and saturated thickness of the aquifer) in the x and y directions, assumed to be the principal directions of anisotropy in the plane, respectively; R $(L T^{-1})$ is an external volumetric flux per horizontal unit area entering or exiting the aquifer; in practice, it represents recharge, or sometimes evapotranspiration from a shallow aquifer, or else leakage fluxes between aquifers; S $(-)$ is the storage coefficient (product of S_s, the specific storage coefficient, by the thickness of the aquifer). In the case of unconfined conditions, Equation (11.1) can be integrated over the vertical and simplified as follows:

$$\frac{\partial h}{\partial x}\left[K_x h\frac{\partial h}{\partial x}\right] + \frac{\partial h}{\partial y}\left[K_y h\frac{\partial h}{\partial y}\right] = \pm R + S_y\frac{\partial h}{\partial t},$$

(11.3)

where $S_y(-)$ indicates the specific yield of the water-table aquifer. In this case, the hydraulic head is expressed taking the bottom of the aquifer (assumed horizontal) as the $z = 0$ reference plane, so that the head h is also the saturated thickness of the aquifer. Since this equation is non-linear, and difficult to deal with, in many instances, Equation (11.2) is used as an approximation for unconfined aquifers as well, making the simplifying assumption that the product Kh of the saturated thickness h of the aquifer by its permeability K is equal to a constant, the transmissivity T, even if the saturated thickness varies with time. In that case, the storage coefficient S in Equation (11.2) is replaced by the specific yield S_y.

These equations, with the appropriate boundary and initial conditions, can be solved by standard numerical methods such as finite differences or finite elements. A large body of literature is available on the analytical or numerical formulation, development, and solution of the groundwater flow equations. Some useful references are Freeze (1971), Remson *et al.* (1971), Prickett (1975), Trescott *et al.* (1976), Pinder and Gray (1977), Cooley (1977, 1979, 1982), Brebbia (1978), Mercer and Faust (1981), Wang and Anderson (1982), Townley and Wilson (1985), Marsily (1986).

11.3.1 Application to hard-rock aquifers

For modelling hard-rock aquifers, the weathered zone, which is a true porous medium, can be represented by this continuum model. For the weathered-fractured zone, it is usual to assume that the flow can be represented, by the same equation, as an equivalent continuous, possibly anisotropic, porous medium. This assumption is only valid at a certain scale, for a volume of rock including a sufficient number of fractures, so that the individual properties of each fracture are averaged out within the large-scale volume. The size of the averaging volume is never precisely defined, it is a function of the average distance between fractures and of the spatial variability of their properties. To give an order of magnitude, it may vary between tens and hundreds of meters. Measurements made at a smaller scale than that of the averaging volume may not easily be interpreted with this equivalent medium approach. This has to be kept in mind for practical applications.

With these assumptions, a hard-rock aquifer is, in general, represented by a two-layer model, one for the weathered zone, and one for the weathered-fractured one. These layers exchange water by vertical leakage, which can be upward or downward, depending on the difference in hydraulic heads between the two layers. In practice, infiltration initially reaches the upper layer, and then leaks into the lower one. In local discharge zones, the leakage may be upward from the lower layer to the upper one, to reach the outlet. In the model, the weathered-fractured zone will be taken as a confined aquifer, and the weathered zone as an unconfined layer. Only in the case where the weathered zone becomes dry will the weathered-fractured zone become an unconfined layer as well. This situation can be transient, if the water-table fluctuates between the upper and lower layer. Although the two layers of the model represent the weathered zone and the fissured zone as two different entities, they constitute in reality one single aquifer, as the two layers exchange water vertically by leakage, and therefore generally have approximately the same hydraulic head over the vertical. Only in very rare cases can two distinct aquifers be observed in hard rocks: when the transition zone between the weathered layer and the weathered-fractured one is impervious. This has been observed in some granitic terrains e.g., in

France, with an accumulation of clay in this transition zone (Pistre, 1993).

Another common feature of hard-rock aquifers is the presence of numerous dykes (e.g., dolerites) or quartz veins that intersect the rock. These can be almost contemporary with the formation of the rock or they may have intruded much later. They have to be represented in the model as localized features that may act either as barriers to flow (they are generally much less weathered) or as high-permeability zones, if they are intensely fractured. Engerrand (2002) found in a granitic aquifer near Hyderabad (Andhra Pradesh, India) that the dolerite dykes were mostly impervious and partitioned the aquifer into quasi-independent blocks.

Let us finally briefly mention the existence of another class of models for the fissured-fractured medium: the so-called "discrete fracture network" model. In this case, each fracture is represented as a two-dimensional plane in which the flow is confined, the "blocks" between fractures are assumed totally impervious. In the model, a general flow can only take place if the fractures intersect; the model includes thousands or tens of thousands of fractures, each with their prescribed geometry and orientation, which constitute the network. In general, the geometry of the fractures is defined by a statistical distribution, as it is impossible to precisely describe the exact geometry of every fracture in the system. See for example, Cacas *et al.* (1990a,b), Chiles and Marsily (1993). In each fracture plane, laminar flow is assumed to occur and Darcy's law is used, so that the continuous porous medium approximation (Equations 11.1 or 11.2) can be applied within each fracture plane. Such models can only represent a very limited portion of a hard-rock aquifer and are not suited to water-resource management. They may, however, be used to interpret a local pumping test, as described later, see for example Bruel *et al.* (2002), Maréchal *et al.* (2004).

11.4 REQUIREMENTS FOR A GROUNDWATER MODEL

The requirements for a regional groundwater model concern the key elements of the conceptual model of the aquifer system, the anticipated future flow conditions and the spatial and temporal distribution of the parameters. The requirements for regional groundwater models were outlined by Mann and Myers (1998) in order to develop technical conditions for selecting a computer code to be used in the implementation of a comprehensive model. A brief discussion of the rationale is provided with each requirement.

11.4.1 Major hydrogeologic units

Major hydrogeologic units should be identified. For hard-rock aquifers, these units are, in general, the weathered layer, its geom-

etry and thickness, and the fissured-fractured zone, in particular, the depth below which the fractures are no longer pervious. The nature and permeability of the transition zone between the two layers must be determined. These data can be obtained by boreholes and geophysical surveys. Furthermore, all intrusive dykes or quartz veins should be identified, from field surveys, aerial photographs, or remote-sensing images. Finally, all surface-water bodies should be taken into account, e.g., streams, dams, small tanks, as they may act as recharge zones or drains for the aquifer.

11.4.2 Hydraulic properties

The regional groundwater model should represent the spatial variability of hydraulic properties in the major hydrogeologic units that have been inferred from the initial survey. These hydraulic properties have to be determined by hydraulic tests performed *in situ* (e.g., pumping tests in wells). Such tests can provide the transmissivity (the product of the horizontal hydraulic conductivity and aquifer thickness) and storage coefficient for a confined aquifer, or the specific yield for an unconfined aquifer. If borehole flowmeter logs are recorded in the wells during the test, they can also detect the position of the major fractures that provide water to the wells. Generally, all these properties are variable in space. The key features of this variability must be considered in order to accurately represent past, present, and future groundwater flow. As long as the aquifer remains confined, it is easy to use the transmissivity (T) as it does not change with time. However, if the aquifer conditions change and it becomes unconfined, it may be necessary to work with the permeability K as T is no longer constant in time. In fractured rocks, this will depend on the depth of the major fractures providing water to the wells. The vertical permeability of the two layers is much more difficult to assess. It is in general calibrated, i.e., the model is fitted to the observations. However, the vertical permeability of the upper layer can be measured locally at the surface, for example by double-ring infiltrometer tests.

11.4.3 Hydrologic boundaries

Various types of boundary conditions have to be defined (see for example, Marsily, 1986) for the selected limits of the area to be modelled, to accommodate real field situations. In general, these limits are chosen at a reasonable distance from the major withdrawal zones, to minimize the influence on the model results of the choice of these limits, which are often imprecisely defined. If perennial streams are present in the system, they are selected as boundaries, and prescribed heads are imposed on them. Between streams, a superficial water-divide line is selected as a limit, where a no-flow boundary condition is prescribed. When no perennial streams exist, the limits are often set at the water-divide line of the superficial watershed overlying the hard-rock aquifer. On these boundaries, a no-flow condition is prescribed.

11.4.4 Rainfall recharge and other inputs

A regional groundwater model should consider all sources of significant recharge to the aquifer system, including:

- natural recharge from direct infiltration of precipitation;
- recharge from runoff that infiltrates into the aquifer, e.g., along streams or in ponds or tanks;
- return flow from irrigation;
- artificial recharge to the aquifer system from past and current operations, e.g., injection in boreholes or dug wells.

Natural recharge can be estimated from a detailed water balance of rainfall and potential evapotranspiration, assuming that the upper soil layer constitutes a reservoir where rainwater is stored and can be used for evapotranspiration by plants. The storage capacity of this soil reservoir is generally comprised between 30 and 150 mm. In India, water-balance modelling has shown that this value is 100 mm as an average in the soils resulting from hard-rock weathering in semi-arid areas. The water that is not evapotranspired can flow as runoff or infiltrate into the aquifer. In this case, the water level in the aquifer will eventually rise, and, if the porosity is known, the water-level rise is directly linked to the recharge. A soil model can help to calibrate the recharge, if the rainfall, potential evapotranspiration, and water-level fluctuations are known, as well as the runoff in streams, if any. Such a model is GARDENIA, developed by BRGM (see below).

In India, natural recharge can be measured by tritium injection experiments. In other countries, using tritium as a tracer in soil is severely restricted, and rarely permitted, not for safety reasons, but to keep tritium as a natural tracer. The principle is to inject a slug of tritium into a borehole, at a shallow depth. This slug is displaced by recharge during the rainy season. By measuring the vertical migration in the soil of the tritium slug after the end of the recharge season, the amount of recharge can be estimated. This is, of course, a local measurement, and it is desirable to repeat the experiment at several locations over several years. Alternatively, use of chloride profiles or natural isotopic tracers can provide insights into recharge fluxes (Edmunds, 1999; Gaye and Edmunds, 1996).

Recharge from streams and ponds can be estimated by calculating rigorous water budgets, and measuring the incoming and outgoing flows and the amount lost by direct evaporation. Return flow from irrigation is generally estimated by the difference between the crop needs and the amount of water pumped into the fields.

Artificial recharge to the aquifer system can have a significant impact on water-table conditions. It is generally measured by the implementing agencies.

11.4.5 Water level and other hydrogeologic records for calibration and validation

To develop a regional groundwater model, historical records are needed, at the very minimum for one hydrological year, but preferably for a longer period, e.g., ten years. The data needed are water-level fluctuations in piezometers, measured at a regular interval (weekly, or monthly at the minimum), daily rainfall and temperature data, withdrawals from wells, and natural discharge from the aquifer to the streams or various outlets, if any. The withdrawal data are often difficult to assess; they can be indirectly estimated from the number of wells, their average flow rates, the average number of hours of operation per day or alternatively, from the electric consumption (in the case of electric pumps), which indicates the amount pumped. If no such data are available, the irrigated surface area and the type of crop can help estimate the withdrawal.

All these data are used to run the model over the available time period and to compare the calculated water level fluctuations with the observed ones; if the match is not satisfactory, the poorly known parameters of the model (transmissivity, storativity, recharge, boundary conditions) and their spatial and temporal variations are changed manually and the model is run again, until a good match is obtained. This is called the calibration (or inverse modelling) phase. The validation is done by saving some parts of the historical record, using the other parts for calibration. When the model is calibrated, the remaining part of the historical data is used to check that the calibrated model is able to reproduce the observed response for periods of time which have not been used in the calibration. Automatic calibration refers to the use of a program that can by itself modify the parameters of the model, until a good match is reached. Detailed discussion is, however, beyond the scope of the present chapter – see for example, Carrera and Neuman (1986a,b,c) and Marsily *et al.* (2000) for a review.

11.4.6 Anticipated future flow conditions

A regional groundwater model should be able to evaluate future transient and steady-state flow conditions in the aquifer system. The future pumping requirements for irrigation, domestic, and industrial purposes must be assessed and operational plans defined so that future withdrawal scenarios can be tested with the model and their sustainability assessed (no drying out of the aquifer, no irreversible damage to its quality). Usually, the predictions concern the conditions in the next 20–50 years, sometimes longer.

However, the meaningful period for predicting aquifer response depends on the number of years or hydrological cycles on which the model is calibrated.

11.4.7 Modelling documentation and database

The regional groundwater model, including the database supporting the conceptual model and its numerical implementation, must be maintained under configuration control, i.e., detailed records must be maintained concerning model configuration, parameters, and databases.

Since a regional groundwater model provides the framework for all groundwater modelling analyses performed on the site, a common site-wise groundwater model database containing all the necessary information should be maintained. It should contain:

- the basic geologic and hydrologic information that provides the basis for the conceptual model;
- results of all pumping tests and hydraulic tests performed in the aquifer;
- the key interpretations of geologic and hydrologic data and all relevant information, including descriptions of methods and approaches used to make the interpretations;
- the historical records (climate data, water levels in piezometers and wells, withdrawals, natural discharge, water quality data, etc.).

The database and data interpretations are updated, as new data become available, both at the local and regional scale. The modelling database should be stored in a form independent of the computer code used or the assumptions made for a particular modelling study. By storing high-resolution, regularly gridded information, it is possible to use the model information at different scales (e.g., in submodels) or with different groundwater computer codes. Thus, the most appropriate numerical representation and computer code can be chosen to simulate the problem at hand. The database should include all information necessary to develop parameter distributions based on geologic data (e.g., geometry of the main hydrogeologic units), hydraulic property estimates, boundary conditions, initial conditions, locations and volumes of sources and sinks, and natural recharge estimates.

The regional groundwater model must be a flexible and evolving platform for analyzing groundwater flow and contaminant transport. As more data are collected, the site-wise groundwater model needs to be updated, as the conceptual model may change, and the parameters may become better calibrated, thus improving the predictive capabilities. The adopted model framework must be one in which new concepts can be tested and improvements readily included. The data used in the site groundwater model are stored in a geographic information system (GIS), which allows easy data retrieval, display, and update. Collections of raw data (measurements) are described as databases and interpretations as information bases. Because data are gathered continuously and new information does not always fit the existing conceptual model, a continuous effort is required to evaluate the data and refine the geologic and hydrogeologic conceptual models (see for example, Bates *et al.*, 2000; Gogu *et al.*, 2001; Mogheir and Singh, 2002; Fert *et al.*, 2005).

Any modelling application that makes simplifications in the conceptual model and modelling database for use in specific analyses should include adequate documentation to demonstrate the consistency of the modelling assessment with the accepted conceptual model. Such documentation may include a list of assumptions made, their justification, and comparisons with simulation results based on the most complete conceptual model.

11.4.8 Model uncertainty

A regional groundwater model should provide explicit acknowledgements and estimates of uncertainty. More specific requirements will be formulated after an additional evaluation in the next section of the alternatives and methodologies available to address uncertainty. Considerable resources and model development efforts are required to establish an uncertainty framework for the databases, model, and code.

11.5 APPLICATION OF GEOSTATISTICS TO AQUIFER MODELLING TO ASSESS UNCERTAINTY

We have seen that many data that are required to build a model are measured at local points (e.g., a well, a geophysical log), but they need to be known in each mesh of the entire domain to be modelled. There is therefore a need for a method to extrapolate the local measurements to the entire study area. Many such methods exist, and are available routinely in contouring packages, but one method, known as *kriging* and forming part of what is called *geostatistics*, has proved particularly successful, especially for hydrogeological processes. It has the additional advantage of providing not only the estimated value, but also the uncertainty of this estimation. And this is precisely what is needed to estimate the uncertainty of the model predictions, as discussed above and below. The method is briefly presented below, and subsequently used in the case study application that follows.

Geostatistics is based on the theory of regionalized variables (Matheron, 1965, 1971). A regionalized variable is a quantity that varies over one-, two- or three-dimensional space and possible time. Geostatistics has found many applications in mining, geology, hydrology, and in almost all domains of hydrogeology, from parameter estimation to predictive modelling in groundwater management, e.g., designing an optimal groundwater monitoring network, estimating parameters at unmeasured locations, constructing groundwater models (optimal discretization), unbiased model calibration using estimation errors, and choosing the best predictive models (Ahmed, 2004)

Kriging is a "best linear unbiased estimator" (a BLUE). It estimates, at an unmeasured location, the "optimal" value of a variable (meaning the value with the smallest uncertainty) which has been measured at a set of measurement points, and also produces the variance of the estimation error. Kriging is a two-step procedure: first, the "structure" of the variable of interest is identified from the measurements. This "structure" describes the spatial correlation of the variable and its spatial variability. It is characterized by a function, the variogram, which is specific to each variable and each study site (see below). The second step is to use the identified variogram to produce the estimation of the variable at any unmeasured location. Kriging shares with other variants of BLUE techniques the following features:

- The estimator is a weighted linear sum of the measured values with weights calculated according to the condition of unbiasedness and minimum variance. Unbiasedness means that, on average, the error of estimation is zero. Minimum variance means that, again on average, the sum of the squared estimation error is as small as possible.

- The weights are determined by solving a system of linear equations with coefficients that depend only on the variogram (see below) describing the structure of the variable in space.

- In the absence of measurement errors, it is an exact interpolator at measurement points, meaning that the kriged surface is continuous and exact at all measurement points (whereas a polynomial approximation, for instance, would not be exact at the measurement points).

11.5.1 Estimation of the variogram

The variogram of a "regionalized variable" is a function $\gamma(h)$ defined by:

$$\gamma(h) = 1/2E\{[Z(x+h) - Z(x)]^2\}. \qquad (11.4)$$

Here, h is the lag, i.e., the distance between two points, $Z(x)$ is the value of the variable Z taken at point x (x represents here the 1, 2, or 3 coordinates of the point in one-, two- or three-dimensional space), $Z(x+h)$ is the value of the variable Z taken at a second point $(x+h)$ at a lag h from point x. E represents here the expected value, i.e., the average of the squared differences $[Z(x+h) - Z(x)]^2$ taken for all pairs of points x and $x+h$ in the domain lying with a lag distance h from each other. In other

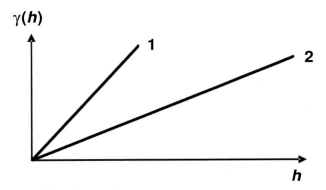

Figure 11.7 Example of two linear variograms

words, h is fixed, and x takes all possible values within the domain of interest, and the average E is taken over all these positions of x. If the variable Z displays anisotropy in its spatial variability, then the lag h is not just a distance, but a vector.

The variogram is a measure of the correlation of the value of Z taken at two different points. If the lag h is small, and if the variable Z is well correlated in space, then $Z(x)$ and $Z(x+h)$ will be very close (on average), and the variogram $\gamma(h)$ will take a small value. If h is large, $Z(x)$ and $Z(x+h)$ will be more different (on average), and $\gamma(h)$ will take a larger value. The shape of a variogram is thus generally increasing from the origin (one sees from Equation (11.4) that $\gamma(h) = 0$ if $h = 0$). Figure 11.7 is an example of two variograms:

It is clear from Fig. 11.7 that the variable with variogram (2) is better correlated in space than that with variogram (1): for a given lag h, the difference between $Z(x)$ and $Z(x+h)$ is (on average) larger for (1) than for (2). In this example, the functional form of the variograms is linear, but other types of variograms can be used, e.g., exponential, Gaussian, spherical, etc.

Note that some conditions must be met by the variable Z to be able to estimate the variogram as explained briefly above, the major one is stationarity of the mean, i.e., that the variable Z does not have a definite trend in the domain (like an average slope in one direction). If this is not the case, other forms of kriging are available to deal with it.

The variogram is the "tool" used in geostatistics to characterize the "structure" of a variable in space, i.e., the type of spatial variability of this variable, which is a characteristic of the underlying process (e.g., very regular, or erratic, or correlated over a given distance, etc.).

To determine the variogram of a given variable in a study area, where n measurements at n different locations are available, one forms all the pairs of points that can be made with n points, i.e., $n(n-1)/2$ pairs. These pairs are grouped into classes of distances h. For each class, the average of $1/2[Z(x+h) - Z(x)]^2$ is calculated and plotted on the $\gamma(h)$ versus h graph. A functional form of the variogram (linear, exponential, etc.) is then fitted on the experimental plot. The variogram can also be validated by using it to estimate (by kriging, see below) the values of the variable Z at the measurement points, one by one. This is called cross-validation. For more details, see for example, Matheron (1971), Journel and Huijbregts (1978), Marsily (1986), Isaaks and Srivastava (1989), Samper and Carrera (1990), Deutsch and Journel (1992), Wackernagel (1995), Kitanidis (1997), and Chiles and Delfiner (1999), among others.

11.5.2 Estimation of a regionalized variable by kriging

The kriging technique was originally developed by Matheron (1965, 1971) to estimate regionalized variables such as the grade of an ore body at unknown locations in space, given a set of observed data. Its application to groundwater hydrology has been described by a number of authors, namely Delhomme (1974, 1976,1978,1979), Delfiner and Delhomme (1973), Gambolati and Volpi (1979), Marsily et al. (1984), Marsily (1986), Aboufirassi and Marino (1983, 1984), Chiles and Delfiner (1999), to name a few.

If $Z(x)$ represents any variable (for instance, the thickness of an aquifer) with values measured at n locations in space $(z(x_i), i, = 1, \ldots, n)$ and if the value of the function Z has to be estimated at a point x_0, which has not been measured, the kriging estimate is defined as:

$$Z^*(x_0) = \sum_{i=1}^{n} \lambda_i z(x_i), \qquad (11.5)$$

where $Z^*(x_0)$ is the estimation of function $Z(x)$ at the points x_0 and the λ_i are the weighting factors of the linear combination of the measured values $Z(x_i)$ at the n measurement points x_i. Two conditions are imposed on Equation (11.5), namely the unbiasedness condition and the condition of optimality. The unknowns λ_i can then be determined by the following set of linear algebraic equations, based on these two conditions, which can be shown to yield:

$$\begin{cases} \sum_{j=1}^{n} \lambda_j \gamma(x_i, x_j) + \mu = \gamma(x_i, x_0), \qquad i = 1, \ldots, n, \\ \sum_{j=1}^{n} \lambda_j = 1, \end{cases} \qquad (11.6)$$

where μ is a Lagrange multiplier, and $\gamma(x_i, x_j)$ is the value of the variogram for the lag $(x_i - x_j)$. The variance of the estimation error is given by:

$$\sigma_k^2 = \sum_{i=1}^{n} \lambda_i \gamma(x_i, x_0) + \mu. \qquad (11.7)$$

To estimate the value of the variable Z in all meshes of a model, the point x_0 in Equation (11.6) is taken successively at each node

of the grid, and the same equation is solved successively for each grid point. Note that if the number n of measurement points is not too large, the linear system (11.6) can be inverted only once, since the location x_0 only intervenes on the right-hand side of (11.6). If n is too large, only a subset of all the measurements is used to krige each point x_0; it is called "moving neighborhood;" in this case, a new linear system must be solved for each point x_0. Note that it is also possible to directly estimate the average of the variable Z over a given surface, e.g., the area of the mesh of a model, rather than at a node. The kriging equation uses a spatial integral of the variogram, see for example, Marsily (1986), Chiles and Delfiner (1999).

In hydrogeology, kriging can be used directly on the variable, or in some cases on the logarithm of the variable. This is the case, in particular, of permeability and tansmissivity, which are always kriged on their logarithm. The reason is that the best estimate of an average transmissivity, in two dimensions, is the geometric mean, not the arithmetic mean. The weighted sum of the logarithm is thus a better estimator of the natural value.

Kriging has many advantages, particularly in groundwater modelling, over more classical deterministic extrapolation methods (Teles et al., 2002; Ahmed et al., 2002; Michalak and Kitanidis, 2002; Ahmed, 2004):

(1) The closer the values of the aquifer parameters are to reality, the faster the model is calibrated. Better estimated values (with lower estimation variances) are initially assigned to the nodes of the aquifer when kriging is used.

(2) In aquifer modelling, the assumption is made that a single value of a system parameter represents the average over the entire mesh (assumed small). Estimating the true average is possible with a special form of kriging, called block-kriging.

(3) An optimal mesh size and number of nodes to discretize the aquifer system can be obtained through the estimation error variance, and the best location of new control points can be predicted.

(4) A confidence interval given by the standard deviation of the estimation error provides a useful guide to transmissivity modifications in each mesh during model calibration and to checking that the calculated heads fall inside the confidence interval of the observed heads. The following criterion can be used:

$$\left| \frac{h_i^m - h_i^*}{\sigma_i} \right| < 2.0 \qquad \forall i, \qquad (11.8)$$

where h_i^m, h_i^*, σ_i are the model calculated head, the kriged estimated head, and the standard deviation of the kriging estimation error in the ith mesh. With these constraints, the model is calibrated in all the meshes and free from the bias of the aquifer modeller. The model parameters should then be changed in or around the meshes where the above criterion is not satisfied. A faster and unbiased model calibration can thus be achieved.

The large number of applications of geostatistics to groundwater hydrology includes the first one by Delhomme (1974) followed by many others, e.g., Delhomme (1978), Gambolati and Volpi (1979), Mizell (1980), Darricau-Beucher (1981), Aboufirassi and Marino (1983, 1984), Neuman (1984), Hoeksema and Kitanidis (1984), Marsily and Ahmed (1987), Ahmed and Marsily (1988), Ahmed and Gupta (1989), Thangarajan and Ahmed (1989), Dong *et al.* (1990), Parriaux and Bensimon (1990), Krasny (1990), Jaquet *et al.* (1990), Ahmed and Murali (1992), Ahmed (1995), Roth (1995), Roth *et al.* (1996), Wen (2001), Lyon *et al.* (2005), Patriarche *et al.* (2005) and several others who have demonstrated the use of geostatistics in groundwater hydrology. As a result of these studies, it has become clear that multi-variate and non-stationary geostatistics are comparatively more important in groundwater hydrology than some of the other geostatistical techniques. Furthermore, some of the conventional techniques have to be modified, and special procedures developed, to make them applicable to the field of hydrogeology. These modified techniques include kriging with linear regression and kriging in the presence of faults, developed by Delhomme (1976). Conditional simulation has also been used in aquifer modelling (Delhomme, 1979). Galli and Meunier (1987) and Ahmed (1987) worked on kriging with an external drift which is extremely useful, in practice, for geohydrological parameters. Ahmed and Marsily (1987) compared a number of multi-variate geostatistical methods for estimating transmissivity from data on transmissivity and specific capacity. Ahmed (1987) developed a special antisymmetric and anisotropic cross-covariance between residuals of hydraulic head and transmissivity based on the work by Mizell (1980) and used the coherent nature of various covariances to cokrige transmissivity and hydraulic head to solve the inverse problem (Ahmed and Marsily, 1987, 1989, 1993 and Ahmed *et al.*, 1990). Bardossy *et al.* (1986), Ahmed *et al.* (1988), Kupfersberger and Bloschl (1995) combined electrical and hydraulic parameters in geostatistical analyses of transmissivity.

11.6   A CASE STUDY OF AQUIFER MODELLING IN A SEMI-ARID REGION

There are a number of examples of groundwater modelling in hard-rock systems including many case studies in arid and semi-arid regions, e.g., Handa (1975), Gupta *et al.* (1979), Huyakorn *et al.* (1983), Gupta *et al.* (1985), Witherspoon *et al.* (1987), Barker (1988), Zhu *et al.* (1990), Resele and Job (1990), Qian and Cai (1990), Milanovic (1990), Kovalevsky (1990), Jackson

and Porter (1990), Das Gupta *et al.* (1990), Civita *et al.* (1990), Chen and Chen (1990), Carrera (1990), Cacas *et al.* (1990a,b), Bredenkamp (1990), Basabe and Bieler (1990), Amadi and Fontes (1990), Ahmed *et al.* (1990), Aghassi (1990), Geier and Axelsson (1991), Anderson and Woessner (1991), Thangarajan (1999a,b, 2000), Thangarajan *et al.* (2000). Here, a detailed description is given of a recent case study of flow simulation in a granitic aquifer in a semi-arid region of India, in an overexploited area from a recent project (Ahmed *et al.*, 2003).

11.6.1   The Maheshwaram watershed: a typical aquifer representing the groundwater conditions in a semi-arid region and hard-rock terrain

The Maheshwaram watershed of about 53 km^2 in the Ranga Reddy district (Fig. 11.8) of Andhra Pradesh, India, is underlain by granitic rocks. This watershed is a representative southern India catchment in terms of overexploitation of its weathered hard-rock aquifer (more than 700 borewells in use), its cropping pattern, rural socio-economy (based mainly on traditional agriculture), agricultural practices, and semi-arid climate. The declining groundwater level is monitored regularly (15 minutes time-step on ten observation wells and twice a year on 155 wells) to study its fluctuations due to rainfall, spatio-temporal variability, and local pumping. The objective of this study is to develop and test well-suited modelling approaches in order to propose physically based decision-support tools at suitable management scale.

The granite outcrops in and around Maheshwaram form part of the largest of all granite bodies recorded in Peninsular India. Alcaline intrusions, aplite, pegmatite, epidote, quartz veins, and dolerite dykes traverse the granite. There are three types of fracture patterns (Fig. 11.9) in the area, namely: (i) mineralized fractures, (ii) fractures traversed by dykes, and (iii) late-stage fractures represented by joints. The vertical fracture pattern is partly responsible for the development of the weathered zone and the horizontal fractures are the result of the weathering. Hydrogeologically, the aquifer occurs both in the weathered zone and in the underlying weathered-fractured zone. However, due to deep drilling and heavy groundwater withdrawal, the weathered zone has now become dry. About 150 dug wells were examined and the nature of the weathering was studied. The weathered-zone profiles range in thickness from 1 to 5 m below ground level (bgl). They are followed by semi-weathered and fractured zones that reach down to 20 m bgl. The weathering of the granite has occurred in different phases and the granitic batholith appears to be a composite body that has emerged in different places and not as a single body. One set of pegmatite veins displacing another set of pegmatitic veins has been observed in some well sections. Joints are well developed in the main directions – N 0° – 15° E, NE-SW, and NW-SE that vary slightly from place to place.

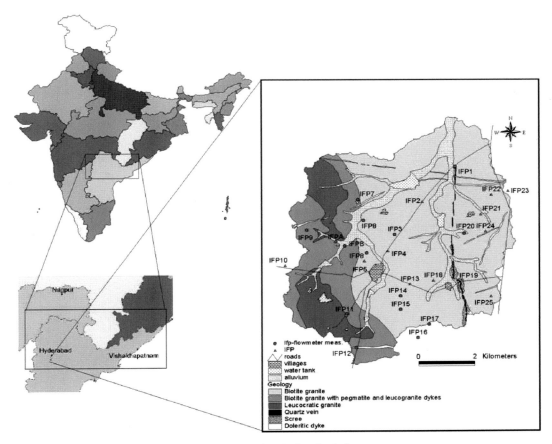

Figure 11.8 Maheshwaram hard-rock watershed in Ranga Reddy District, Andhra Pradesh.

The groundwater flow system is local, i.e., with its recharge area at a topographic high and its discharge area at a topographic low adjacent to each other. Intermediate and regional groundwater-flow systems also exist since there is significant hydraulic conductivity at depth. Aquifers occur in the permeable saprolite (weathered) layer, as well as in the weathered-fractured zone of the bedrock and the quartz pegmatite intrusive veins when they are jointed and fractured. Thus only the development of the saprolite zone and the fracturing and inter-connectivity between the various fractures allow a potential aquifer to develop, provided that a recharge zone is connected to the groundwater system.

Mean annual precipitation (P) is about 750 mm, of which more than 90 % falls during the Monsoon season. The mean annual temperature is about 26 °C, although in summer ("Rabi" season from March to May), the maximum temperature can reach 45 °C. The resulting potential evaporation from the soil plus transpiration by plants (PET) is 1800 mm/year. Therefore, the aridity index (AI = P/PET = 0.42) is 0.2 < AI < 0.5, typical of semi-arid areas (UNEP 1992). Although the annual rainfall is around 750 mm and the recharge around 10–15 %, the water levels have been lowered by about 10 m during the last two decades due to intensive exploitation. The transmissivity of the fractured aquifer was measured by seven pumping tests in wells and it varies considerably from about 1.7×10^{-5} m^2/s to about 1.7×10^{-3} m^2/s, and a low storage coefficient (0.6 %) indicates a weak storage potential of the aquifer (Maréchal *et al.*, 2004). The withdrawal which increases year by year has to be controlled to allow recharge by rainfall to maintain or restore the productive capacity of the depleted aquifer.

The information acquired in the study area was used to define a conceptual hydrodynamic model for the weathered-fractured layer of the hard-rock aquifer (Fig. 11.10).

Our data show that the weathered-fractured layer is conductive mainly from the surface down to a depth of 35 m, the range within which conductive fracture zones with transmissivities greater than 5×10^{-6} m^2/s are observed (Maréchal *et al.*, 2004). The lower limit corresponds to the top of the unweathered basement, which contains few or poorly conductive fractures ($T < 5 \times 10^{-6}$ m^2/s) and where only local deep tectonic fractures are assumed to be significantly conductive at great depths, as observed by studies in Sweden (Talbot and Sirat, 2001), Finland (Elo, 1992) and the United States (Stuckless and Dudley, 2002). An unpublished study

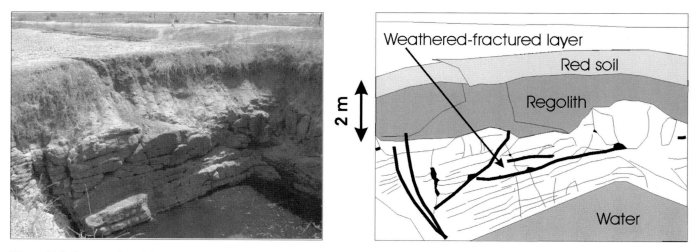

Figure 11.9 Cross-section of the typical substratum in a dug well, after Maréchal *et al.* (2003b)

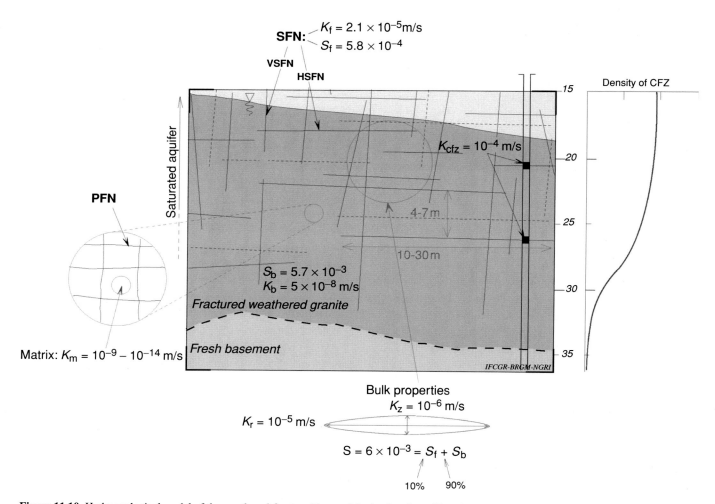

Figure 11.10 Hydrogeological model of the weathered-fractured layer of the hard-rock aquifer, after Maréchal *et al.* (2004). The dashed lines represent poorly conductive fractures unidentified by the flowmeter tests; solid lines: conductive fractures identified by the flow meter

of the same area has statistically confirmed this result by the analyses of airlift flow rates in 288 boreholes, 10 to 90 m deep, where the cumulative airlift flow-rate ranges from 1.5 m^3/h to 45 m^3/h, and the mean value increases drastically in the weathered-fractured layer at depths between 20 and 30 m. Below 30 m, the flow rate is constant and does not increase with depth. In practice, drilling deeper than the bottom of the weathered-fractured layer (30–35 m) does not increase the probability of improving the well discharge. The data confirm that the weathered-fractured layer is the most productive part of the hard-rock aquifer, as already shown elsewhere by other authors (e.g., Houston and Lewis, 1988; Taylor and Howard, 2000).

Two different scales of fracture networks are identified and characterized by hydraulic tests (Maréchal *et al.*, 2004): a primary fracture network (PFN) which affects the matrix at the decimeter scale, and a secondary fracture network (SFN) affecting the blocks at the borehole scale. The latter is the first one described below.

The secondary network is composed of two conductive fractures sets – a horizontal one (HSFN) and a subvertical one (VSFN) as observed on outcrops. They are the main contributors to the permeability of the weathered-fractured layer. The average vertical density of the horizontal conductive set ranges from 0.15 m to 0.24 m, with a fracture length of a few tens of meters (10–30 m in diameter for the only available data). This corresponds to a mean vertical thickness of the blocks ranging from $\approx$ 4 m to $\approx$ 7 m. The strong dependence of the permeability on the density of the conductive fractures indicates that individual fractures contribute more or less equally to the bulk horizontal conductivity ($K_r = 10^{-5}$ m/s) of the aquifer. No strong heterogeneity is detected in the distribution of the hydraulic conductivities of the fractures, and therefore no scale effect was inferred at the borehole scale. The subvertical conductive fracture set connects the horizontal network, ensuring a vertical permeability ($K_z = 10^{-6}$ m/s) and a good connectivity in the aquifer. Nevertheless, the subvertical set of fractures is less permeable than the horizontal one, introducing a horizontal-to-vertical anisotropy ratio for the permeability close to ten due to the preponderance of horizontal fractures.

As discussed earlier, the horizontal fracture set is due to the weathering processes, through the expansion of the micaceous minerals, which induces cracks in the rock. These fractures are mostly subparallel to the contemporaneous weathering surface, as in the flat Maheshwaram watershed where they are mostly horizontal (Maréchal *et al.*, 2003b). In the field, the bore wells drilled with a fairly homogeneous spacing throughout the watershed confirm this conclusion: of 288 wells, 257 (89 %) were drilled deeper than 20 m and 98 % of these are productive. In any case, the probability of a vertical well crossing a horizontal fracture induced by such wide-scale weathering is very high.

The primary fracture network (PFN), operating at the block scale, increases the original matrix permeability of $K_m =$

10^{-14} –10^{-9} m/s to $K_b = 4 \times 10^{-8}$ m/s. Regarding the matrix storage, it should also contribute to the storage coefficient in the blocks of $S_b = 5.7 \times 10^{-3}$. The storage in the blocks represents 91 % of the total specific yield ($S_y = 6.3 \times 10^{-3}$) of the aquifer; storage in the secondary fracture network accounts for the rest. This high storage in the blocks (in both the matrix and the PFN) and the generation of the PFN would result from the first stage of the weathering process itself. The development of the secondary fracture network (SFN) is the second stage in the weathering process: that is why the words "primary" and "secondary" have been chosen to qualify the different levels of fracture networks. The obtained storage values are compatible with those of typical unconfined aquifers in low-permeability sedimentary layers.

The universal character of granite weathering and its worldwide distribution underline the importance of understanding its impact on the hydrodynamic properties of the aquifers in these environments.

11.6.2 Numerical groundwater modelling in the Maheshwaram watershed

The Maheshwaram watershed was modelled with the MARTHE software developed at the BRGM (French Geological Survey: Thiery, 1993a and 1993b). MARTHE is a transient hydrodynamic modelling code, representing three-dimensional and/or multi-layer flow in aquifers. The solution method uses finite differences with a rectangular grid and offers the possibility of having a free surface in a mesh of any layer.

11.6.3 Conceptualization of the hydogeological system

MODEL DISCRETIZATION

The studied aquifer was represented by a two-layer aquifer flow system. The upper layer is located in the weathered rocks, the lower one in the weathered-fractured granite represented as an equivalent porous medium. Each layer was divided into 5272 square meshes with a 100 m sides (Fig 11.11). Layer 1 is unconfined, and layer 2 is confined, but may become unconfined when layer 1 has become dry. The MARTHE code was used with its coupled climatic-balance model (GARDENIA: Thiéry, 1988, 1993a,b; Thiéry and Boisson, 1991; Thiéry *et al.*, 1993; Thiéry and Amraoui, 2001). The groundwater flow in the Maheshwaram watershed was simulated in transient regime in order to represent the piezometric variations observed in the wells in the studied area from January 2001 to July 2003.

TOPOGRAPHY

The DEM values (Fig. 11.12) were averaged to fit the 100 m $\times$ 100 m grid (Fig. 11.11). For the meshes containing the wells

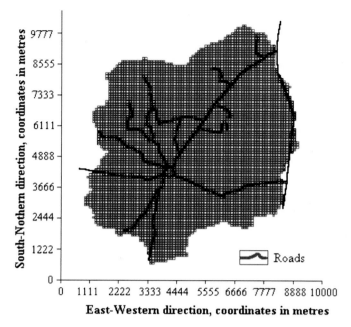

Figure 11.11 Grid of the watershed, layer 1 is similar to layer 2

AQUIFER GEOMETRY

The thickness of the weathered layer (Fig. 11.13) was estimated by kriging from the measurements made in the 25 existing lithologs and the vertical electrical sounding (VES) interpretations (Krishnamurthy *et al.*, 2000). The geometry of the weathered-fractured granite layer (Fig. 11.13) was deduced jointly from the total depth of the 900 inventoried farm wells of the watershed after removing the wells with a total depth of more than 70 m and from the result of the VES.

BOUNDARY CONDITIONS

The topographical limits of the watershed were taken as the groundwater divides (no-flow boundaries), except at the northern limit where a non-perennial stream at the outlet of the watershed can evacuate the surface water. At this location, the hydraulic heads were prescribed and set at the average of the 2000–2 measured field values.

The streams are ephemeral, they flow only during heavy rainfall a few hours (or maximum one or two days) a year due to very rare runoff. That is why, as a first approximation, the role of the hydrographic network in recharging the aquifer is assumed to be negligible. The hydrographic network in the study area includes a few low order ephemeral streams connected to a surface storage tank. The assumption of a negligible recharge is justified in this study as the flow in the streams are rare and fast as well as the vertical hydraulic conductivity of the tank beds (measured using the double infiltrometer) are almost zero due to thick silting. However, the groundwater is allowed to overflow into the rivers as base flow during and following high floods.

drilled in the Maheshwaram project, where some precise elevation values were estimated with a differential global positioning system (Lebert, 2001), the topographic values in these meshes were taken as the DGPS ones. For the meshes containing the other project wells, the adopted topography values were the nearest local values of the DEM (interpolated every 10 m).

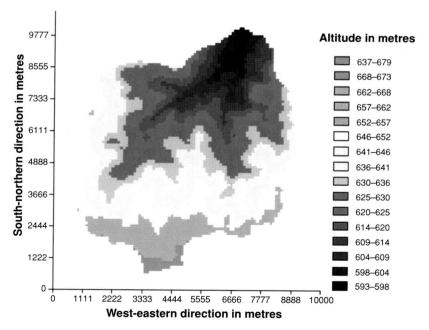

Figure 11.12 Topography of Maheshwaram watershed

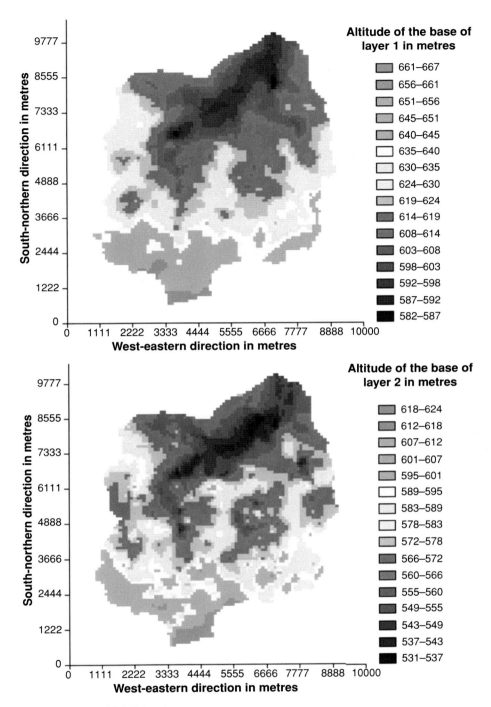

Figure 11.13 Aquifer bottom of layer 1 and 2 (asl, in m)

HYDRAULIC CONDUCTIVITY

No hydraulic tests could be carried out in the first layer because, during the project period (1999 to 2003), the water table was constantly deeper than the bottom of the weathered zone. Thus the hydraulic parameters for that layer were estimated from the literature. The hydraulic conductivities chosen for the weathered layer were selected according to two assumptions:

(1) The hydraulic conductivity values must be in the range of the ones found in the literature for the same type of geology.

(2) The hydraulic conductivity variability in the weathered rocks is based on the conceptual model of the hydrogeological functioning of weathered-rock/hard-rock aquifers in Africa proposed by Chilton and Foster (1995). This model assumes that, in the general case, the weathered rocks have a weak

hydraulic conductivity which increases with depth. In the Maheshwaram watershed, if it is assumed that the thickness of the weathered-rock layer is linked to its degree of erosion, the thin, eroded weathered rock layer is, in reality, the base of the weathered horizon and, consequently, it is composed of coarser grains than the thick layers where the weathered rocks show the entire profile. Thus, the thicker the weathered-rock layer, the weaker its average hydraulic conductivity.

Figure 11.14 shows the distribution of the calibrated hydraulic conductivities for the weathered layer. They range between 8.10^{-7} m/s and 5.10^{-6} m/s.

In the weathered-fissured granite layer, the transmissivities were calculated from the results of 34 aquifer tests. But the high variability observed in the field and the difficulties of matching the simulated heads with the measured ones led to additional manual calibration of the hydraulic conductivities in the model.

Few of the calibrated transmissivities match the values obtained by the interpretation of the hydraulic tests. However, the average hydraulic conductivity obtained by calibration was compared with the equivalent horizontal hydraulic conductivity calculated by the FRACAS model (Bruel *et al.*, 2002), a "discrete fracture network" model, briefly described above, which was used to interpret 25 slug tests. Table 11.1 shows the comparison and suggests that the model fits well with the discrete-fracture network study results obtained for the weathered-fractured granite.

The hydraulic conductivities in the weathered-fissured granite layer are shown in Fig. 11.15. They vary greatly from one place

Table 11.1 *Hydraulic conductivities: comparison with the discrete-fracture network study interpretations*

	Mean hydraulic conductivity value, calibrated by MARTHE (m/s)	Equivalent horizontal hydraulic conductivity from FRACAS (m/s)
Layer 1	2.61×10^{-6}	No interpretation
Layer 2	7.67×10^{-6}	$7.70.10^{-6}$

to another because of the heterogeneity of the rocks: 1.10^{-7} to 3.10^{-5} m/s.

In the weathered-fissured granite layer, linear heterogeneities (impervious vertical barriers) had to be introduced into the model because this was the only way to take into account the observed piezometric data. These heterogeneities were attributed to the dykes that crop out in the studied area, to a south–north quartz reef crossing the watershed, and to some assumed extensions of dykes (Fig. 11.16).

SPECIFIC YIELD AND STORAGE COEFFICIENTS
The specific yield was taken to be 2.4 % in the weathered rocks. This value is an average of the specific yields observed in the weathered rocks of similar watersheds (Rangarajan and Prasada Rao, 2001). The specific yield of the weathered rocks was also used for the calculation of preferential recharge. This value lies within the orders of magnitude given by the proton magnetic

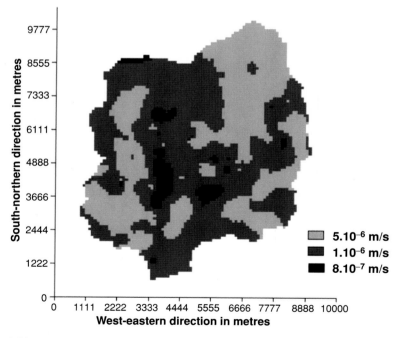

Figure 11.14 Hydraulic conductivities in the weathered layer, after calibration

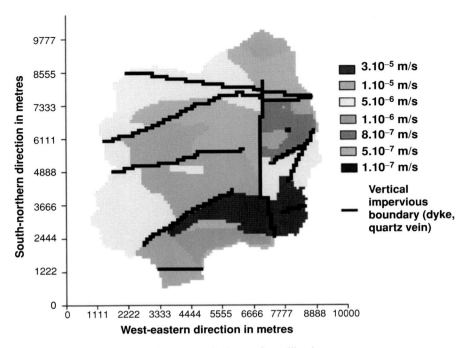

Figure 11.15 Hydraulic conductivities in the weathered-fissured granite layer, after calibration

resonance (PMR) measurements carried out in the watershed (Legchenko and Baltassat, 1999). In the fissured granite, this specific yield is assumed constant at a value of 1 % which is close to the value deduced from the discrete-fracture network interpretation (0.8 %: Bruel *et al.*, 2002), from pumping-test interpretations (0.63 %: Maréchal *et al.*, 2004) and water-table fluctuation interpretations (1.4 %: Maréchal *et al.*, 2006). Some sensitivity tests were made by decreasing the specific yield in layer 2 by one order of magnitude. This led to a lowering of the simulated water levels due to the effect of pumping. This effect was amplified by the fact that layer 1 was almost dry and could not play its role of reservoir which might have hidden the influence of the second-layer specific yield, as observed in other watersheds (Engerrand, 2002). Storativity of the weathered layer (as a confined aquifer) plays a role in the flow calculations only when this layer becomes saturated (Engerrand, 2002). Its value was taken as 8×10^{-5}. For layer 2, the storativity value was set at 1×10^{-5}.

RECHARGE

Direct recharge

Total recharge can be divided into three main components (Lerner *et al.*, 1990): direct recharge (by direct vertical percolation through the vadose zone – saprolite), indirect recharge (percolation to the water table through the beds of surface-water courses, close to nil in the study area due to quasi absence of water in surface streams) and localized recharge (various-scale pathways such as those due

to shrinkage cracks, roots and burrowing animals, trenches, dug wells, brick factories, and major landscape features).

The direct recharge was calculated on the basis of tritium injection tests that were carried out in 1999 and 2000. The interpretation by piston flow of the tests indicated that between the end of July 1999 and November 1999, the direct recharge was 22.2 mm (Rangarajan and Prasada Rao, 2001) and that during the 2000 Monsoon, the direct recharge was 42 mm (Rangarajan, personal communication, 2002). The GARDENIA code (see above), which is a soil hydrologic balance code, was used at a daily time-step to calculate the balance between:

- Rainfall (daily data from a rain gauge located in Maheshwaram village).
- Potential evapotranspiration, calculated from the evaporation in a Pan A and multiplied by a factor of 0.9 (Monteith *et al.*, 1989). The data are daily when collected at Maheshwaram village and weekly, from an average of six years of monitoring at the CRIDA farm (CRIDA, 1991–2000), when they are not available at Maheshwaram.
- Runoff. For the Musi basin, the Central Water Commission calculated a runoff/rainfall ratio of 11.2 % for the period 1989–94 (Maréchal, personal communication, 2001).
- Available water capacity of the soil. The most common soils in the area have an available water capacity ranging from 75 to 110 mm (CRIDA, 1990).

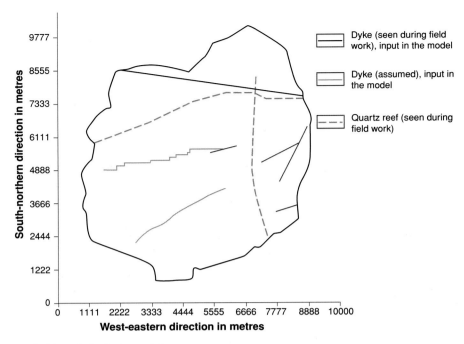

Figure 11.16 Impervious vertical boundaries in the model

The parameters of the GARDENIA code (runoff and available water capacity) were calibrated for 1999 and 2000 to obtain the measured 22.2 mm and 42 mm of recharge, respectively, for these periods, as obtained by tritium injection. The same parameters were then used in the code to estimate the recharge for the years 2001 and 2002.

Indirect and localized recharge
Indirect and localized recharge were calculated from the groundwater level rise in the weathered rocks during the tritium experiments of 1999. In 2000, most of the dug wells located near the tritium injection points were dry, preventing us from estimating the preferential recharge. It was thus calculated only for 1999 with the specific yield value of the weathered layer that was known for similar watersheds (Rangarajan and Prasada Rao, 2001) and the groundwater level fluctuations measured in the dug wells within this layer.

In 1999, the water level rose by 1.5 m on average, during the time of the tritium experiment, near the injection points. The specific yield of the weathered layer is assumed to be 2.4 % (Rangarajan and Prasada Rao, 2001). This means that the total recharge was around 36 mm in 1999 while the amount of 22.2 mm measured with the tritium experiment constituted direct recharge. The extra recharge that takes into account whatever is not measured by a piston flow recharge model is attributed to indirect (including recharge from streams, negligible) and localized recharge.

Table 11.2 *Annual recharge calculated for 1999, 2000, and 2001*

	Direct recharge (mm)	Indirect and localized recharge (mm)	Total recharge (mm)
1999	22.2	13.8	36
2000	42	25	67
2001	83	35	118

If the ratio between indirect as well as localized recharge and total recharge is assumed constant every year, then $(36–22.2) \times 100/350\%$ of the total annual rainfall recharges the aquifer as indirect and localized recharge. Table 11.2 shows the calculated direct recharge and indirect and localized recharge for the years 1999, 2000, and 2001.

DISCHARGE RATES AND RETURN FLOW
The major part of the groundwater withdrawal from the system is due to pumping from the wells and this should be evaluated with precision. However, in the absence of a good database on the wells and because of other practical limitations, the estimates of groundwater withdrawal are often poor and erroneous. Therefore, various means were used to cross-check the estimates and, finally, an average was taken in the model. Two quite independent

Table 11.3 *Annual water balance in Maheshwaram watershed*

	1/1/2000–31/12/2000	1/1/2001–31/12/2001	1/1/2002–31/12/2002
Recharge (m^3/y)	3 521 000	6 213 000	2 372 000
Return flow (m^3/y)	5 214 000	6 228 000	6 720 000
Outlet from prescribed heads (m^3/y)	− 46 000	− 110 000	− 74 000
Withdrawals (m^3/y)	− 9 469 000	− 11 293 000	− 11 927 000
Groundwater overflows (m^3/y)	0	− 13 000	− 5 000
Storage variation (m^3/y)	− 780 000	1 026 000	− 2 912 000
Balance deviation (%)	− 0.02	− 0.02	0.42

methodologies were applied, namely, a direct one based upon the well inventory including the location of the wells, their discharge and pumping duration, etc., and an indirect one based upon the demand using data on irrigated areas, cropping pattern and crop water requirements. The two results were cross-checked.

The discharge rates due to pumping are evaluated from the well-inventory survey. This is a direct method of estimation and also provides grid-wise information as the wells pumping the groundwater can be located on the grids (Fig. 11.17). For the years 2000, 2001, and 2002, all the wells drilled up to 2000, 2001, and 2002 respectively were taken into account. The wells were assumed to be pumped in layer 2 only as there was no water in the weathered layer. The total yearly amount of groundwater withdrawal from the aquifer is given in Table 11.3.

Due to water level declines, the groundwater pumping is mainly from the 2nd layer but the return flow from the irrigated fields

occurs in the 1st layer. It was deduced from the pumping according to:

• Andhra Pradesh Groundwater Department (APGWD) (personal communication, 2000) report on Maheshwaram land use.

• APGWD (1977) report on hydrologic parameters of groundwater recharge in Andhra Pradesh.

In the ideal case, 1200 mm (V_{mmrice}) of water is required for one rice crop and 350 mm ($V_{mmother}$) per crop for other cultures (APGWD, personal communication, 2000). There are, on average, two rice crops and two vegetable crops in one year. It is assumed that the rainfall amount is 700 mm per year. In 2000, the cultivated areas of rice and vegetables are known respectively for winter and Monsoon ($S_{ricewinter}$, $S_{ricemonsoon}$, $S_{otherwinter}$, $S_{othermonsoon}$). The total

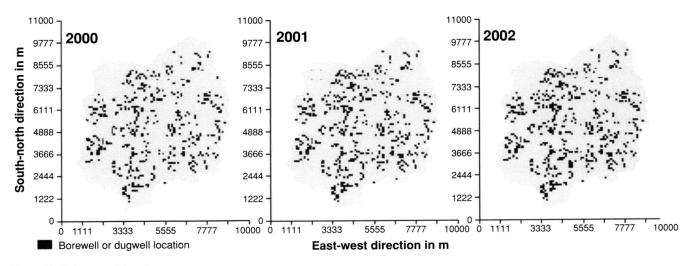

Figure 11.17 Farm-well locations

amount of groundwater required for irrigation, V_{tot}, can be written as:

$$V_{tot} \, (m^3) = V_{rice} \, (m^3) + V_{other} \, (m^3)$$

$$V_{tot} \, (m^3) = V_{mmrice} \, (mm)/1000^* S_{rice} \, (m^2)$$
$$+ V_{mmother} \, (mm)/1000^* S_{other} \, (m^2)$$

$$V_{tot} \, (m^3) = [(V_{mmricemonsoon} \, (mm) - 700)^* S_{ricemonsoon} \, (m^2)$$
$$+ (V_{mmricewinter} \, (mm)^* S_{ricewinter} \, (m^2)]/1000$$
$$+ [(V_{mmothermonsoon} \, (mm) - 700)^* S_{othermonsoon}$$
$$(m^2) + V_{mmotherwinter} \, (mm)^* S_{otherwinter} \, (m^2)]/1000.$$

$$(11.9)$$

If $V_{mmricemonsoon} \, (mm) - 700 < 0$, we take $V_{mmricemonsoon} \, (mm) - 700 = 0$, and if $V_{mmothermonsoon} \, (mm) - 700 < 0$, we take $V_{mmothermonsoon} \, (mm) - 700 = 0$.

$$V_{tot} \, (m^3) = (1200 - 700) \times 362.56 \times 10^4 + 1200 \times 217.38 \times 10^4)$$
$$/1000 + (0 + 350 \times 198.10 \times 10^4)/1000 V_{tot} \, (m^3)$$
$$= 5\,114\,650 \, m^3/yr, \qquad (11.10)$$

With

$$V_{rice} \, (m^3) = 4\,421\,300 \, m^3/yr \qquad (11.11)$$

and

$$V_{other} \, (m^3) = 693\,350 \, m^3/yr, \qquad (11.12)$$

i.e., 86 % of the pumped water is used for the rice and 14 % for other crops.

In the APGWD report (1977), experiments on the return flow from the rice in a nearby watershed showed that 55 to 88 % of the irrigation water returned to the aquifer. After calibration, 60 % of return flow from the rice and 20 % from the other crops were assumed. In the model, we do not distinguish between the locations of the rice and the other crops, because they are not known. The average return flow of one crop (rice/other) is then taken as $0.86 \times 60 + 0.14 \times 20 = 54.4 \approx 55$ % of return flow from the withdrawals.

TIME-STEPS

The time-steps of the hydrodynamic calculation in transient flow are 14 days during the dry season and one week in the rainy season (they are daily when it rains). The MARTHE code uses GARDENIA to calculate the hydroclimatic balance; this balance is calculated with a weekly time-step during the dry season and a daily one during rainy days throughout the year.

11.6.4 Model calibration and fitting criteria

The fitting of the model was carried out by calibrating the following parameters:

- the hydraulic conductivities of the two layers;
- the return flow from withdrawals.

The fitting criteria were based on:

- probable orders of magnitude of the fitting parameters;
- the similarity between the hydraulic heads observed in the project wells in the field and the simulated ones.

11.6.5 Balance

In 2000, the balance (storage variation) was negative (Table 11.3). The withdrawals were greater than the recharge and the storage decreased. The pumping wells were in operation at 99.9 % of normal capacity. The discharge from the prescribed head-boundaries was 1.3 % of the recharge, which was negligible. The overflows were nil.

For 2001, the balance was positive. The recharge was greater than the withdrawals and the stock increased. The pumping wells were in operation at 99.7 %. The discharge from the prescribed heads was 1.8 % of the recharge, which again was negligible. The overflows were 0.2 % of the recharge, which was also negligible. The aquifer overflows were located near the draining meshes, which seems logical.

In 2002, the balance was strongly negative. The recharge was weaker compared to the other years and the withdrawals increased. The pumping wells were in operation at 97.6 %. But the return flow was calculated on the basis of the maximum demand and not on the basis of the actual withdrawals; in this case, because the quantity of water that cannot be pumped was not negligible compared to the demand for water, the prescribed return flow was slightly over-estimated and the balance was optimistic. In fact, the return flows ranged between $6\,720\,000 \, m^3$ and $6\,560\,000 \, m^3$. The discharge from prescribed heads was 3.1 % of the recharge. The overflows were 0.2 % of the recharge which was negligible and also located near the draining meshes.

11.6.6 Discussion of the simulated heads

Figure 11.18 shows the comparison between the simulated heads and the measured ones for January 2001 and January 2002. Most of the water levels are well simulated. Some wells show lower simulated water levels than those observed. Figure 11.19 presents the comparison between the simulated groundwater levels as a function of time and the field data for all the project wells located within the watershed.

Figure 11.20 shows the area-wise comparison. The majority of the watershed area has water level differences within the tolerance limit of 5 m decided arbitrarily. The value is not very

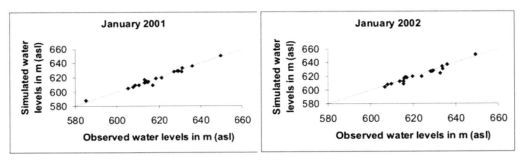

Figure 11.18 Comparison between simulated and calculated heads for January 2001 and January 2002

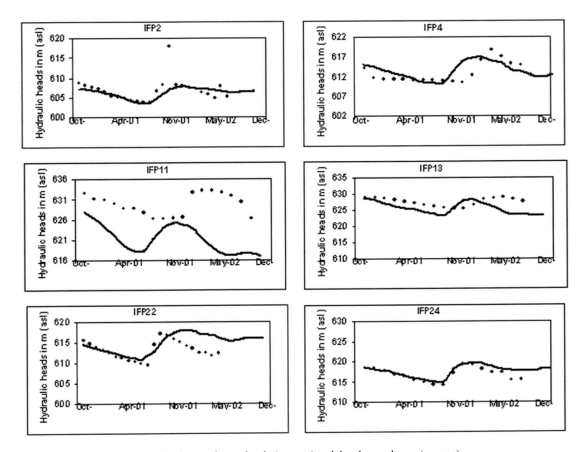

Figure 11.19 Comparison between the simulated groundwater levels (squares) and the observed ones (crosses)

different from the 2σ (σ being the standard deviation of water level data and the average σ^2 for the simulation period was $12.0\,\mathrm{m}^2$).

The wells located in the western part of the watershed show lower groundwater levels in the simulations than those observed in the field. Some tests were made by reducing the pumping rate values, i.e., assuming that the farmers were not pumping during the summer; the results show that the water levels were then better simulated in this area. Varying the pumping rates based on more accurate data may lead to a better fitting. In the north-eastern part of the watershed, the water levels show a more complicated pattern: within a few hundred metres, the simulated water levels are either too high, or too low, or else fit the field observations. This is an illustration of the complexity of the system; the variations may stem from a lack of accuracy in the pumping data or from local geological heterogeneities; this area is highly pumped and

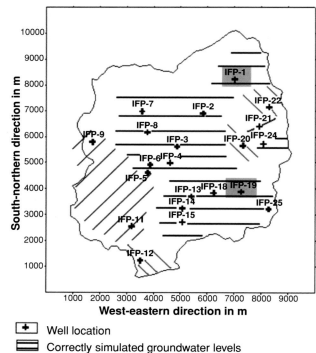

Well location

Correctly simulated groundwater levels

Under-estimated simulated groundwater levels

Slightly over-estimated simulated groundwater levels

Smaller simulated amplitude in the groundwater levels

Figure 11.20 Results of the simulation

the fitting problem may also be caused by the proximity of a farm well to the observation wells: the observation wells can be influenced by the pumping rates at a very small scale that is not adapted to the mesh size of the model. In this case, a smaller mesh size at this location could give a better result and guide the development of the model. Even after varying the hydraulic conductivities and storativities near the project wells IFP1 and IFP19, the simulated water levels in these wells show smaller amplitude in the simulations than in reality. They are the only wells in the watershed to show such a large amplitude in the water level fluctuations and these two wells are also aligned with the quartz vein that crosses the watershed from north to south. This high amplitude in the water-level fluctuation may be due to the low storativity induced by the presence of the quartz vein that is only a few tens of meters wide and cannot be represented at the model mesh scale. Here again a smaller mesh size near these wells might produce better results.

Figure 11.21 presents some simulated water-level maps for both layers. There is not much difference in the water levels between layer 1 and layer 2 except in the zone of large groundwater withdrawals where layer 2 shows lower hydraulic heads than layer 1.

Table 11.4 *Water storage in both layers*

	Layer 1	Layer 2
Water storage capacity (m^3)	14 491 474	19 803 453
Minimal water storage (m^3)	724 574	1 980 345
Storage in 01/2001 (m^3)	1 182 864 (8 %)	16 561 160 (84 %)
Storage in 01/2002 (m^3)	1 896 799 (13 %)	16 944 939 (86 %)
Storage in 01/2003 (m^3)	967 094 (7 %)	14 837 332 (75 %)

11.6.7 Discussion of the water storage capacity of the watershed and the probable evolution of its groundwater resources

The storage capacity of each layer was calculated (Table 11.4) as follows:

$$\sum_{i=1}^{n} H_i \times S \times Sy, \qquad (11.13)$$

with:

> n, the total number of meshes in the layer,
> H_i, the thickness of the aquifer in the mesh, in m,
> S, the surface area of the mesh (100 m × 100 m),
> S_y, the specific yield of the layer (0.024 for layer 1 and 0.01 for layer 2).

In the same way, the water storage of each layer was also calculated for each year with the same formula by replacing H_i by the hydraulic head of the month of January calculated for each mesh in the layer, as all water levels at that time (minimum) are very close to the saturated thickness of the aquifer.

The minimum water storage is a parameter that the user can prescribe in the model to help the convergence when the layers become unsaturated or saturated. It is also an input through a limitation of the withdrawals in layer 2.

The term expressed in % in Table 11.4 is the real storage of the layer compared to the total water storage capacity. It appears that layer 1 is almost totally unsaturated. Around 80 % of layer 2 is filled in the month of January. In 2002, the water demand reached 12 200 000 m^3, which represents around 65 % of the total water storage available at the beginning of the year in both layers. Table 11.5 presents the main components that control the balance.

Water that cannot be pumped is simply the difference between the demand based on crop requirement and the actual withdrawal. The cropping pattern requires this amount of water but, due to desaturation of the layer, the water is not actually withdrawn. Table 11.5 shows that even if the water demand is much higher than the recharge (up to five times higher in 2002), the balance is not alarming because of the high rate of return flow (55 % of the withdrawals). Another parameter that controls the state of the aquifer is the "water that cannot be pumped" due to the

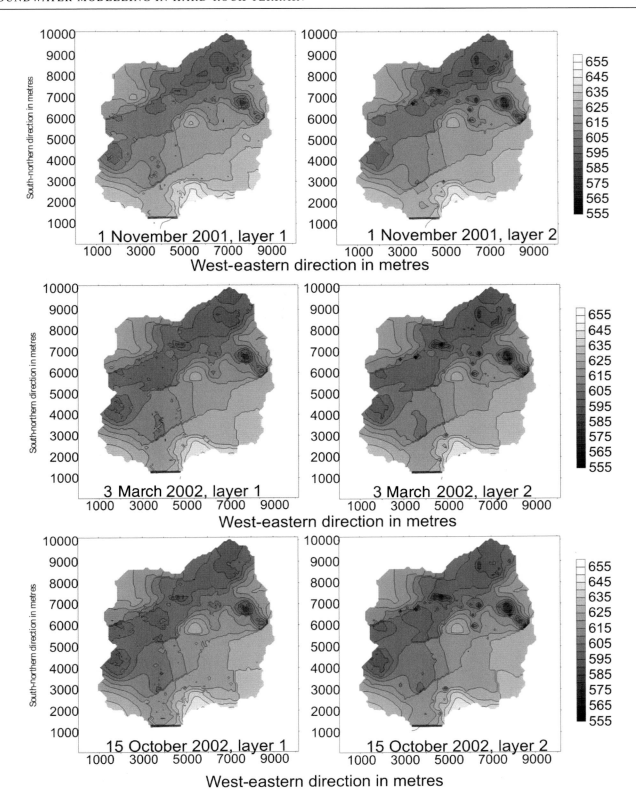

Figure 11.21 Simulated water levels for layers 1 and 2

Table 11.5 *Main parameters controlling the balance.*

	Rainfall (mm)	Water demand (m³)	Water that cannot be pumped (m³)	Recharge (m³)	Return flow (m³)	Groundwater storage (m³)
2000	630	−9 479 540	10 399	3 521 310	5 213 780	*17 744 024*
2001	886	−11 324 400	30 901	6 212 840	6 228 460	*18 841 738*
2002	584	−12 218 600	291 152	2 372 280	6 720 260	*15 804 426*

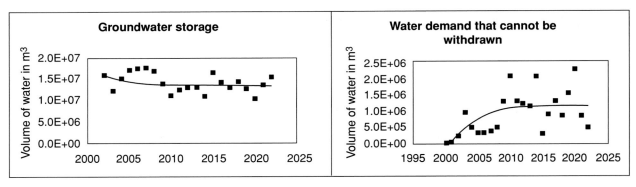

Figure 11.22 Predictive scenarios: evolution of the groundwater storage and of the water demand that cannot be pumped. Squares, the recharge varies; continuous line, the recharge is set constant

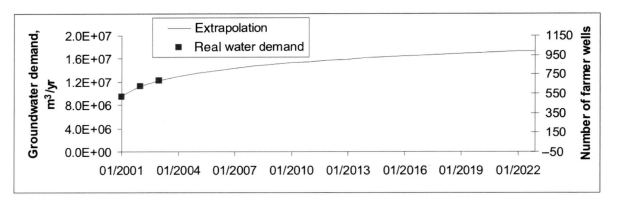

Figure 11.23 Estimate of the increase in the water demand for the next 20 years

desaturation of the second layer in some meshes of the watershed. This value may regulate the balance in the long term.

Some predictive scenarios were tested until the year 2023 assuming:

- that the recharge was set at a constant value equivalent to the average of the recharge calculated between 1986 and 2002;
- that the recharge varied as it had done between 1986 and 2002.

In both scenarios, the water demand was set constant and equal to the one in 2002.

Figure 11.22 shows that the groundwater system reaches a steady-state regime. This regime is reached when the Recharge = (Water demand – Water demand that cannot be pumped) – Return flow. In 2023, 69 farm wells of a total of 709 will not be able to provide the water demanded due to the desaturation of the aquifer.

If the water demand increases, the scheme is different. Figure 11.23 shows an extrapolation of the water demand over the next 20 years. This figure is only an example. It proposes a logarithmic increase in the water demand justified by the assumption that the yearly increase of the number of wells will decrease due to a saturation of the land use. With this figure, the number of farm

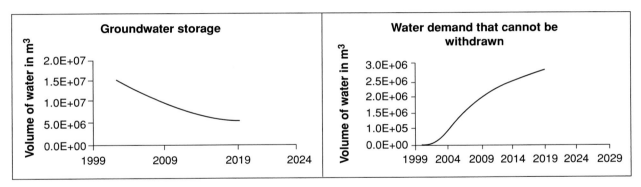

Figure 11.24 Predictive scenario: evolution of the groundwater storage and of the water demand that cannot be pumped. Recharge is set constant and the water demand increases

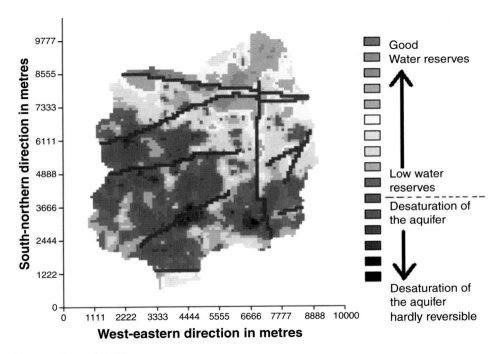

Figure 11.25 State of the groundwater in 2023

wells located in the watershed will have increased from 709 to 1004 in January 2023.

Predictions were made with this new scenario. Figure 11.24 shows that the global groundwater storage decreases rapidly and, as a consequence, the "water demand that cannot be pumped" due to the desaturation of the aquifer increases (Fig. 11.25) and touches 149 wells of 1004 in 2023. In the longer term (a few more years), the groundwater storage will be directly equivalent to the recharge. The Monsoon will allow the aquifer to recharge but the irrigation will, within one year or less, empty the aquifer until the next Monsoon season. In these conditions, the farmers will be more vulnerable to dry years than at present and may not be

able to grow two or even one crop in these years. This will lead to economic instability and hardship for the farmers.

11.6.8 Conclusion

Groundwater flow in the Maheshwaram watershed was simulated taking into account the withdrawals for agriculture. A two-layer aquifer model was chosen: the upper layer being the weathered rocks and the lower layer the weathered-fractured rocks. For a majority of wells, hydraulic head simulations are in accordance with the water levels observed in the project wells for the years 2001 and 2002. Hydraulic heads that are not in accordance with the field observations are located in areas where the withdrawals have to be verified. The average hydraulic conductivity and the specific yield of the weathered-fissured layer are in accordance with those found with the "discrete fracture network" model used to interpret the pumping tests in the aquifer (Bruel *et al.*, 2002). The other hydraulic parameters are in accordance with those in the literature. The model has to be further improved with:

- the input of a more accurate geometry of the layers;
- a more accurate estimate of the recharge;
- the verification of some withdrawal values.

The return flow from the irrigation water is very difficult to validate. It is an approximate parameter that can also be calibrated with the model, but that must stay within the range of values found in the literature.

Some predictive scenarios were also tested and showed that:

- If the water demand remains equal to the one in 2002, the regime will reach a steady-state flow where the mean water level in the wells will be the same as the present one in some areas, and up to 30 m lower in others.
- If the water demand increases, the water levels in the watershed will decrease drastically everywhere. More and more wells will become dry and the groundwater storage at the beginning of a given year will be entirely dependent on the recharge of that year. In other words, there will not be any more reserves in the aquifer to even out the temporal variability of recharge from year to year. This will lead to difficulties for the farmers and economic instability.

11.7 DISCUSSION AND GENERAL CONCLUSIONS

This chapter has provided an overview of groundwater resources and their management in hard-rock areas, with emphasis on arid and semi-arid zones and the Indian experience. The focus has also been on groundwater modelling, which is a necessary step in an attempt to quantify the resource and test management scenarios to forecast the behavior of the aquifer under various withdrawal schemes. Here, only the quantity of water is considered, not its quality, although the exploitation of an aquifer may change the quality of its water. This is particularly true in coastal aquifers, where over-exploitation can cause salt water intrusion, but salt intrusion can also occur in continental aquifers, e.g., in southern Madagascar, where in a semi-arid zone, surface water and shallow groundwater are subjected to strong evaporation, which creates salt crusts that dissolve during the following rainy season. Salt infiltrates into the aquifer, creating highly concentrated brines, of up to several grams of salt per litre in certain areas (see e.g.,

Rabemanana *et al.*, 2005). This has also been observed in the arid zone of the Bolivian Altiplano (Coudrain *et al.*, 2001). Man-made pollution can also degrade the groundwater quality, requiring special remediation schemes or preventing withdrawals in the affected areas.

Beyond describing the main steps in groundwater modelling, we have shown here that one of the key factors that determine groundwater exploitation sustainability is the long-term recharge rate. Contrary to expectations, hydrogeologists are still poorly equipped to accurately estimate the total recharge rate of aquifers. This is due to the diversity of recharge mechanisms (direct recharge, indirect and localized recharge: infiltration in stream beds and ponds, irrigation return flow, etc., see for example, Marsily, 2003) but also to the strong temporal variability of recharge, as a function of climatic variations. In desert areas, in particular, recharge to the aquifers can occur only every 10 to 30 years, requiring long-term monitoring of the aquifer behavior to evaluate the resources, see for example, Besbes *et al.* (1978). The method able to estimate recharge with the least uncertainty is still the one where a groundwater flow model is fitted on the piezometric records and withdrawal/discharge data, even if this fitting is uncertain and can be done by calibrating parameters other than the recharge. More work is required to improve recharge estimates. A concern is the influence of the predicted climate changes on rainfall and infiltration. Although, on average, the global warming due to greenhouse gases is likely to increase rainfall across the globe, some arid areas will receive less rainfall and recharge, which are still difficult to predict with the current atmospheric global circulation models.

Recharge estimates by groundwater modelling, as described in this chapter, cannot be made for every aquifer in a country as large as India. We have seen how data-intensive and time-consuming this work is. It is therefore clear that only a few "representative watersheds," of a given class of geology and climate, can be studied in this way. The results obtained in these watersheds must then be extended to all similar systems in their class and adapted to the local situation. The methods by which these classifications, extensions, and adaptations can be achieved are still to be determined.

Finally, we would like to draw attention to an attitude very common in the hydrogeologic community, which is to say that an aquifer is "overexploited." Again and again, observing that the water levels in an aquifer are dropping regularly, hydrogeologists will say that the aquifer is "overexploited." This word should be used more carefully. The decline of the water level is not enough to qualify an aquifer as overexploited. In purely natural conditions, an aquifer receives recharge from precipitation and discharges this water at its natural outlets: a spring, a river, a lake, or the sea. As soon as exploitation by artificial withdrawal starts, the water levels begin to decline, generally indicating that less water is lost to the natural discharge zones and that the water stored in the aquifer is put to work. Using the storage is one of the benefits of groundwater exploitation: it is available all year round, and in particular during the summer, when water is needed, and recharge is nil. It is therefore very important to precisely define what "overexploitation" means; declining water levels are simply an indication that the aquifer is exploited, but not necessarily that it is "overexploited."

Three clear signs indicate overexploitation:

(i) When the low water levels in the aquifers cause a deterioration of the water quality, as for example, seawater intrusion in coastal aquifers.

(ii) When the water stored in the system is too low to average out recharge fluctuations over several years: it is then no longer possible to keep a quasi-constant rate of withdrawal, regardless of the value of a given year recharge, and the water demand has to be regulated by the annual recharge, not the long-term average recharge. This makes the economic and social system much more fragile and prone to crisis.

(iii) When the water level is so low that some parts of the aquifer are dry (or the aquifer is not permeable enough at these depths to be exploitable by wells), creating new social inequalities between users, some farmers can still exploit the resource, whereas others are deprived of its benefit. Alternatively, the water level is so low for all users that it is no longer economical to pay the pumping costs for irrigation. This economic limit is considered today to be reached when the dynamic water level in wells is deeper than 100 to 200 m.

At the present time, it seems that the hard-rock aquifers in semi-arid zones in India are approaching real overexploitation, mainly due to reasons (ii) and (iii), i.e., insufficient storage to cope with annual recharge variations, and unequal access to water. These reasons argue for regulation of the resource exploitation before the onset of crisis situations.

11.8 ACKNOWLEDGEMENTS

The authors would like to thank all those involved in running the G-WADI program who have given us the opportunity to include these results in this chapter on groundwater modelling. However, the authors wish to extend special thanks to Professor Howard Wheater, Imperial College, UK, Professor S. Sorooshian, University of California, USA, Dr. Salih Abdin of UNESCO as well as Dr. K. D. Sarma of NIH, India for encouraging our studies and including us in the program. Thanks are due to the Director of NGRI for extending the facilities for carrying out most of the research reported in this chapter and to various funding agencies particularly the IFCPAR, New Delhi, for supporting a major project whose findings are part of this chapter. Several colleagues were actively involved in the studies that are reported and we are grateful in particular to C. Engerrand and D. Bruel of the Paris School of Mines. The authors thank the Indo-French Centre for Groundwater Research and several scientists among which B. Dewandel, J. M. Gandolfi, P. D. Sreedevi, D. Kumar, P. Lachassagne, N. S. K. Murthy, K. Subrahmanyam, and F. Touchard, have dealt with the study.

11.9 REFERENCES

Aboufirassi, M. and Marino, M. A. (1983). Kriging of water levels in Souss aquifer, Morocco, *J. Math. Geol.*, **15(4)**, 537–51.

Aboufirassi, M. and Marino, M. A. (1984). Cokriging of aquifer transmissivities from field measurements of transmissivity and specific capacity, *J. Math. Geol.*, **16(1)**, 19–35.

Aghassi, A. V. (1990). Groundwater evaluation in fractured zones with special emphasis on solubility of carbonated rocks. In *Proc. of IAH Congress, Water Resources in Mountainous Regions*, ed. A. Parriaux. IAH Memoirs, Vol. XXII, Part 1. Lausanne, Switzerland: Ecole Polytechnique Federale de Lausanne, 480–5.

Ahmed, S. (1987). Estimation des transmissivités des aquifères par methodes géostatistiques multivariables et résolution du problème inverse. Doctoral thesis, Ecole Nationale Supérieure des Mines, Paris, France.

Ahmed, S. (1995). An interactive software for computing and modeling a variogram. In *Proc. of a Conference on Water Resources Management (WRM'95), August 28–30*. Iran: Isfahan University of Technology, 797–808.

Ahmed, S. (2004). Contribution of geostatistics to the aquifer modeling for groundwater management, *Proc. of International Conference on Advanced Modeling Techniques for Sustainable Management of Water Resources, Jan 28–30, 2004, Warangal, India*. Hyderabad: Allied Publishers Pvt. Limited, 264–272.

Ahmed, S. and Gupta, C. P. (1989). Stochastic spatial prediction of hydrogeologic parameters: role of cross-validation in kriging. *International Workshop on Appropriate Methodologies for Development and Management of Groundwater Resources in Developing Countries" Hyderabad, India, February 28–March 4, 1989*, Vol. III. New Delhi: Oxford and IBH Publishing Co. Pvt. Ltd., IGW, 77–90.

Ahmed, S. and Marsily, G. de (1987). Comparison of geostatistical methods for estimating transmissivity using data on transmissivity and specific capacity, *Water Resour. Res.*, **23(9)**, 1717–37.

Ahmed, S., and Marsily, G. de (1988). Some application of multivariate kriging in groundwater hydrology. *Science de la Terre, Série Informatique*, **28**, 1–25.

Ahmed, S., and Marsily, G. de (1989). Cokriged estimates of transmissivity using jointly water levels data. In *Geostatistics*, ed. M. Armstrong. Dordrecht: Kluwer Academic Publishers, 615–28.

Ahmed, S., Marsily, G. de, and Talbot, A. (1988). Combined use of hydraulic and electrical properties of an aquifer in a geostatistical estimation of transmissivity. *Groundwater*, **26(1)**, 78–86.

Ahmed, S., Marsily, G. de, and Gupta, C. P. (1990). Coherent structural models in cokriging aquifer parameters: estimation of transmissivity and water levels. In *Proc. of IAH Congress, Water Resources in Mountainous Regions*, ed. A. Parriaux. IAH Memoirs, Vol XXII, Part 1. Lausanne, Switzerland: Ecole Polytechnique Federale de Lausanne, 123–31.

Ahmed, S. and Murali, G. (1992). Regionalization of fluoride content in an aquifer. *J. Environ. Hydrol.*, **1(1)**, 35–9.

Ahmed, S. and Marsily, G. de (1993). Cokriged estimation of aquifer transmissivity as an indirect solution of the inverse problem: a practical approach. *Water Resour. Res.*, **29(2)**, 521–30.

Ahmed, S., Kumar, D., and Maréchal, J. C. (2002). Geostatistical analyses of water level in a fractured aquifer and optimisation of monitoring network. *Proc. of the International Groundwater Symposium, LBNL, March 25–28, 2002, USA*, ed. A. N. Findikakis. Spain: IAHR publication, 379–82.

Ahmed, S., Engerrand, C., Sreedevi, P. D. *ed al.* (2003). *Geostatistics, aquifer modelling and artificial recharge*, Scientific Report, Vol. 3, Indo-French

Collaborative Project (2013–1), Technical Report No. NGRI-2003-GW-41, Hyderabad, India.

Amadi, U. M. P. and Fontes, J. Ch. (1990). Inland hydrologic sub-basins and their control on the groundwater system from the Benue Trough-Chad basin. In *Proc. of IAH Congress, Water Resources in Mountainous Regions*, ed. A. Parriaux. IAH Memoirs, Vol. XXII, Part 1. Lausanne, Switzerland: Ecole Polytechnique Federale de Lausanne, 550–6.

Anderson, M. P. and Woessner, W. (1991). *Applied Groundwater Modeling: Simulation of Flow and Advective Transport*. London, New York: Academic Press.

Andhra Pradesh Groundwater Department (APGWD) (1977). *Studies on hydrologic parameters of groundwater recharge in water balance computations, Andhra Pradesh*. Research series no.6, Hyderabad: Government of Andhra Pradesh Ground Water Department.

Bardossy, A., Bogardi, I., and Kelly, W. E. (1986). Geostatistical analysis of geoelectric estimates for specific capacity. *J. Hydrol.*, **84**, 81–95.

Barker, J. (1988). A generalized radial flow model for hydrologic tests in a fractured rock. *Water Resour. Res.*, **24(10)**, 1796–804.

Basabe, P. and Bieler, G. (1990). Différenciation des écoulements du Malm inférieur et du Dogger supérieur dans le Jura vaudois, In *Proc. of IAH Congress, Water Resources in Mountainous Regions*, ed. A. Parriaux, IAH Memoirs, Vol. XXII, Part 1. Lausanne, Switzerland: Ecole Polytechnique Federale de Lausanne, 383–91.

Bates, L. E., Otto, C. J., Bartle, G., and Johnston, C. (2000). *Development of a hydrogeological database and groundwater vulnerability assessment using GIS and integrated modelling, Garden Island, WA*. Progress Report: Vulnerability Assessment. CSIRO Land and Water, Final Report 2000.

Bear, J. (1979). *Hydraulics of Groundwater*. New York: McGraw-Hill.

Belden, M. and Osborn, I. (2002). *Hydrogeologic investigation of the Ogallala aquifer in Roger Mills and Beckham counties, Western Oklahoma*. Tech. Report No. GW-2002–2, Oklahoma Water Resources Board, USA.

Besbes, M. (2003). Système Aquifère du Sahara Septentrional, Gestion commune d'un bassin transfrontière. *La Houille Blanche*, **5**, 128–33.

Besbes, M., Delhomme, J. P., and Marsily, G. de (1978). Estimating recharge from ephemeral streams in arid regions: a case study at Kairouan (Tunisia). *Water Resour. Res.*, **14(2)**, 281–90.

Brebbia, C. A. (1978). *The Boundary Element Method for Engineers*. London: Pentech Press.

Bredenkamp, D. B. (1990). Simulation of the flow of Dolomitic springs and of groundwater levels by means of annual recharge estimates. In *Proc. of IAH Congress, Water Resources in Mountainous Regions*, ed. A. Parriaux. IAH Memoirs, Vol. XXII, Part 1. Lausanne, Switzerland: Ecole Polytechnique Federale de Lausanne, 150–57.

Bruel, D., Engerrand, C., Ledoux, E., *et al.* (2002). *The evaluation of aquifer parameters in Maheshwaram Mandal, RR dist., AP., India*. Ecole des Mines de Paris, CIG, technical report LHM/RD/02/28, July 2002, updated November 2002.

Cacas, M. C., Ledoux, E., Marsily, G. de, *et al.* (1990a). Flow and transport in fractured rocks: an in situ experiment in the Fanay-Augeres mine and its interpretation with a discrete fracture network model. In *Proc. IAH Congress, Water Resources in Mountainous Regions*, ed. A. Parriaux. IAH Memoirs, Vol. XXII, Part 1. Lausanne, Switzerland: Ecole Polytechnique Federale de Lausanne, 13–38.

Cacas, M. C., Ledoux E., Marsily G. de, *et al.* (1990b). Modelling fracture flow with a discrete fracture network: calibration and validation. 1: The flow model. *Water Resour. Res.*, **26(1)**, 479–89.

Carrera, J. and Neuman, S. P. (1986a). Estimation of aquifer parameter under transient and steady state conditions. 1: Maximum likelihood method incorporating prior information. *Water Resour. Res.*, **22(2)**, 199–210.

Carrera, J. and Neuman, S. P. (1986b). Estimation of aquifer parameter under transient and steady state conditions. 2: Uniqueness, stability, and solution algorithms. *Water Resour. Res.*, **22(2)**, 211–27.

Carrera, J. and Neuman, S. P. (1986c). Estimation of aquifer parameter under transient and steady state conditions 3: Application to synthetic and field data. *Water Resour. Res.*, **22(2)**, 228–42.

Carrera, J. (1990). A modeling approach incorporating quantitative uncertainty estimates. In *Proc. of IAH Congress, Water Resources in Mountainous Regions*, ed. A. Parriaux. IAH Memoirs, Vol. XXII, Part 1. Lausanne, Switzerland: Ecole Polytechnique Federale de Lausanne, 67–78.

Chen, B. and Chen, J. (1990). The application of groundwater chemical constituent cluster analysis method for groundwater resources elevation in

Maxian County. In *Proc. of IAH Congress, Water Resources in Mountainous Regions*, ed. A. Parriaux. IAH Memoirs, Vol. XXII, Part 1. Lausanne, Switzerland: Ecole Polytechnique Federale de Lausanne, 460–7.

Chiles, J. P. and Marsily, G. de (1993). Flow in fractured rocks. In *Flow and Transport in Fractured Rocks*, ed. J. Bear, G. de Marsily, and C. F. Tsang, Orlando, FL: Academic Press.

Chiles, J. P. and Delfiner, P. (1999). *Geostatistics: Modeling Spatial Uncertainty*. New York: Wiley.

Chilton, P. J. and Foster, S. S. D. (1995). Hydrogeological characteristics and water-supply potential of basement aquifers in tropical Africa. *Hydrogeo. J.*, **3(1)**, 36–49.

Civita, M., Olivaro, G., Vigna, B., and Pavia, R. (1990). Hydrodynamic and chemical features of a high altitude karstic system in the maritime Alps (Italy). In *Proc. of IAH Congress, Water Resources in Mountainous Regions*, ed. A. Parriaux. IAH Memoirs, Vol. XXII, Part 1. Lausanne, Switzerland: Ecole Polytechnique Federale de Lausanne, 444–51.

Cooley, R. L. (1977). A method of estimating parameters and assessing reliability for models of steady state groundwater flow. 1: Theory and numerical property. *Water Resour. Res.*, **13(2)**, 318–24.

Cooley, R. L. (1979). A method of estimating parameters and assessing reliability for models of steady state groundwater flow. 2: Application of statistical analysis. *Water Resour. Res.*, **15(3)**, 603–17.

Cooley, R. L. (1982). Incorporation of prior information into non-linear regression groundwater flow models. 1: Theory. *Water Resour. Res.*, **18(4)**, 965–76.

Coudrain, A., Talbi, A., Ledoux, E. *et al.* (2001). Subsurface transfer of chloride after a lake retreat in the central Andes. *Groundwater*, **39(5)**, 1–9.

CRIDA (1990). *Soil Map of Hayathnagar Farm and Appendix*. Ranga Reddy District, Andhra Pradesh, India: Center for Research In Dry-land Agriculture.

CRIDA, (1991–92; 1993–94; 1994–95; 1997–98; 1998–99; 1999–2000). Annual Report. Center for Research in Dry-land Agriculture, Ranga Reddy District, Andhra Pradesh, India.

Custodio, E. (2002). Aquifer overexploitation: what does it mean? *Hydrogeol. J.*, **10(2)**, 254–77.

Darricau-Beucher, H. (1981). Approche géostatistique du passage des données de terrain aux paramètres des modèles en hydrogéologie. Doctoral thesis, Ecole Nationale Supérieure des Mines, Paris, France.

Das Gupta, A., Paudyal, G. N., and Aryal, S. K. (1990). Optimization of pumping and recharge pattern for a depleted aquifer. In *Proc. of IAH Congress, Water Resources in Mountainous Regions*, ed. A. Parriaux. IAH Memoirs, Vol. XXII, Part 1. Lausanne, Switzerland: Ecole Polytechnique Federale de Lausanne, 982–90.

Delfiner, P. and Delhomme, J. P. (1973). Optimum interpolation by kriging. In *Display and Analysis of Spatial Data*, ed. J. C. Davis and M. J. McCullough. London: Wiley, 96–114.

Delhomme, J. P. (1974). La cartographie d'une grandeur physique à partir des données de différentes qualités. In *Proc. of IAH Congress, Montpelier, France*, Vol. X, Part 1. Lausanne, Switzerland: Ecole Polytechnique Federale de Lausanne, 185–194.

Delhomme, J. P. (1976). Application de la théorie des variables régionalisées dans les sciences de l'eau, Doctoral Thesis, Ecole des Mines de Paris, Fontainebleau, France.

Delhomme, J. P. (1978). Kriging in the hydrosciences. *Adv. Water Resour.*, **1(5)**, 252–66.

Delhomme, J. P. (1979). Spatial variability and uncertainty in groundwater flow parameters: a geostatistical approach. *Water Resour. Res.*, **15(2)**, 269–80.

Deutsch, C. V. and Journel, A. G. (1992). *GSLIB, Geostatistical Software Library and User's Guide*. New York: Oxford University Press.

Dong, A., Ahmed, S., and Marsily, G de (1990). Development of geostatistical methods dealing with boundary condition problems encountered in fluid mechanics. In *Proceedings, 2nd European Conference on the Mathematics of Oil Recovery, Arles, France, September 11–14*, ed. Guerillot and Guillon. Paris: Technip, 21–30.

Edmunds, W. M. (1999). Unsaturated zone chemical and isotopic tracers to estimate groundwater recharge, recharge history and past environments in North Africa. *Abstr. Progr., Geolog. Soc. Am.*, **31(7)**, 87.

Elo, S. (1992). Geophysical indications of deep fractures in the Narankavaara-Syote and Kandalaksha-Puolanka zones. Geolog. Surv. Finland, **13**, 43–50.

El-Shibini, F. and El-Kady, M. (2002). Coping with water scarcity: the future challenges. In *Water Resources Development and Management*, Vol. 4, ed. Al-Rashed, Singh and Sherif. The Netherlands: Swets & Zeitilinger B.V., 43–54.

Engerrand, C. (2002). Hydrogeology of the weathered-fissured hard-rock aquifers located in Monsoon areas: hydrogeological study of two water-sheds in Andhra Pradesh (India). Doctoral Thesis, Université Pierre et Marie Curie, Paris, France.

Fert, M., Mordzonek, G., and Wêglarz, D. (2005). The management and data distribution system of the hydrogeological map of Poland 1: 50 000. *Przegl Geologiczny*, **53(10/2)**, 940–1.

Freeze, R. A. (1971). Three-dimensional, transient, saturated-unsaturated flow in a groundwater basin. *Water Resour. Res.* **7(2)**, 347–66.

Galli, A. and Meunier, G. (1987). Study of a gas reservoir using the external drift method. In *Geostatistical Case Studies*, ed. G. Matheron and M. Armstrong. Dordrecht: D. Reidel Publ. Co., 105–20.

Gambolati, G. and Volpi, G. (1979). A conceptual deterministic analysis of the kriging technique in hydrology. *Water Resour. Res.*, **15(3)**, 625–9.

Gaye, C. B. and Edmunds, W. M. (1996). Groundwater recharge estimation using chloride, stable isotopes and tritium profiles in the sands of north-western Senegal, Environ. Geol., **27(3)**, 246–51.

Geier, J. E. and Axelsson, C. L. (1991). *Discrete fracture modeling of Finnsjön rock mass. Phase I: Feasibility study*. Swedish Nuclear Fuel and Waste Management Co., Technical Report No. SKB 91–13, Stockholm, Sweden.

Gogu, R., Carabin, G., Hallet, V., Peters, V., and Dassargues, A. (2001). GIS based hydrogeological databases and groundwater modelling. Hydrogeol. J., **9(4)**, 555–69.

Gupta, C. P., Thangarajan, M., and Gurunadha Rao, V. V. S. (1979). Electric analog model study of aquifer in Krishni-Hindon interstream region, Uttar Pradesh, India. *Groundwater*, **17(3)**, 284–90.

Gupta, C. P., Ahmed, S., and Gurunadha Rao, V. V. S. (1985). Conjunctive utilization of surface and ground water to arrest the water-level decline in an alluvial aquifer. *J. Hydrol.*, **76(3/4)**, 351–61.

Handa, B. K. (1975). Geochemistry and genesis of fluoride contains groundwater in India. *Groundwater*, **13**, 275–81.

Havlík, M. and Krásný, J. (1998). Transmissivity distribution in southern part of the Bohemian Massif: regional trends and local anomalies, hardrock hydrogeology of the Bohemian Massif. Proc. 3rd Internat. Workshop 1998, Windischeschenbach., *Münchner Geol. Hefte*, **B8**: 11–18.

Hoeksema, R. J. and Kitanidis, P. K. (1984). An application of the geostastical approach to the inverse problem in two-dimensional ground water modeling. *Water Resour. Res.*, **20(7)**, 1003–20.

Houston, J. F. T. and Lewis, R. T. (1988). The Victoria Province drought relief project. II: Borehole yield relationships. *Groundwater*, **26(4)**, 418–26.

Huyakorn, P. S., Lester B. H., and Faust, C. R. (1983). Finite element techniques for modeling groundwater flow in fractured aquifers. *Water Resour. Res.*, **19(4)**, 1019–31.

Isaaks, M. and Srivastava, R. M. (1989). *An Introduction to Applied Geostatistics*. New York: Oxford University Press.

Jackson, C. P. and Porter, J. D. (1990). Uncertainty in groundwater flow and transport calculations for repository. In *Proc. of IAH Congress, Water Resources in Mountainous Regions*, ed. A. Parriaux. IAH Memoirs, Vol. XXII, Part 1. Lausanne, Switzerland: Ecole Polytechnique Federale de Lausanne, 87–96.

Jaquet, O., Thompson, B., Vomvoris, S., and Hufschmied, P. (1990). Geostatistical methods applied to inverse modelling in the macro-permeability experiment at the Grimsel test site, Switzerland. In *Proc. of IAH Congress, Water Resources in Mountainous Regions*, ed. A. Parriaux. IAH Memoirs, Vol. XXII, Part 1. Lausanne, Switzerland: Ecole Polytechnique Federale de Lausanne, 142–9.

Journel, A. G., and Huijbregts, Ch. J. (1978). *Mining Geostatistics*. London: Academic Press.

Kitanidis, P. K. (1997). *Introduction to Geostatistics: Application to Hydrogeology*. Cambridge: Cambridge University Press.

Kovalevsky, V. S. (1990). Irregularity of groundwater recharge and its consideration in planning rational groundwater development. In *Proc. of IAH Congress, Water Resources in Mountainous Regions*, ed. A. Parriaux. IAH Memoirs, Vol. XXII, Part 2, 941–9.

Krasny, J, (1990). Regionalization of transmissivity data: hard rocks of the Bohemian Massif (Czechoslovakia). In *Proc. of IAH Congress, Water Resources in Mountainous Regions*, ed. A. Parriaux. IAH Memoirs,

Vol. XXII, Part 1. Lausanne, Switzerland: Ecole Polytechnique Federale de Lausanne, 98–105.

Krásný, J. (1999). Hard-rock hydrogeology in the Czech Republic. *Hydrogéologie*, **2**, 25–38.

Krisnamurthy, N. S., Kumar, D., Sankaran, S., *et al.* (2000). *Magnetic investigations across some lineaments in Maheshwaram watershed, Andhra Pradesh, India*. Technical Report National, Geophysical Research Institute of India, NGRI-2000-GW-288, September 2000.

Kromm, D. E. and White, S. E. (1992). *Groundwater Exploitation in the High Plains*. Lawrence, Kansas: University of Kansas Press.

Kupfersberger, H. and Bloschl, G. (1995). Estimating aquifer transmissivities – on value of auxiliary data. *J. Hydrol.*, **165**, 85–99.

Lebert, F. (2001). *Mesure de topographie par DGPS sur le bassin de Maheshwaram, (Andhra Pradesh, Inde)*. Report BRGM/RP-50728-FR; February 2001.

Le Borgne, T., Bour, O., Paillet, F. L., and Caudal, J. P. (2006). Assessment of preferential flow path connectivity and hydraulic properties at single-borehole and cross-borehole scales in a fractured aquifer. *J. Hydrol.*, **328**, 347–59.

Legchenko, A. and Baltassat, J. M. (1999). *Application of the "Numis" proton magnetic resonance equipment for groundwater exploration in a fractured granite environment 30 km south of Hyderabad, Inida*. BRGM-Internal report – December 1999 – R40925.

Lerner, D. N., Issar, A., and I. Simmers. (1990). A guide to understanding and estimating natural recharge. *Int. Contribution to Hydrogeology*, IAH. Publ., 8. The Netherlands: Verlag Heinz Heisse.

Lyon, S. W., Seibert, J., Lembo, A. J., Walter, M. T., and Steenhuis, T. S. (2005). Geostatistical investigation into the temporal evaluation of spatial structure in a shallow water table. *Hydrol. Earth Sys. Sci. Discuss.*, **2**, 1683–716.

Mann, F. M. and Myers, D. A. (1998). *Computer Code Selection Criteria For Flow and Transport Code(s) to be Used in Undisturbed Vadose Zone Calculations for TWRS Environmental Analyses*. (HNF-1839, Rev. B). Richland, Washington: Lockheed-Martin Hanford Company.

Maréchal, J. C., Dewandel, B., Subrahmanyam, K., and Torri, R. (2003a). Review of specific methods for the evaluation of hydraulic properties in fractured hard-rock aquifers. *Curr. Sci. India*, **85(4)**, 511–16.

Maréchal, J. C., Wyns, R., Lachassagne, P., Subrahmanyam, K., and Touchard, F. (2003b). Anisotropie verticale de la perméabilité de l'horizon fissuré des aquifères de socle: concordance avec la structure géologique des profils d'altération, *C. R. Geosci.*, **335**, 451–60.

Maréchal, J. C., Dewandel, B., and Subrahmanyam, K. (2004). Use of hydraulic tests at different scales to characterize fracture network properties in the weathered-fractured layer of a hard-rock aquifer, *Water Resour. Res.*, **40**, W11508.

Maréchal, J. C., Dewandel, B., Ahmed, S., Galeazzi, L., and Zaidi, F. K. (2006). Combined estimation of specific yield and natural recharge in a semi-arid groundwater basin with irrigated agriculture, *J. Hydrol.*, **329**, 281–93.

Marsily, G. de (1986). *Quantitative Hydrogeology: Groundwater Hydrology for Engineers*. Orlando, FL: Academic Press.

Marsily, G. de., Lavedan, G., Boucher, M., and Fasanino, G. (1984). Interpretation of interference tests in a well field using geostatistical techniques to fit the permeability distribution in a reservoir model. In *Geostatistics for Natural Resources Characterization*, Part 2, ed. G. Verly *et al.*, Proc. NATO-ASI, Ser. C., 182. Hingham, MA: D. Reidel, 831–49.

Marsily, G. de, and Ahmed, S. (1987). Application of kriging techniques in groundwater hydrology, *J. Geol. Soc. India*, **29(1)**, 47–69.

Marsily, G. de, Delhomme, J. P., Coudrain-Ribstein, A., and Lavenue, M. A. (2000). Four decades of inverse problems in hydrogeology. In *Theory, Modeling, and Field Investigation in Hydrogeology: A Special Volume in Honor of Shlomo P. Neuman's 60th Birthday*, ed. D. Zhang and C. L. Winter, Boulder, CO: Geological Society of America, Special Paper 348, 1–17.

Marsily, G. de (2003). Importance of temporary ponds in arid climates for the recharge of groundwater. *Académie des Sciences, CR Géosciences* **335**, 933–4.

Matheron, G. (1965). *Les Variables Régionalisées et Leur Estimation*. Paris: Masson.

Matheron, G. (1971). *The theory of regionalized variables and its application*. Paris School of Mines, Cah. Cent. Morphologie Math., 5, Fontainebleau, France.

Mercer, J. W. and Faust, Ch. R. (1981). *Groundwater Modeling*. Columbus, Ohio: National Water Well Assoc.

Michalak, M. and Kitanidis, P. K. (2002). Application of geostatistical inverse modeling to contaminant source identification at Dover AFB, Delaware. *Proc. of the International Groundwater Symposium, LBNL, March 25–28, 2002 USA*, ed. A. N. Findikakis. Spain: IAHR Publication, 137–9.

Milanovic, P. (1990). The optimal management of groundwater for meeting seasonally fluctuating demands. In *Proc. of IAH Congress, Water Resources in Mountainous Regions*, ed. A. Parriaux. IAH Memoirs, Vol. XXII, Part 2. Lausanne, Switzerland: Ecole Polytechnique Federale de Lausanne, 933–40.

Mizell, S. A. (1980). Stochastic analysis of spatial variability in two-dimensional groundwater flow with implications for observation-well-network design. Doctoral thesis, New Mexico Institute of Mining and Technology, Socorro, New Mexico.

Mogheir, Y. and Singh, V. P. (2002). Specification of information needs for groundwater resources management and planning in developing country: Gaza Strip case study. *Groundwater Hydrology*, ed. Sherif, Singh and Al-Rashed, Vol. 2, the Netherlands: Swets & Zeitilinger B.V., 3–20.

Monteith, J. L., Huda, A. K. S., and Midya, D., (1989). Modelling Sorghum and Pearl Millet – Rescap: A resource capture model for Sorghum and Pearl Millet. In *Modeling the Growth and Development of Sorghum and Millet*, Research Bulletin no. 12. Pantacheru, India: International Crops Research Institute for the Semi-Arid Tropics, 30–4.

Neuman, S. P. (1984). Role of geostatistics in subsurface hydrology. In *Geostatistics for Natural Resources Characterization, Proc. NATO-ASI*, ed. G. Verly, M. David, A. G. Journel, and A. Maréchal, Dordrecht, the Netherlands: D. Reidel Publ. Co., 787–816.

Oroz, L. (2001). *Modelo Conceptual Hidrogeológico e Hidrogeoquímico de la Costa de Hermosillo*. Doctoral Thesis, UNISON.

Parriaux, A. and Bensimon, M. (1990). Some rules for the design and the management of observation networks for groundwater resources. In *Proc. of IAH Congress, Water Resources in Mountainous Regions*, ed. A. Parriaux. IAH Memoirs, Vol. XXII, Part 1. 719–27.

Patriarche, D., Castro, M. C., and Goovaerts, P. (2005). Estimating regional hydraulic conductivity fields – a comparative study of geostatistical methods. *J. Math. Geol.*, **37(6)**, 587–613.

Pinder, G. F. and Gray, W. G. (1977). *Finite Element Simulation in Surface and Subsurface Hydrology*. London: Academic Press.

Pistre, S. (1993). Rôle de la fracturation dans les circulations souterraines du massif granitique de Millas (Pyr. Or., France). *C. R. Acad. Sci., Paris*, **317, série II (11)**, 1417–24.

Prickett, T. A. (1975). Modeling techniques for groundwater evaluation. *Adv. Hydrosci.*, **10**, 1–143.

Pye, K. (1986). Mineralogical and textural controls on the weathering of granitoids rocks, *Catena*, **13**, 47–57.

Qian, X. X. and Cai, S. H. (1990). Groundwater modeling in massif of complex geological structure. In *Proc. of IAH Congress, Water Resources in Mountainous Regions*, ed. A. Parriaux. IAH Memoirs, Vol. XXII, Part 1. Lausanne, Switzerland: Ecole Polytechnique Federale de Lausanne, 39–45.

Rabemanana, V., Violette, S., Marsily, G. de, *et al.* (2005). Origin of the high variability of the water mineral content in the bedrock aquifers of southern Madagascar. *J. Hydrol.*, **310(1–4)**, 143–56.

Radhakrishna, B. P. (1970). Problems confronting the occurrence of groundwater in hard rocks. In *Proc. of Seminar on Groundwater Potential in Hard Rocks of India*, Bangalore: Geological Society of India, 27–44.

Radhakrishna, B. P. (2004). *Groundwater in hard rock aquifers of South India: some facts which every one should know*. Bangalore: Geological Society of India Public.

Rangarajan, R. and Prasada Rao, N. T. V. (2001). *Natural recharge measurements in Maheshwaram granitic watershed, Ranga Reddy district, Andhra Pradesh, India – 1999 Monsoon*. Technical Report No. NGRI-20001-GW-298, National Geophysical Research Institute, Hyderabad, India.

Rangarajan R. D., Muralidharan, G. K., Hodlur, S. D., Deshmukh, N. T. V., Prasad Rao, U., Satyanarayana, G. B. K., Sankar, S. and Athavale, R. N. (2002). *Natural recharge rates in granites, basalt, sedimentary and alluvium formations of Andhra Pradesh using injected tritium tracer*. Technical Report No. NGRI-20001-GW-298, National Geophysical Research Institute, Hyderabad, India, 14 p.

Remson, I., Hornberger, G. M., and Molz, F. J. (1971). *Numerical Methods in Subsurface Hydrology with an Introduction to the Finite Element Method*. New York: Wiley (Intersciences).

Republic of Botswana (1997). *Mathematical model of the Shashe River valley wellfield and aquifer system* (Appendix H). Prepared by Eastend Investments, Gaborne, Botswana, 22 p.

Resele, G. and Job, D. (1990). Calibration validation and uncertainty analysis of a numerical groundwater model. In *Proc. of IAH Congress, Water Resources in Mountainous Regions*, ed. A. Parriaux. IAH Memoirs, Vol. XXII, Part 1. Lausanne, Switzerland: Ecole Polytechnique Federale de Lausanne, 79–86.

Roth, C. (1995). Contribution de la géostatistique à la résolution du problème inverse en hydrogéologie. Doctoral thesis, Ecole Nationale Supérieure des Mines, Paris, France.

Roth, C., Chiles, J. P., and de Fouquet, C. (1996). Adapting geostatistical transmissivity simulations to finite difference flow simulators. *Water Resour. Res.*, **32(10)**, 3237–42.

Samper, F. J. and Carrera, J. (1990). *Geoestadística – Aplicaciones la Hidrología Subterránea*. Barcelona: Barcelona University.

Sauty, J. P. (1977). Contribution à l'identification des paramètres de dispersion dans les aquifères par interprétation des expériences de traçage. Doctoral Thesis, Univ. Grenoble, France.

Sauty, J. P. (1978). Identification des paramètres du transport hydrodispersif dans les aquifères par interprétation de traçages en écoulement cylindrique convergent ou divergent. *J. Hydrol.* **39**, 69–103.

Simpson, E. S. and Duckstein, L. (1975). Finite state mixing cells models. *Proc. US-Yugoslavian Symp. Karst Hydrology Water Resour.*, Dubrovnik, (1975). 489–508.

Strang, G., and Fix, G. T. (1973). *An Analysis of the Finite Element Method*. Prentice Hall, Englewood Cliff, New Jersey.

Stuckless, J. S. and Dudley, W. W. (2002). The geohydrologic setting of Yucca Mountain, Nevada. *Appl. Geochem.*, **17(6)**, 659–82.

Subrahmanyam, K., Ahmed, S., and Dhar, R. L. (2000). *Geological and hydrogeological investigations in the Maheshwaram watershed, A.P., India*. Technical Report No. NGRI 2000-GW-292, Hyderabad, India.

Summers, W. K. (1972). Specific capacities of wells in crystalline rocks, *Ground Water*, **10**, 37–47.

Talbot, A. (1979). The accurate numerical inversion of Laplace transforms, *J. Inst. Math. Appl.* **23**, 97–120.

Talbot, C. J. and Sirat, M. (2001). Stress control of hydraulic conductivity in fracture-saturated Swedish bedrock. *Eng. Geol.*, **61(2–3)**, 145–53.

Taylor, R. and Howard, K. (2000). A tectono-geomorphic model of the hydrogeology of deeply weathered crystalline rock: evidence from Uganda. *Hydrogeol. J.*, **8(3)**, 279–94.

Teles, V., Perrier, E., Delay, F., and Marsily, G. de (2002). Generation of alluvial aquifers with a new genetic/stochastic sedimentation model: Comparison with geostatistical approaches by means of groundwater flow simulations. *Proc. of the International Groundwater Symposium, LBNL, March 25–28, (2002), USA*, ed. A. N. Findikakis. Spain: IAHR publication, 29–35.

Thangarajan, M. (1999a). Modeling multi-layer aquifer system to evolve pre-development management scheemes. *Environ. Geol.*, **38(4)**, 285–95.

Thangarajan, M. (1999b). Numerical simulation of groundwater flow regime in a weathered hard rock aquifer. *J. Geolog. Soc. India*, **53(5)**, 561–70.

Thangarajan, M. (2000). Approaches for modeling the hard rock aquifer system. *J. Geolog. Soc. India*, **56**, 123–38.

Thangarajan, M. and Ahmed, S. (1989). Kriged estimates of water levels from the sparse measurements in a hard rock aquifer. In *Proc. of Internat. Groundwater Workshop (IGW-89), Hyderabad, India, Feb 28 to March 4*, ed. C. P. Gupta, *et al.*, Vol. I. New Delhi: Oxford and IBH Pub. Co., 287–302.

Thangarajan, M., Masie, M., Rana, T., *et al.* (2000). Simulation of arid multi-layer aquifer system to evolve optimal management schemes: a case study in Shashe River Valley, Okavango Delta, Botswana. *J. Geolog. Soc. India*, **55**, 623–48.

Thiéry, D. (1988). Analysis of long-duration piezometric records from Burkina Faso used to determine aquifer recharge. In *Estimation of Natural Groundwater Recharge*, ed. Simmers. The Netherlands: Springer 477–89.

Thiéry, D. (1993a). *Logiciel Marthe: Modélisation d'Aquifères par un maillage Rectangulaire en régime Transitoire pour le calcul Hydrodynamique des Ecoulements Version 4.3*. Rapport BRGM 4S/EAU n. R32210.

Thiéry, D. (1993b). Modélisation des aquifères complexes – Prise en compte de la zone non saturée et de la salinité. Calcul des intervalles de confiance. *Hydrogéologie*, **4**, 325–36.

Thiéry, D. and Boisson, M. (1991). *Logiciel GARDENIA modèle Global A Réservoirs pour la simulation des DEbits et des NIveaux Aquifères*. Guide d'utilisation (version 3.2). BRGM, R32209.

Thiéry, D., Dilucas, C., and Diagana, B. (1993). Modelling the aquifer recovery after a long duration drought in Burkina Faso. In *Proceedings of the IAHS/IAMAR International Symposium on Extreme Hydrological Events: Precipitation, Floods and Drought, Yokohama, Japan, July 1993*. IAHS Publ. 213. Wallingford, UK: IAHS Press, 43–50.

Thiéry, D. and Amraoui, N. (2001). Hydrological modelling of the Saône basin. Sensitivity to the soil model. *Phys. Chem. Earth J., Part B*, **26**, 467–2.

Thomas, R. G. (1973). Groundwater models. Irrigation and Drainage. Spec. Pap. Food and Agricultural Organ. No.21, U. N., Rome, Italy.

Townley, L. R. and Wilson, J. L. (1985). Computationally efficient algorithms for parameter estimation and uncertainty propagation in numerical models of groundwater flow. *Water Resour. Res.*, **21(12)**, 1851–60.

Trescott, P. C., Pinder, G. F., and Carson, S. P. (1976). Finite difference model for aquifer simulation in two dimensions with results of numerical experiments. In *Technique of Water Resources Investigations of the USGS*. Reston, VA: USGS, Chapter C1.

United Nations (UN) (1987). *Non-Conventional Water Resources Use in Developing Countries*. UN Pub. No. E.87.11.A.20.

United Nations Environmental Programme (UNEP) (1992). *World Atlas of Desertification*. Sevenoaks, UK: Edward Arnold.

Varga, R. S. (1962). Matrix Iterative Analysis. Prentice Hall, Englewood Cliffs, New Jersey.

Wackernagel, H. (1995). *Multivariate Geostatistics: An Introduction With Applications*. New York: Springer.

Wang, H. F. and Anderson, M. P. (1982). *Introduction to Groundwater Modeling: Finite Difference and Finite Element Methods*. San Francisco, CA: Freeman.

Wen, J.-C. (2001). A study of mean areal precipitation and spatial structure of rainfall distribution in the Tsen-Wen river basin. *J. Chinese Inst. Eng.*, **24(5)**, 649–58.

Witherspoon, P. A., Long, J. C. S., and Majer, L. R. (1987). A new seismic-hydraulic approach modeling flow in fractured rocks. In *Proc. NWWA/IGWMC Conference on solving groundwater problems with models, Denver, CO, February, 10–12*, Dublin, Ohio: National Water Well Assn.

World Bank (1999). *Groundwater in rural development: facing the challenges of supply and resource sustainability*. World Bank Technical Paper No. 463. Washington, DC.

Wyns, R., Baltassat, J. M., Lachassagne, P., Legchenko, A., Vairon, J., and Mathieu, F. (2004). Application of SNMR soundings for groundwater reserves mapping in weathered basement rocks (Brittany, France). *Bull. Société Géologique de France*, **175(1)**, 21–34.

Younes, F. and Razack, M. (2003). Hydrodynamic characterization of a Sahelian coastal aquifer using the ocean tide effect (Dridrate Aquifer, Morocco). *Hydrolog. Sci. J.*, **48(3)**, 441–54.

Zhu, X. Y., Zhu, G. R., Liu, Y. H., and Xie, X. Y. (1990). Modeling of karst-fracture groundwater resource in Xindian area, China. In *Proc. of IAH Congress, Water Resources in Mountainous Regions*, ed. A. Parriaux. IAH Memoirs, Vol. XXII, Part 1. 45–56.

Appendix   Access to software and data products

The data products and models referred to in the book are available as described below. Additional workshop material, including lecture/tutorial presentations, are available from the G-WADI web-site www.g-wadi.org.

DATA PRODUCTS

The satellite-based precipitation measurement algorithm, PERSIANN (Precipitation Estimation from Remotely Sensed Information using Artificial Neural Networks), continuously provides near global rainfall estimates at hourly $0.25° \times 0.25°$ scale from geostationary satellite longwave infrared imagery. An adaptive training feature enables model parameters to be constantly adjusted whenever independent sources of precipitation observations from low-orbital satellite sensors are available. The current PERSIANN algorithm has been used to generate multiple years of research quality precipitation data. PERSIANN data is available through HyDIS (Hydrologic and Data Information System: http://hydis8.eng.uci.edu/persiann/) at CHRS (Center for Hydrometeorology and Remote Sensing), University of California at Irvine. See Chapter 2 for further details.

MODELLING TOOLS

The Institute for Water Research, Rhodes University, South Africa, has developed the SPATSIM (Spatial and Time Series Information Modelling) system for Windows. This package makes use of an ESRI Map Objects spatial front end, linked to a database table structure, for data storage and access and includes relatively seamless links to a range of hydrological and water-resource estimation models. It also includes a wide range of data preparation and analysis facilities typically required in hydrological modelling studies. The SPATSIM approach has been adopted as the core modelling environment to be used for the update of the South African water-resource information system. Further details about the SPATSIM package can be accessed through the IWR web site at http://www.ru.ac.za/institutes/iwr and looking for the "Hydrological Models and Software" link. See Chapter 3 for further details.

The IHACRES software is available free of charge from the Catchment Modelling Toolkit website http://www.toolkit.net.au/. Users will need to register as members (free) in order to be able to download software from the site (registration allows users to be notified of updates to the software they download). There is also an e-group for asking questions included on the IHACRES product page that users can register with (this also includes an archive of questions and answers). See Chapter 4 for further details.

The KINEROS (KINematic runoff and EROSion) model was originally released in 1990. The latest version of the model, KINEROS2 is open-source software, available freely, along with associated model documentation, from www.tucson.ars.ag.gov/kineros.

A GIS interface, AGWA, has been developed for watershed-based analysis in which landscape information is used for both deriving model input, and for visualization of the environment and modelling results. This is an extension for the Environmental Systems Research Institute's ArcView versions 3.X. AGWA is distributed freely as a modular, open-source program suite (www.tucson.ars.ag.gov/agwa or www.epa.gov/nerlesd1/land-sci/agwa). See Chapter 5 for further details.

The US Geological Survey (USGS) Modular Modelling System (MMS) is an integrated system of computer software that provides a research and operational framework to support the development and integration of a wide variety of hydrologic and ecosystem models, and their application to water- and environmental-resources management. MMS provides a toolbox approach to model and system development that supports model integration and the development of decision support systems. Open-source software design means that individual modules, models, and tools can be developed by those with the relevant expertise and added to the common toolbox for use by others. MMS, the GIS support system, GISWeasel, selected models and tools, and additional information on MMS can be downloaded at:

> http://www.brr.cr.usgs.gov/mms/
> http://www.brr.cr.usgs.gov/weasel/
> http://www.brr.cr.usgs.gov/warsmp/
> See Chapter 7 for further details.

Imperial College has developed a set of tools for rainfall-runoff modelling and stochastic analysis of model structure and performance in the MATLAB modelling environment. The Rainfall-Runoff Modelling Toolbox (RRMT) allows a wide range of lumped rainfall-runoff models to be created and compared using a variety of loss and routing modules. The Monte Carlo analysis toolbox (MCAT) provides a range of analysis tools to display and analyze parameter identifiability and model uncertainty, including the use of multiple objective functions and generalized likelihood uncertainty analysis (GLUE). An extension of RRMT has been developed with the University of Lancaster, with UNESCO support, to include a time-series analysis capability and an arid-zone tutorial example. These can be downloaded free for research users from: http://ewre.cv.imperial.ac.uk. Please note that MATLAB software is required for their implementation. See Chapter 8 for further details.

In Chapter 9, Young presents data-based mechanistic (DBM) time-series models for flow (or river level) forecasts. Both the model structure and the associated parameters are inferred from the data and can be embedded easily within a state-estimation algorithm, such as the Kalman Filter, to provide for state updating, data assimilation, and real-time forecasting. The statistical tools for the identification, estimation and validation of DBM models are all available in Lancaster University's CAPTAIN Toolbox, a novel collection of computational algorithms designed for use in the MATLAB/Simulink™ software environment (see http://www.es.lancs.ac.uk/cres/captain/).

Index

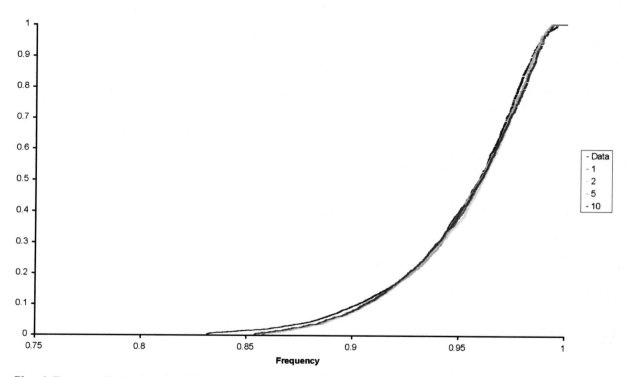

Plate 1 Frequency distribution of spatial coverage of Walnut Gulch rainfall. Observed versus alternative simulations

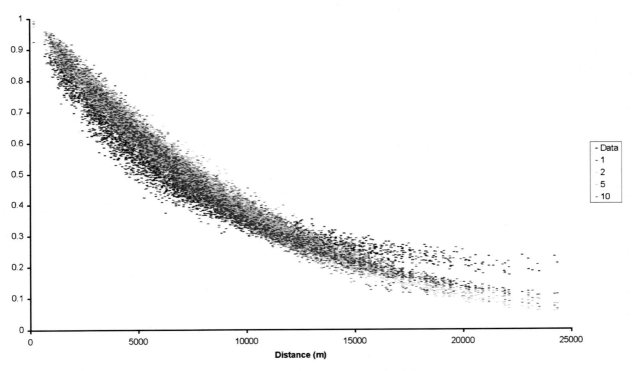

Plate 2 Spatial correlation of Walnut Gulch rainfall. Observed versus alternative simulations

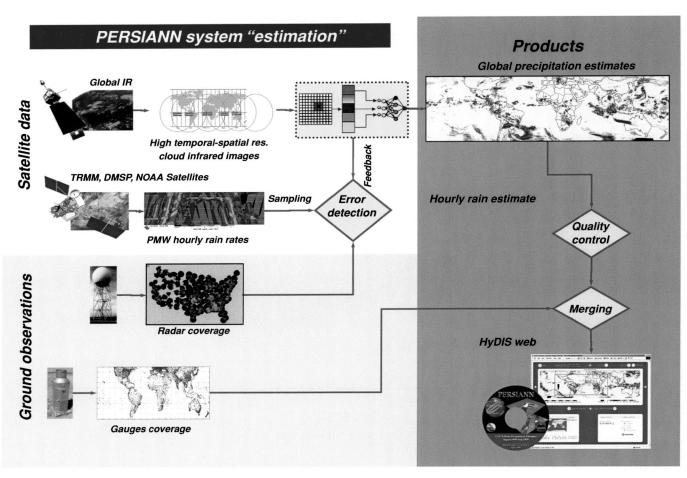

Plate 3 Current operational implementation of the PERSIANN system produces and distributes near-real-time global precipitation products at 0.25° six-hourly resolution.

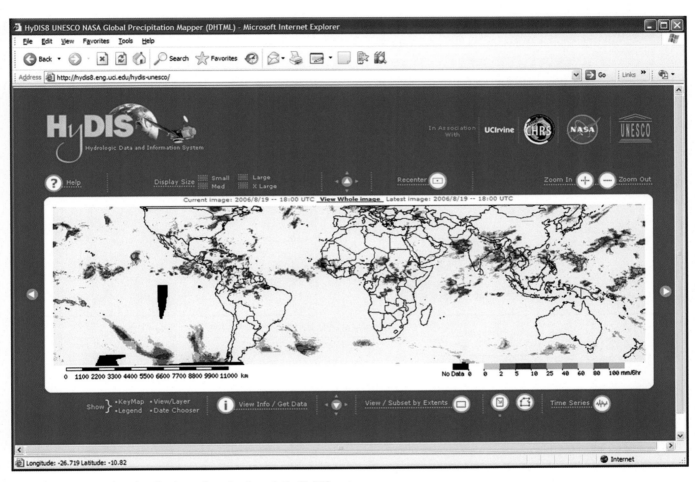

Plate 4 PERSIANN data visualization and service through the HyDIS system

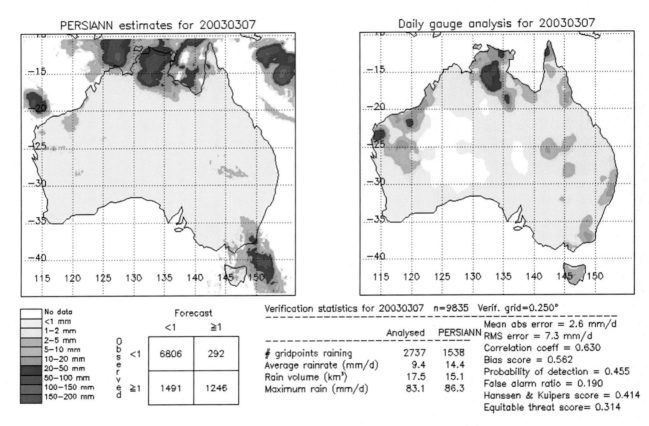

Plate 5 Evaluation of PERSIANN daily precipitation data over Australia region (this figure adapted from: http://www.bom.gov.au/bmrc/SatRainVal/dailyval.html)

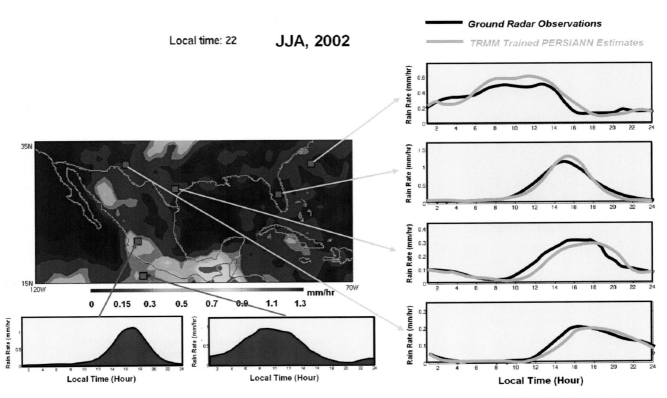

Plate 6 Evaluation of diurnal rainfall pattern of PERSIANN estimates in the summer season (JJA) 2002 using NCEP WSR-88 radar data

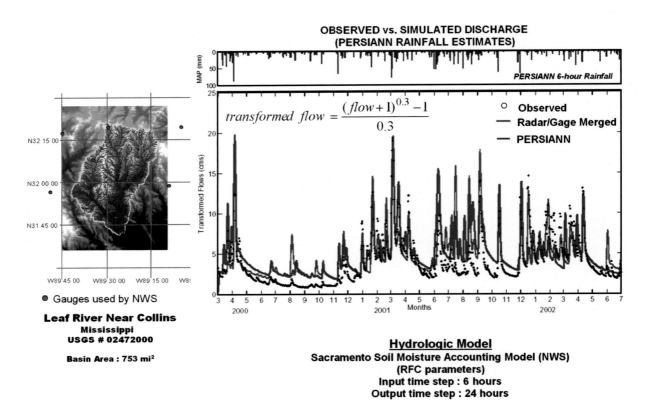

Plate 7 Simulation of daily streamflow using six-hourly accumulated rainfall from (1) satellite-based PERSIANN data and (2) radar and gauge merged data

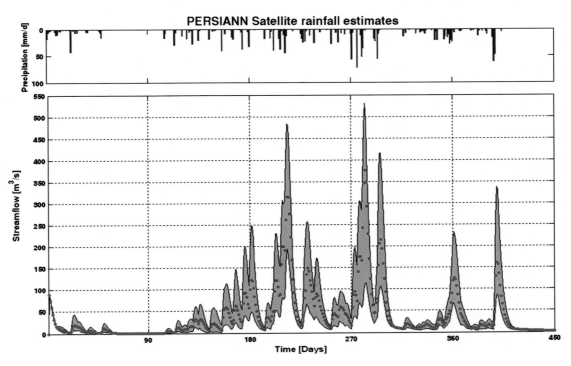

Plate 8 Confidence interval (95 %) of daily streamflow is generated based on the uncertainty of PERSIANN precipitation estimates

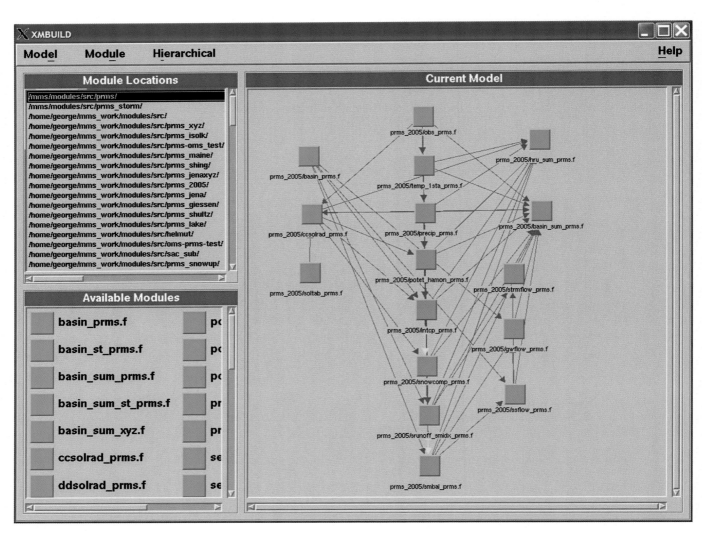

Plate 9 The model builder interface Xmbuild

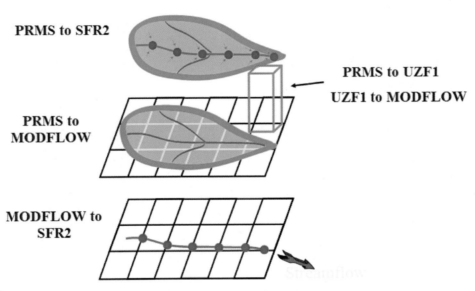

Plate 10 Fully coupled surface-water/groundwater model

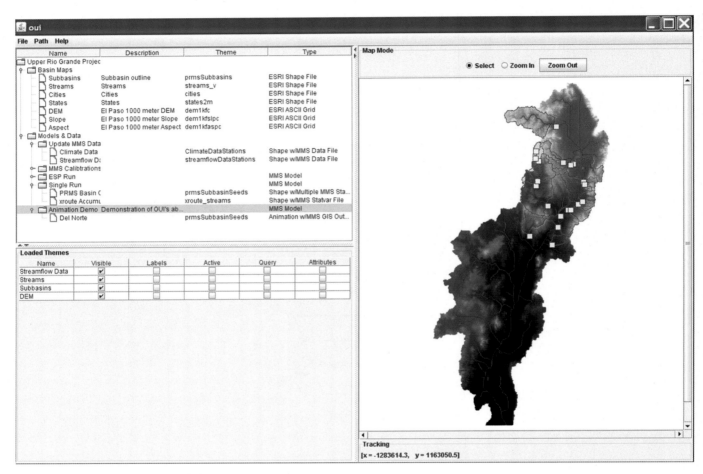

Plate 11 The Object User Interface (OUI)

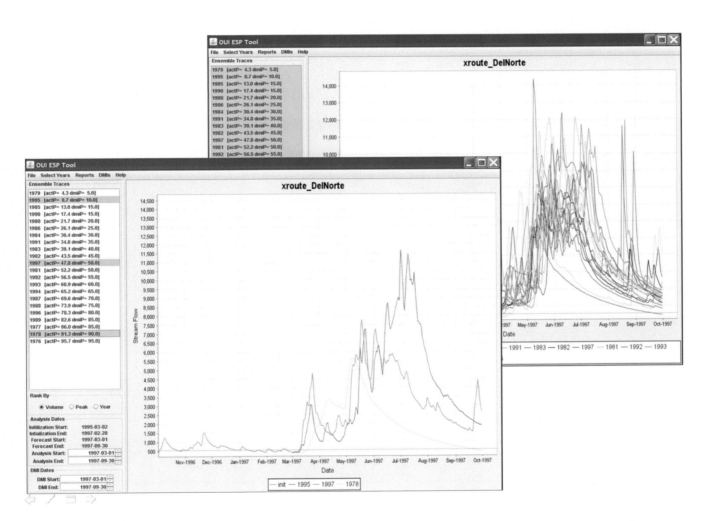

Plate 12 The ESP tool

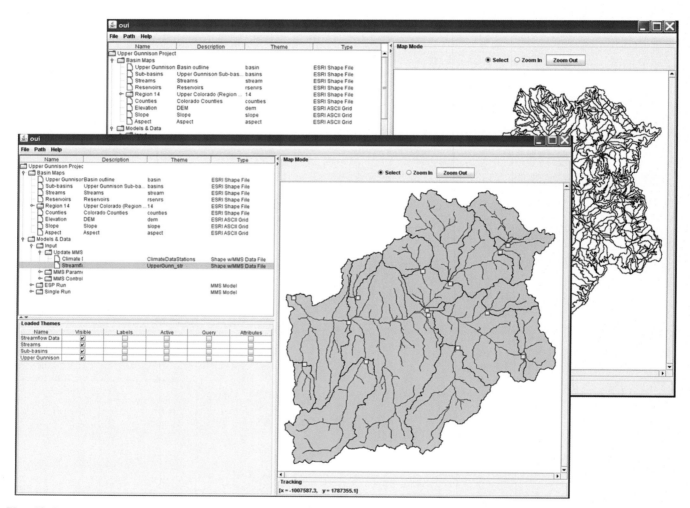

Plate 13 Gunnison River basin delineation into HRUs, subbasins, and forecast nodes

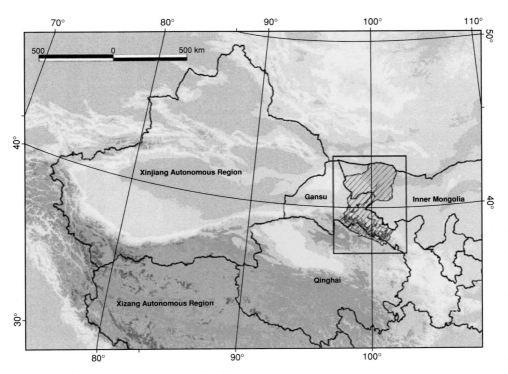

Plate 14 Location of the Heihe River basin, China

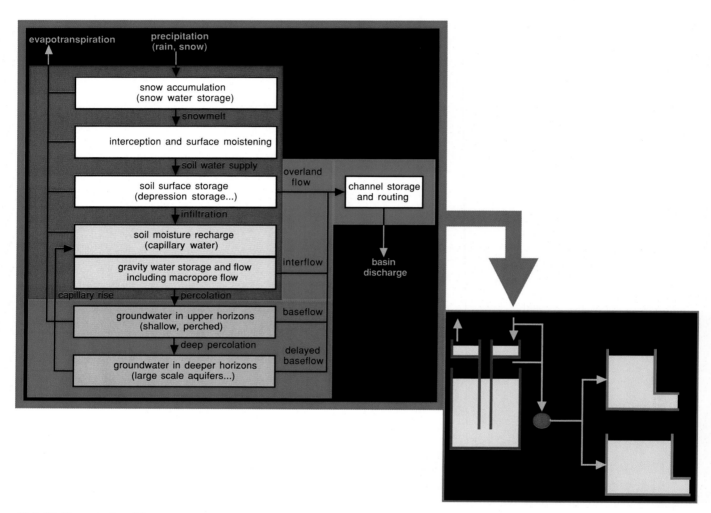

Plate 15 Conceptual model

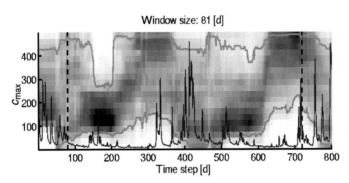

Plate 16 DYNamic Identifiability Analysis (DYNIA)

Printed in the United States
By Bookmasters